Chitosan-Based Nanocomposite Materials

Shikha Gulati
Editor

Chitosan-Based Nanocomposite Materials

Fabrication, Characterization and Biomedical Applications

 Springer

Editor
Shikha Gulati
Department of Chemistry
Sri Venkateswara College
University of Delhi
New Delhi, Delhi, India

ISBN 978-981-19-5337-8   ISBN 978-981-19-5338-5 (eBook)
https://doi.org/10.1007/978-981-19-5338-5

This Springer imprint is published by the registered company Springer Nature Singapore Pte Ltd.
The registered company address is: 152 Beach Road, #21-01/04 Gateway East, Singapore 189721, Singapore

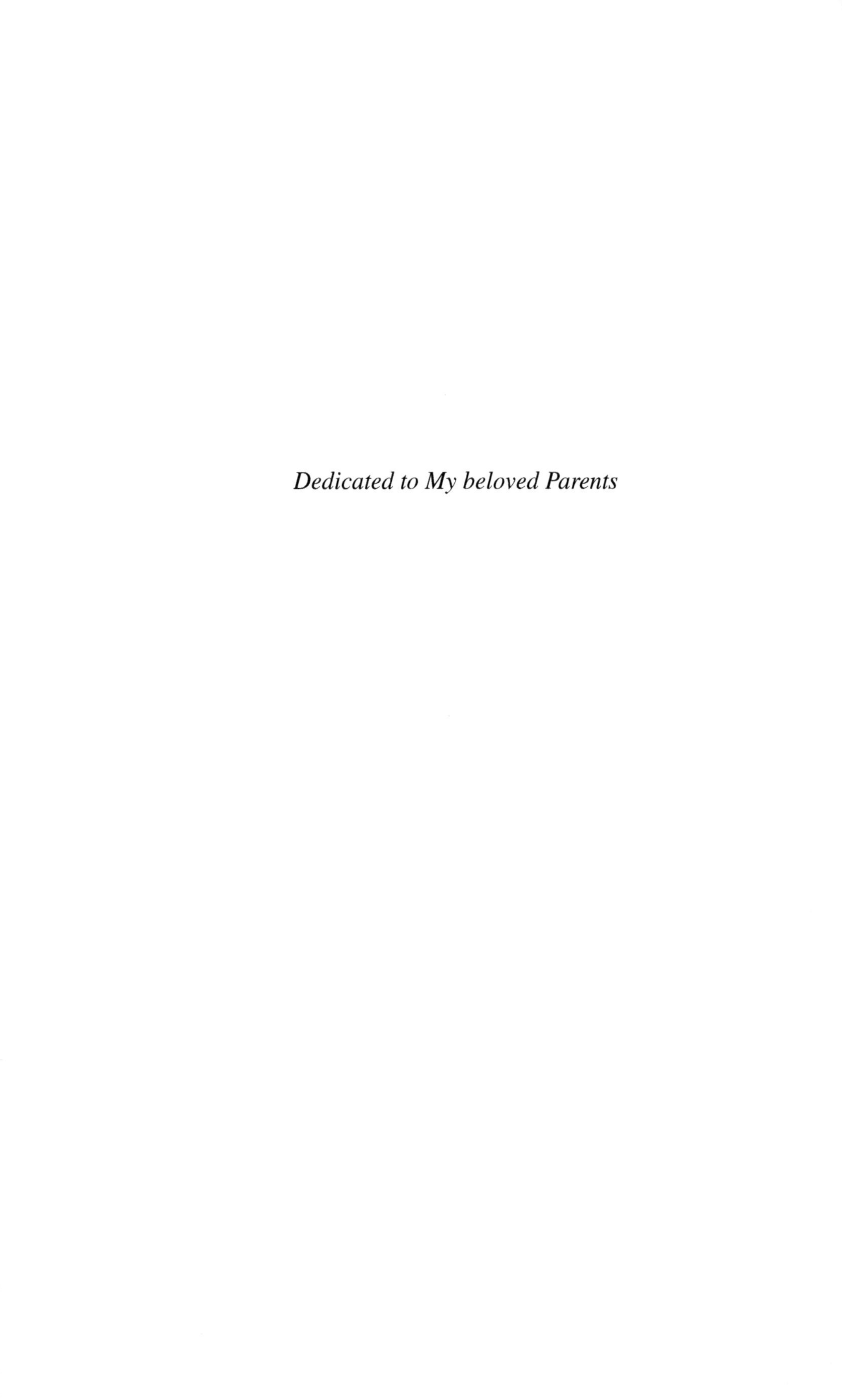

Dedicated to My beloved Parents

Preface

Chitosan is an intriguing second most abundant natural biopolymer that has been widely used in diverse fields owing to its biodegradability, biocompatibility and non-toxicity, predominantly in the biomedical field due to its wide range of structural possibilities for chemical and mechanical modifications to produce novel properties and applications. Chitosan contains several therapeutic qualities, including antibacterial, antioxidant and low immunogenicity, all of which boost its utility in biomedical applications. Chitosan can be easily fabricated into a variety of nanomaterials or nanocomposites, including nanosponges, nanofibers, nanoparticles, membranes, gels, micelles and liposomes, to meet a variety of purposes. In the recent decade, the number of studies employing chitosan has risen dramatically. As a result, many barriers have been broken down, and new research fields have emerged. Based on the facts given above, the classic and current trends in chitosan-based materials are grouped in a book: *Chitosan-Based Nanocomposite Materials: Fabrication, Characterizations and Biomedical Applications*. This book delves into the full chemistry of chitosan, as well as synthesis and chemical modification methodologies, characterization and applications of chitosan-based nanomaterials in biomedical and healthcare domains. The purpose of this book is to give readers of both academia and industry a comprehensive grasp of a topic where novel research is emerging that is of interest to wider scientific researchers.

As a result, a multi-author book with outstanding fourteen contributions from leading research groups working on chitosan-based nanocomposite materials is compiled. The fourteen chapters explain crucial discoveries and substantial advancements that have resulted from these possibilities. Chapter 1 provides an overview of chitosan and chitosan-based nanocomposite materials, highlighting their state-of-the-art potential in biomedical applications. Chapter 2 then focuses on the different methods of synthesis of chitosan-based nanocomposites, as well as their chemical modification strategies, which allow organic groups to be added to chitosan-based nanocomposites for a wide range of biological applications. Moreover, to understand the properties of the produced chitosan-based nanomaterials, several characterization techniques have been employed. Hence, Chap. 3 briefly describes various characterization techniques for chitosan and its based materials. Chapters 4–9 are focused on the biomedical applications of chitosan-based nanomaterials such as drug delivery, gene therapy, engineering, regenerative medicine, cancer diagnosis and treatment, bioimaging and wound healing agents. Further, the use of chitosan-based nanocomposites as antibacterial agents, antifungal agents and antiviral agents is discussed comprehensively in Chaps. 10–12. Chapter 13 discusses the application of chitosan-based nanocomposites in orthopaedics and dentistry. Finally, Chap. 14 ends with conclusions and future outlooks on chitosan-based nanocomposites, permitting researchers interested in this subject to design their future work and development. The contributing authors have wisely selected references to guide the reader through the extensive literature, allowing a wide spectrum of people to learn about the subject. As an editor, it will give me immense pleasure if this book assists individuals doing research on chitosan-based nanocomposites.

New Delhi, India   Dr. Shikha Gulati

Acknowledgments

First and foremost, I bow down to the heavenly almighty for providing me inspiration and constant strength to achieve milestones in life that can add meaning to it. Despite having one name on the cover, this book would not have been possible devoid of the support and assistance of a huge list of people without whom this task would have been difficult to realize. Firstly, I am highly indebted to Prof. C. Sheela Reddy, Principal, Sri Venkateswara College, the University of Delhi, for facilitating a well-equipped ICT Laboratory and library in the college. Above all and the most needed, she provided me unflinching encouragement and support in various ways. I would like to express my very sincere gratitude to well-known contributing authors without whose hard work this book could not have been written efficiently. I express my sincere thanks to all my respected teachers and mentors for their reassurance and inspiration. My deepest gratitude goes to my adored parents and my darling daughter for their unflagging love and support throughout my life; this book is simply impossible without them. I would also like to extend my thanks to the staff members of Springer, for their help and support. I feel great pleasure in acknowledging a word of appreciation to Springer Nature, Singapore, for publishing this book.

Dr. Shikha Gulati

About This Book

The emergent biomedical applications of chitosan-based nanocomposites encouraged me to write this book that brings together an international and interdisciplinary group of renowned scientists in the field of nanotechnology and biomedical sciences. While several books are existing about the synthetic strategies, significant properties and structures of chitosan-based nanocomposites, there is a dearth of an all-purpose book that not only covers the fundamentals of chitosan-based nanocomposites but also focuses on their biomedical applications. In this framework, the innovative contribution *Chitosan-Based Nanocomposite Materials: Fabrication, Characterizations and Biomedical Applications* in the Polymers and Composite Materials Book Series published by Springer Nature is highly beneficial for the readers. This book focuses on the cutting-edge research and findings on the use of chitosan-based nanomaterials in biomedical applications, emphasizing the potential that researchers have taken advantage of by using these creative bionanocomposites. It provides unparalleled insight into chitosan synthesis and chemical modifications, characterization methods, their use as anticancer, antimicrobial, antiviral and antifungal agents and their role in the biomedical field, as well as applications in gene therapy, drug delivery, dentistry, orthopaedics and other fields. This book will also highlight the difficulties in the light of prior advancements and strategies for additional study, as well as specifics on current groundbreaking technology and future views using a multidisciplinary approach. The book will undoubtedly be fascinating to researchers and scientists who are interested in nanocomposites made of chitosan.

Contents

1 Introduction to Chitosan and Chitosan-Based Nanocomposites 1
Rajender S. Varma, Arikta Baul, Lakshita Chhabra,
and Shikha Gulati

**2 Strategies for Synthesis and Chemical Modifications
of Chitosan-Based Nanocomposites: A Versatile Material
with Extraordinary Potential for Diverse Applications** 53
Mansi, Shikha Gulati, and Anoushka Amar

**3 Characterization Techniques for Chitosan and Its Based
Nanocomposites** 79
Gunjan Purohit and Diwan S. Rawat

**4 Implementation of Chitosan-Based Nanocomposites for Drug
Delivery System** 103
Gyanendra Kumar, Mohd Ehtesham, and Dhanraj T. Masram

**5 Potential of Chitosan-Based Nanocomposites for Biomedical
Application in Gene Therapy** 121
Manoj Trivedi and Sanjay Kumar

**6 Recent Advancements in the Application of Chitosan-Based
Nanocomposites in Tissue Engineering and Regenerative
Medicine** 145
Tailin Rieg, Angelo Oliveira Silva, Ricardo Sousa Cunha,
Karina Luzia Andrade, Dachamir Hotza,
and Ricardo Antonio Francisco Machado

**7 Emerging Applications of Chitosan-Based Nanocomposites
in Multifarious Cancer Diagnosis and Therapeutics** 165
Nandini Sharma, Shikha Gulati, and Jeevika Bhat

8 An Overview of the Application of Chitosan-Based Nanocomposites in Bioimaging 189
Ishita Chakraborty, Sharmila Sajankila Nadumane, Rajib Biswas, and Nirmal Mazumder

9 Chitosan-Based Nanocomposites as Remarkably Effectual Wound Healing Agents 199
Sneha Vijayan, Shikha Gulati, Tanu Sahu, Meenakshi, and Sanjay Kumar

10 Role of Antibacterial Agents Derived from Chitosan-Based Nanocomposites 221
Neha Dhingra, Anubhuti Mathur, Nishaka, and Kanchan Batra

11 Biomedical Application of Chitosan-Based Nanocomposites as Antifungal Agents 251
Richa Arora and Upasana Issar

12 Antiviral Potency of Chitosan, Its Derivatives, and Nanocomposites 273
Upasana Issar and Richa Arora

13 Modifications and Applications of Chitosan-Based Nanocomposites in Orthopaedics and Dentistry 291
Taruna Singh, Parul Pant, and Sarthak Kaushik

14 Conclusion and Future Prospects of Chitosan-Based Nanocomposites 305
Sanjay Kumar, Abhigyan Sarmah Gogoi, Shefali Shukla, Manoj Trivedi, and Shikha Gulati

About the Editor

Dr. Shikha Gulati (M.Sc., Ph.D.) is working as Assistant Professor of Chemistry at Sri Venkateswara College, University of Delhi. She is Expert in analytical chemistry, green chemistry, catalysis, inorganic chemistry and nanomaterials. Dr. Shikha's multiple publications, as well as chapters in numerous other books, speak to her research prowess and strong writing abilities. She has also published several research papers in reputable international journals. In numerous universities throughout India, her writings are cited for various undergraduate and postgraduate courses. She has served as Editor for the book entitled *Metal-organic Frameworks as Catalysts* published by Springer Nature. The Young Researcher Award 2020 was also given to Dr. Gulati for her efforts in the area of nanotechnology. Her expertise in inorganic chemistry, polymers and nanomaterials has greatly benefited this work.

Abbreviations

3D	Three dimensional
5-FU	5-Fluorouracil
A549	Adenocarcinomic human alveolar basal epithelial cell line
AAS	Atomic absorption spectroscopy
ABC	Acrylic bone cements
ACE-2	Angiotensin-converting enzyme-2
AES	Atomic emission spectroscopy
AFM	Atomic force microscopy
AIDS	Acquired immune deficiency syndrome
AMV	Alfalfa mosaic virus
APDI	Antimicrobial photodynamic inactivation
API	Active pharmaceutical ingredient
Ber	Berberine
BET	Brunauer–Emmett–Teller
BG	Active glass
BGC	Bioactive glass ceramic
BMP-2	Bone morphogenetic protein 2
BNNT	Boron nitride nanotube
BYMV	Bean yellow mosaic virus
CA	5-cholanic acid
CAV	Chitosan aloe vera
CG	Gelatin
CHI/GO	Chitosan conjugated graphene oxide
ChNP	Chitosan nanoparticles
CHO	Chinese hamster ovary
CLSM	Confocal laser electron microscopy
CM	Cell membrane
CMA	Confocal microscopic analysis
CMC	Carboxymethyl chitin
CMV	Cucumber mosaic virus
CNM	Carbon nanomaterials

CNP	Chitosan nanoparticles
CNT	Carbon nanotubes
COS	Chito-oligosaccharide
COVID-19	Coronavirus disease of 2019
CPT	Camptothecin
Cryo-SEM/TEM	Cryogenic-scanning electron microscopy and transmission electron microscopy
CS	Chitosan
CSF-Chi-NPs	Colony-stimulating factor chitosan-based nanoparticles
CS-NPs	Chitosan nanoparticles
CTAB	Cetyltrimethylammonium bromide
CTMS	Chlorotrimethylsilane
CU	Curcumin
DA	Degree of acetylation
DC	Direct current
DCS	Differential centrifugal sedimentation
DD	Deacetylation degree
DLS	Dynamic light scattering
DMC/TPP	N,N-dimethyl chitosan/tripolyphosphate
DNA	Deoxyribonucleic acid
DS	Degree of substitution
EA	Elemental analysis
Echo-CNPs	Echogenic CNPs
ECM	Extracellular matrix
EDTA	Ethylenediaminetetraacetic Acid
EDX	Energy-dispersive X-ray spectroscopy
EMR	Electromagnetic radiation
EOs	Essential oils
ESCA	Electron spectroscopy for chemical analysis
ETAs	Electrothermal atomizers
EVA	Ethylene-vinyl acetate
FDA	Food and Drug Administration
FESEM	Field emitter scanning electron microscopy
FTIR	Fourier transform infrared spectroscopy
GA	Glutaraldehyde
GB	Genipin
GC-CA	Glycol chitosan-5-cholanic acid conjugates
GM-CSF	Granulocyte-macrophage colony-stimulating factor
GO	Graphene oxide
GRAS	Generally recognized as safe
GTA	Glutaraldehyde
GTR	Guided tissue regeneration
HA	Hemagglutinin
HAp	Hydroxyapatite
HCL	Hollow cathode lamp

HCMV	Human cytomegalovirus
HCoV-NL63	Human coronavirus NL63
HCT 15	Human colorectal carcinoma cell line
HCV-4a	Hepatitis C virus genotype 4a
HEK	Human embryonic kidney
HeLa cells	Henrietta lacks cell line
HepG2	Liver hepatocellular carcinoma
HIV	Human immunodeficiency virus
HMW	High molecular weight
HPLC	High-performance liquid chromatography
HPV	Human papillomavirus
HSV	Herpes simplex virus
HT 29	Human colorectal adenocarcinoma cell line
HTCC	N-(2-hydroxypropyl)-3-trimethylammonium chitosan chloride
IARC	International Agency for Research on Cancer
ICDD	International Centre for Diffraction Data
ICP-MS	Inductively coupled plasma mass spectroscopy
ID	Iprodione
IFN-γ	Interferon gamma
IL-1	Interleukin 1
IL-4	Interleukin 4
IL-6	Interleukin 6
IL-10	Interleukin 10
IL-13	Interleukin 13
ISO	International Organization for Standardization
JCPDS	Joint Committee on Powder Diffraction Standards
LbL	Layer by layer
LCC	N lauryl-carboxymethyl chitosan
LDPE	Low-density polyethylene
LMW	Low molecular weight
LPS	Lipopolysaccharide
MB	Methylene blue
MBC	Minimum bacterial concentration
mBG	Microparticles of active glass
MCF 7	Michigan Cancer Foundation
MD	Molecular dynamics
MIC	Minimum inhibitory concentration
MMT	multifunctional montmorillonite
MNPs	Metal nanoparticles
MOF-5	Metal-organic framework
mPEG	Methoxy polyethylene glycol
MRI	Magnetic resonance imaging
mRNA	Messenger ribonucleic acid
MRSA	Methicillin-resistant Staphylococcus aureus
MTT	3-(4,5-dimethylthiazol-2-yl)-2,5-diphenyltetrazolium bromide

mV	Millivolts
MW	Molecular weight
MWCNTs	Multiwalled carbon nanotubes
N,O-CMC	Nitrogen, oxygen-carboxymethyl chitosan
Nar	Naringin
NBCs	Nanobiocomposites
nBG	Nanoparticles of active glass
NC	Nanocomposite
N-CMC	Nitrogen-carboxymethyl chitosan
NGCs	Neural guidance channels
nHA	Nanohydroxyapatite
NHCS	N-hexanoyl chitosan
NiO	Nickel oxide
NMR	Nuclear magnetic resonance
NO	Nitric oxide
NP	Nanoparticle
NPN	N-phenyl-1-naphthylamine
NTA	Nanoparticle tracking analysis
O-CMC	Oxygen-carboxymethyl chitosan
OCNP	Drug interaction oral contraceptive pill
OM	Outer membrane
OMPA	Outer membrane protein A
PAA	Polyacrylic acid
PAI	Photoacoustic imaging
PCDHN	Physically cross-linked hydrogel double network
PCL	Polycaprolactone
PDMAEA	Poly[2-(dimethylamino) ethyl acrylate]
PDMAEMA	Poly[2-(dimethylamino) ethyl methacrylate]
PEC	Polyelectrolyte complex
PEG	Polyethylene glycol
PEG/PCL	Poly(ethylene glycol)-poly(ε-caprolactone)
PEI	Polyethylenimine
pHPMA	Poly[N-(2-hydroxypropyl methacrylamide]
PLGA	Poly(lactic-co-glycolic) Acid
PLGA-NPs	Poly(lactic-co-glycolicacid) nanoparticles
p-NIPAM	Poly(N-isopropylacrylamide)
PNS	Peripheral nervous system
PPE	Personal protective equipment
PSTV	Potato spindle tuber viroid
PSV	Peanut stunt virus
PTT	Photothermal therapy
PVA	Polyvinyl alcohol
PVX	Potato virus X
QAS	Quaternized ammonium salt
QCh/PVP	Quaternized chitosan/polyvinylpyrrolidone

QNC	Quercetin nanocomposites
RBD	Receptor binding domain
rGO	Reduced graphene oxide rGO
RNA	Ribonucleic acid
RNAi	Ribonucleic acid interference
ROI	Region of interest
ROS	Reactive oxygen species
RR	Reactive dye
RSV	Respiratory syncytial virus
RT	Room temperature
RVFV	Rift valley fever virus
SARS-Cov-2	Severe acute respiratory syndrome coronavirus-2
SAXS	Small-angle X-ray scattering
SEM	Scanning electron microscopy
SEP	Sepiolite
SPM	Scanning probe microscopy
SSA	Specific surface area
SWCNTs	Single-walled carbon nanotubes
TA	Teichoic acid
TCID50	Tissue culture infectious dose
TEM	Transmission electron microscopy
TEMPO	(2,2,6,6-Tetramethylpiperidin-1-yl)oxyl
TGF-β1	Transforming growth factor beta
THI	T helper 1
TMV	Tobacco mosaic virus
TNF-α	Tumour Necrosis Factor alpha
TNV	Tobacco necrosis virus
TPP	Tripolyphosphate
UV	Ultraviolet
VEGF	Vascular endothelial growth factor
XPS	X-ray photoelectron spectroscopy
XRD	X-ray diffraction

Chapter 1
Introduction to Chitosan
and Chitosan-Based Nanocomposites

Rajender S. Varma, Arikta Baul, Lakshita Chhabra, and Shikha Gulati

Abstract Chitosan, produced by the deacetylation of chitin, is the second-most abundant natural polymer. Known for its appealing properties like non-toxicity, chemical versatility, biodegradability, biocompatibility, high adsorption capacity, etc., chitosan finds appliance in a broad range of biomedical applications such as wound dressing, drug delivery, biosensors, and dietary supplement. However, as a pristine material, chitosan may display some limitations like poor thermal stability, low mechanical strength, and poor barrier attributes. With the advent of nanotechnology over the years, chitosan has garnered immense interest as a matrix of nanocomposites that have led to the advancement in its effectiveness, eventually expanding its application areas. Chitosan nanocomposites are promising bio-based polymeric nanocomposites with exceptional biological and physicochemical properties that offer great potential for the delivery along with the controlled release of drugs and active compounds, thus functioning as a superior nanocarrier system. Additionally, chitosan nanocomposites have been widely deployed in not only tissue engineering but also assorted fields of biomedical research such as bioimaging, cancer therapy, dentistry, and wound healing, among others. This chapter offers an introduction to chitosan, its structure, manufacturing process, and its chemical and biological activity. Furthermore, the basics of its cross-linking to form nanocomposites and the related synthesis scheme are presented. A brief discussion on the remarkable properties of chitosan nanocomposites and their biomedical applications is provided, followed by concluding statements on their limitations, current status, and future prospects including an overview of all the biomedical applications covered subsequently in this book.

R. S. Varma (✉)
Regional Centre of Advanced Technologies and Materials, Czech Advanced Technology and Research Institute, Palacký University in Olomouc, Šlechtitelů 27, 783 71 Olomouc, Czech Republic
e-mail: Varma.Rajender@epa.gov

A. Baul · L. Chhabra · S. Gulati (✉)
Department of Chemistry, Sri Venkateswara College, University of Delhi, New Delhi, Delhi 110021, India
e-mail: shikha2gulati@gmail.com

Keywords Chitosan · Nanocomposites · Nanotechnology · Biomedical applications · Drug delivery

Abbreviations

AFM	Atomic force microscopy
BET	Brunauer Emmett Teller
CLSM	Confocal laser electron microscopy
CMA	Confocal microscopic analysis
CNT	Carbon nanotube
CTAB	Cetyltrimethylammonium bromide
DA	Degree of acetylation
DCS	Differential centrifugal sedimentation
DD	Degree of deacetylation
DLS	Dynamic light scattering
EA	Elemental analysis
EMR	Electromagnetic radiation
FTIR	Fourier transform infrared spectroscopy
GA	Glutaraldehyde
GO	Graphene oxide
HPLC	High-performance liquid chromatography
LbL	Layer by layer
MIC	Minimum inhibitory concentration
MTT	3-(4,5-Dimethylthiazol-2-yl)-2,5-diphenyltetrazolium bromide
NC	Nanocomposite
N-CMC	Nitrogen-carboxymethyl chitosan
NiO	Nickel oxide
NMR	Nuclear magnetic resonance
N,O-CMC	Nitrogen, oxygen-carboxymethyl chitosan
NP	Nanoparticle
NTA	Nanoparticle tracking analysis
O-CMC	Oxygen-carboxymethyl chitosan
SAXS	Small-angle X-ray scattering
SEM	Scanning electron microscopy
SPM	Scanning probe microscopy
SSA	Specific surface area
TEM	Transmission electron microscopy
TPP	Tripolyphosphate
XPS	X-ray photoelectron spectroscopy

1 Introduction

Polysaccharides are natural biopolymers that are commonly utilized in the field of biomedicine owing to their unique attributes, such as biodegradability, biocompatibility, and high availability [1]. One such widely employed polysaccharide is chitosan, the second-most abundantly available biopolymer [2]. It is an eco-friendly, economical, and deacetylated form of chitin isolated from marine wastes, insects, and fungi. Chitin, mainly found in shrimps and crabs, the cell walls of fungi and yeasts, and in the exoskeleton of anthropods, can be extracted from crustaceans via acid treatment (to dissolve calcium carbonate) and alkaline extraction (to solubilize proteins). Chitosan after deacetylation can be obtained in many different forms, like fiber, sponges, and powder appearance from the solution and with different degrees of deacetylation and molecular weights. Their morphological, thermal, structural, crystalline, and many other physicochemical properties can then be investigated using standard characterization techniques. Chitosan is a cationic polysaccharide consisting of glucosamine units linked by glycosidic bonds. The increasing popularity of chitosan among researchers worldwide is justified in view of its versatile properties like non-toxicity, controlled release, biodegradability, biomimetics, stability, and the possibility to chemically and structurally modify them to increase their chelating and absorption properties, solubility, porosity, and permeability [5] (Fig. 1). The added advantage is the presence of functional groups in their chemical structure because of which they can react easily with other active compounds. These favorable properties of chitosan make them an ideal highly economical choice for application in a multitude of industries such as biomedical science, material science,

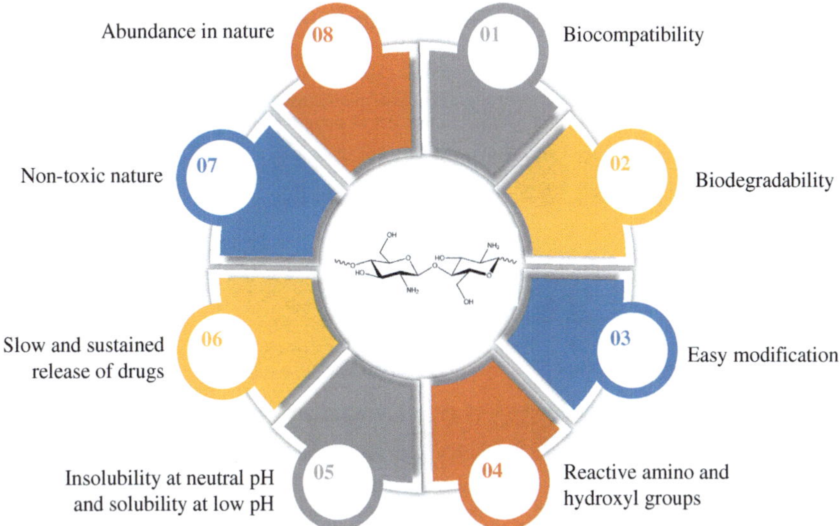

Fig. 1 Exceptional properties of chitosan

bioengineering, and pharmaceuticals, among others. However, chitosan, on its own, may display poor mechanical and thermal properties [6] and may not be as stable in media with variable pH and ionic strength. Hence, nanoscale-sized materials (like nanoparticles (NPs), nanosheets, nanorods, nanocapsules, nanofibers, etc.) can be added to chitosan as fillers, thus embedding into the bulk or by coating on the material surfaces to enhance its biological and physicochemical characteristics. The free amine and hydroxyl groups forming the polymeric backbone of chitosan not only facilitate its solubility but also assist in forming several hydrogen bonds that enable nanoparticles to be embedded as fillers. Chitosan exhibits a strong affinity for metals and its ability to effectively blend with elements like copper, zinc, iron, silver, silicon, zinc oxide, titanium oxide, etc., and results in the formation of metal nanoparticles that exhibit improved absorbency and enhanced antimicrobial, chemical, thermal, and mechanical properties [7]. Nanomaterials offer a plethora of favorable characteristics including pore size, surface-to-volume ratio, and reactive groups on the surface. And justifiably, the potential of using nanoparticles as nanocarriers that can encapsulate drugs, deliver them to the target site, and ensure a regulated release has been extensively exploited in the biomedical arena.

Since its discovery at the beginning of the nineteenth century, chitin, chitosan, and its derivatives have been widely investigated. Figure 2 depicts a timeline of their historical development and progress made over the years.

Chitosan nanoparticles were first characterized in 1994 for the circulatory delivery of 5-fluorouracil by Ohya et al. [8]. Since then, they have been extensively synthesized using many different methods and modified in diverse ways taking into account several factors like size, stability, retention time, and drug loading capacity. The preparation of drug-loaded chitosan nanoparticles is accomplished using two main approaches: nanoencapsulation and chemical modification. Nanoencapsulation, which involves forming nanostructures that contain the drug at the particle

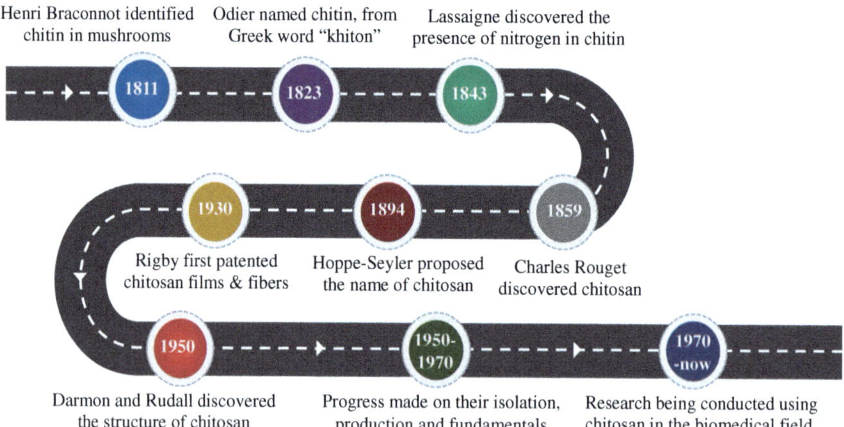

Fig. 2 Timeline of progress pertaining to chitin and chitosan. Based on the information provided in [3, 4]

surface or within [9], helps to improve the efficiency of the drug, and its accessibility to the target site, and lower the toxicity [10]. Chemical modification, like PEGylation, thiolation, carboxylation, quaternization, and alkylation can be attained starting from the particles or the base polymers [11]. Over the years, chitosan nanocomposites (NCs) have been widely studied for use in medicine due to their excellent ability to encapsulate and chain graft drugs and active ingredients, thus preventing the enzymatic degradation of drugs [12], ensuring continuous and controlled drug release [13], and to reduce the damage of non-targeted tissues [14]. Thus, they have profound applications in biomedicine, ranging from wound healing, drug delivery, biosensors, tissue engineering, and food and nutrition, to solving several human health problems like blood clotting, gene therapy, skin burn healing, food allergies, cholesterol control, weight loss, biological imaging and diagnosis, and cancer treatment, among others. Some recent advancements comprise the use of chitosan inside fat-burning supplements [15]; for external skin wound healing by employing the use of nanocomposites containing chitosan and silver nanoparticles [16]; and in fabricating a drug delivery system to prevent multidrug resistance in cancer treatment [17], among others. In continuation to our ongoing studies in the field of nanotechnology and chitosan [18–27], this chapter provides an introduction to chitosan and highlights its exceptional properties that make chitosan and its composite forms an exemplary model for application in the biomedical science arena, while presenting their structural analysis. Furthermore, we bring to attention the preparative techniques for chitosan nanocomposites, their modification approaches, characterization techniques, and their ideal properties that render them novel materials in this research field, along with an overview of the current challenges and future outlooks.

2 Chitosan: Structure and Preparation

2.1 Chitin and Chitosan: Chemical Structures

Chitin is a -(1 4)-linked N-acetyl-D-glucosamine homopolymer that is semicrystalline. The building components -(1 4)-2-amino-D-glucose and -(1 4)-2-acetamido-D-glucose are combined to form this copolymer. Chitosan is a chitin derivative that has been partially deacetylated. Its nitrogen content distinguishes it from other polysaccharides. Figure 3 illustrates the structures of chitin and chitosan. Chitosan organic changes are caused by amino and hydroxyl groups in the structure, which enable the formation of polymeric derivatives of these substances.

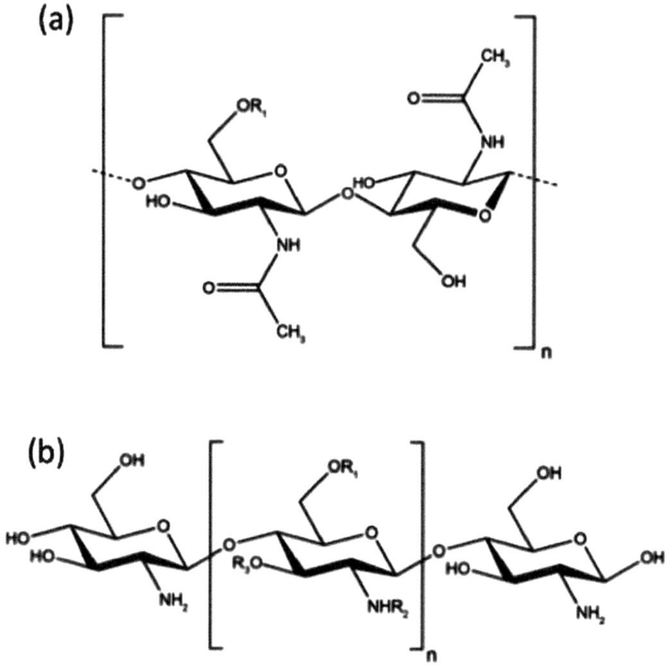

Fig. 3 **a** Structure of chitin and **b** chitosan. Reprinted with permission from [28]. Copyright 2017 Elsevier

2.2 Preparation of Chitin and Chitosan

Crab and shrimp shell exoskeleton wastes are used as a biomass source for the industrial manufacturing of chitin and chitosan. Chitin is found in crustacean exoskeletons, insect cuticles, algae, and fungal cell walls. Vegan needs have recently sparked interest in chitosan derived from mushrooms.

2.2.1 Extraction of Chitin

The chemical method, as well as biological extraction, can both be deployed to recover chitin from the animal exoskeleton. Despite the fact that the chemical approach has a number of drawbacks, including being inefficient and unfriendly to the environment, it is still the most often utilized commercial method in view of its faster processing time [29]. In a study comparing chemical and biological methods for extracting chitin from prawn and shrimp shells, it was discovered that the biological method yielded more chitin [30]. Washing, drying, and crushing the raw material, received from diverse sources, to attain a fine powder form are the basic

Fig. 4 Schematic summary of the synthesis process for chitosan. Reprinted with permission from [1]. Copyright 2021 Elsevier

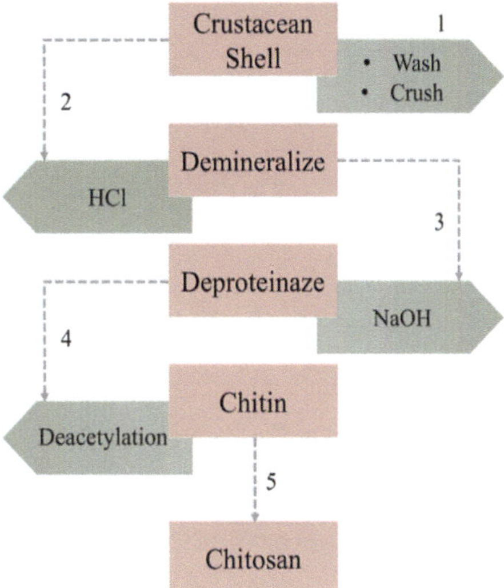

techniques for extracting chitin. The powder is subsequently demineralized, deproteinized, and decolored by a number of processes; both the chemical and biological approaches have these three phases in common [31]. This is usually followed by the important following phases [6, 18, 32]: Raw material shells are washed, crushed, and ground to smaller sizes, with some components, such as calcium carbonate, demineralized by chemical extraction with dilute hydrochloric acid and stirring at room temperature. This is followed by deproteinization with a dilute aqueous sodium hydroxide solution. Proteins can be recovered by reducing the pH to 4.0 and then drying the precipitates. An additional decolorization step is performed to remove the color. Chitin is extracted as the primary raw material for the production of chitosan. Time, temperature, and alkali concentration are the three key reaction parameters. Figure 4 depicts the chitosan manufacturing process.

2.2.2 Conversion of Chitin to Chitosan

Chitin is prepared in the same way as it was previously explained. The method of deacetylation is subsequently used to convert it into chitosan (Fig. 5). The chitin structure is stripped of acetyl groups (COCH$_3$). As the N-deacetylation process is never complete, chitosan with varying degrees of deacetylation is the normal outcome that defines its applications.

Chitin deacetylation is accomplished through two methods: alkaline and enzymatic.

(a) Alkaline method: A 40–50% NaOH solution is applied to the chitin powder. This process causes acetyl groups to be hydrolyzed, as well as converting N-acetyl-D-glucosamine units into D-glucosamine units with free amine groups.

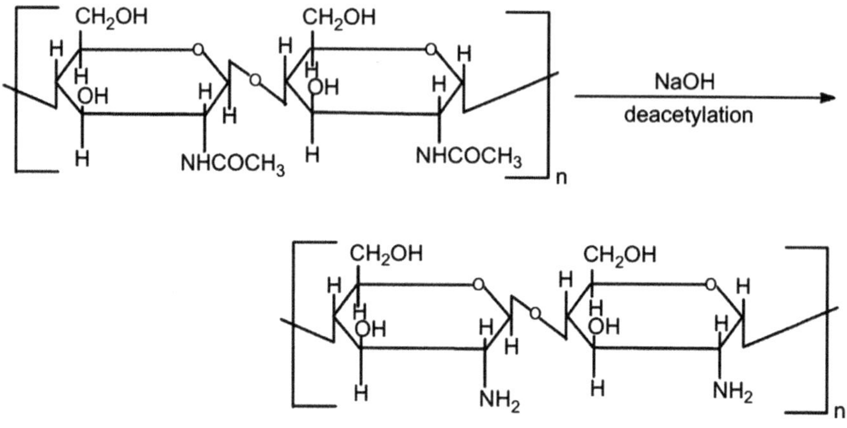

Fig. 5 Schematic showing deacetylation of chitin to produce chitosan. Reprinted with permission from [33]. Copyright 2000 Elsevier

The degree of deacetylation (DD) of the chitosan produced by this technique is determined by temperature, alkali solution concentration, and reaction time. However, this process consumes a lot of energy and pollutes the environment. As a result, an enzymatic process that is more environmentally friendly is often chosen over this procedure.

(b) Enzymatic method: Chitin is converted to chitosan using an enzyme named "chitin deacetylase enzyme"; fungus "*Mucor rouxii*" was responsible for the discovery of this enzyme in 1974. Chitosan was discovered in the cell walls of some fungi, and the chitin deacetylase enzyme was determined to be responsible for the conversion of cell wall chitin to chitosan in those fungal strains. However, the chitin deacetylase-producing fungal strains have a low enzyme yield and entail a difficult fermentation process. Moreover, when compared to enzymes isolated from bacterial strains, the enzyme obtained from a fungal source had lesser activity. Furthermore, bacteria in large-scale fermentation systems multiply faster and more easily.

3 Modifications of Chitosan

Chitosan modifications are made to improve the polymer's performance by introducing helpful features that are tailored to individual demands. Chitosan alterations can be divided into two types: physical changes and chemical modifications [5, 34, 35]:

3.1 Physical Modifications of Chitosan

Blending or physically mixing at least two polymers to generate new material with different, better, and improved physical properties is referred to as physical modification [36]. Blending is primarily intended to be less expensive, easier to accomplish, and less time-consuming. Another advantage of the blending approach is that it allows for the adjustment of the composition of starting materials, resulting in products with a wide variety of qualities tailored to specific applications. The blended membranes have been successful in boosting the wettability and porosity of the absorptive membrane. According to the hypothesis introduced by Pearson in 1963, the best ion to create a complex with the manufactured membrane (182 mol/g) is Zn (II), while Ca (II) and Mg (II) tend to form ionic connections with chitosan NH_2 groups as these two metal ions are hard acids. However, investigations on physical chitosan modifications are dwindling as researchers prefer to use numerous chemical changes to alter the modified chitosan according to certain contaminants [5].

3.2 Chemical Modifications of Chitosan

The introduction of various functional groups to the chitosan's structure, such as photosensitizers, dendrimers, sugars, cyclodextrins, and crown ethers, can be accomplished to enhance the existing properties of chitosan:

(i) Photosensitizer-modified chitosan: A photosensitizer is a chemical substance that absorbs light energy before transferring it to the reactants of interest. When processes require light sources of specific wavelengths that are not readily available, the photosensitizer is typically utilized.

(ii) Chitosan-dendrimer hybrids: Dendrimers are highly branched and symmetrical macromolecules that have lately been recognized as part of the polymer family [37]. The multifunctional features of dendrimers make them an appealing option.

(iii) Carboxymethylation of chitosan: CMC is a carboxymethylated chitosan product with some amino and primary hydroxyl sites of the chitosan glucosamine units substituted by a carboxyl group (COOH). This chemical modification produces three derivatives: nitrogen-carboxymethyl chitosan (N-CMC), oxygen-carboxymethyl chitosan (O-CMC), and nitrogen, oxygen-carboxymethyl chitosan (N-O-CMC) (N,O-CMC). The synthesis of N-CMC, O-CMC, and N,O-CMC from chitosan is depicted schematically in Fig. 6. The modified chitosan possesses hydrophilic and water-soluble qualities, which are sought after not only in membrane technology and wastewater treatment systems, but also in medical engineering, biomedicine, agricultural, and tissue engineering, to name a few fields.

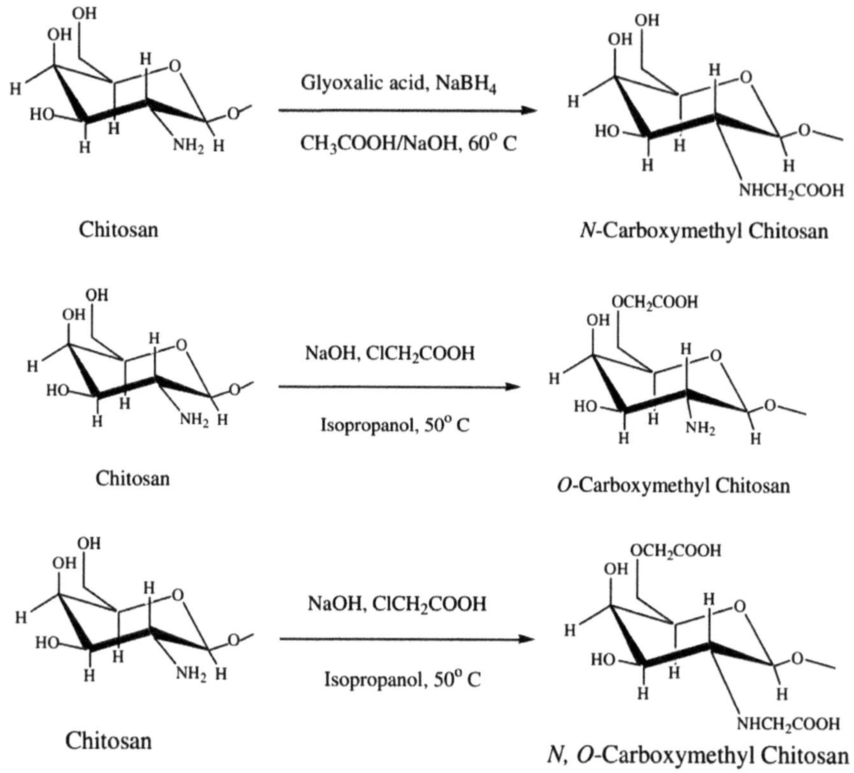

Glyoxalic acid, NaBH$_4$

CH$_3$COOH/NaOH, 60° C

Chitosan

N-Carboxymethyl Chitosan

NaOH, ClCH$_2$COOH

Isopropanol, 50° C

Chitosan

O-Carboxymethyl Chitosan

NaOH, ClCH$_2$COOH

Isopropanol, 50° C

Chitosan

N, O-Carboxymethyl Chitosan

Fig. 6 Synthesis of N-CMC, O-CMC, and N,O-CMC from chitosan. Reprinted with permission from [38]. Copyright 2010 Elsevier

(iv)   Other modifying systems: Hall and Yalpani were the first to describe the sugar modification of chitosan in 1980 [39]. They provided reductive amination processes for connecting a wide variety of carbohydrates as side chains to linear amine-containing polysaccharides that were particularly simple and adaptable. Furthermore, they discovered that the synthetic chitosan derivatives are open to further chemical changes, as demonstrated by the targeted oxidation of the pendant galactose residues with galactose oxidase to obtain a C-6 aldehyde derivative.

4   Properties of Chitosan

(i)   Degree of *N*-deacetylation: Many factors, including the concentration of NaOH, the contact time of NaOH with chitin, temperature, and pH, can influence whether chitosan is partially or completely deacetylated. The source of chitin influences the degree of deacetylation required to produce chitosan. The acetylation degree of chitin (typically 0.90) and chitosan is distinguished by a fine line (the degree of acetylation is less than 0.35).

(ii) Molecular weight: The molecular weight of this biopolymer is another important attribute to consider while performing experiments with it. According to an earlier study [40], membranes created with high molecular weight chitosan had superior tensile strength, elongation, and enthalpy, while membranes prepared with low molecular weight chitosan had better permeability. It was concluded that because chitosan with a higher molecular weight has a higher enthalpy, it has more crystallinity in the membrane and has more intermolecular contacts than chitosan with a lower molecular weight.

(iii) Charge density: The degree of protonation of the amino group determines the charge density of chitosan. The degree of acetylation (DA) of chitosan, as well as ionic strength and pH, governs it. Depending on the type of chitosan and environmental conditions, the dissociation constant of chitosan ranges from 6.2 to 7.0. At constant ionic strength, chitosan samples with various DA values have been examined for mean electrophoretic mobility as a function of pH. When the pH value was increased, the electrophoretic light scattering curve for all of the samples followed the same trend. Chitosans with more DA, and hence lower charge density, had poorer mobility at all pH values, as expected.

(iv) Solubility: The solubility of chitosan in water is determined by three factors: the solvent's ionic strength, which directs the salting-out effect; the pH of the solution; and the ions in the solvent reacting with chitosan and restricting its solubility. Chitosan has a characteristic property of being soluble at acidic and insoluble at basic pH values. All varieties of chitosan are soluble below a pH value of 6. However, chitosans with medium molecular weight and DA around 0.5 can be soluble at pH 7 [41]. Figure 7 depicts the fluctuation in solubility values with pH for chitosans of various DA.

Because of variations in chitosan's availability to enzymes, solubility values can have a big impact on its biological characteristics.

(v) Viscosity: From a technological standpoint, polymer viscosity is a critical characteristic because very viscous solutions are difficult to regulate. Viscometry is also a useful instrument for estimating the molecular weight of chitosan. The Mark–Houwink–Sakurada equation is used to calculate the average molecular weight [42]. As a fall in viscosity is observed during polymer storage due to polymer breakdown, viscosity can be utilized to estimate the stability of the polymer in solution.

(vi) Chemical stability: The glycosidic connections in the chitosan chain structure can be damaged by alkalis and acids, and they can also be oxidized by free radicals. However, because the majority of chitosan's applications are tied to its stability at low pH, the resilience of chitosan against oxidation has received far less attention than its pH stability. The DA value of chitosan appears to be responsible for its acid-catalyzed degradation, and the rate of degradation increases as the DA value rises.

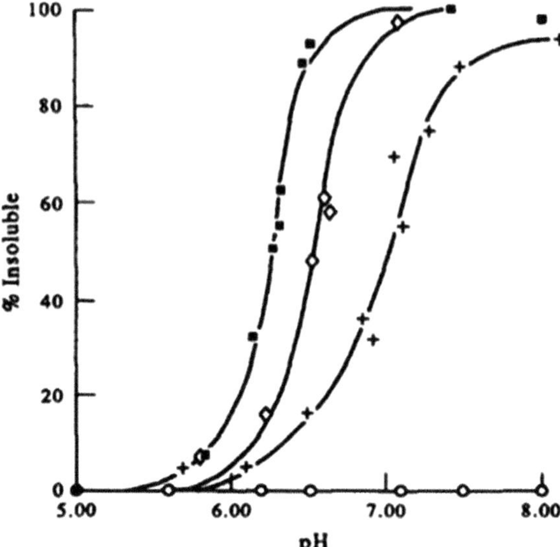

Fig. 7 Solubility versus pH curves for chitosans with DA = 0.01 (filled square), DA = 0.17 (diamond), DA = 0.37 (plus), and DA = 0.60 (circle). Reprinted with permission from [41]. Copyright 1994 Elsevier

(vii) Antimicrobial activity: Antibiotic resistance is a major public health concern, which necessitates finding antibiotic replacements a critical task. The antimicrobial activity of chitosan, chitosan derivatives, and chitooligosaccharides has been demonstrated against a variety of microorganisms, including filamentous fungus, yeast, and bacteria. Food, cosmetics, the textile industry, and a variety of other areas benefit from this antibacterial activity.

(viii) Antioxidant activity: Because of the link between oxidative stress and diseases like Alzheimer's disease, Huntington's disease, Parkinson's disease, cancer, and amyotrophic lateral sclerosis, antioxidants are attaining a more significant stature. It's also linked to complications from other illnesses like diabetes. Chitosan contains amino and hydroxyl groups that can react with free radicals and act as scavengers. Some chitosan derivatives, such as chitosan sulfates or N-2 carboxyethyl chitosan, have more antioxidant activity than others [43–45]. Chitooligosaccharides have also been chemically modified to increase their antioxidant activity, by treating the polymers with gallic acid or phenolic chemicals.

(ix) Anti-inflammatory properties: The inflammatory response is the body's automatic physiological response to tissue injury. The primary goal of the inflammatory response is to direct circulating leukocytes and plasma proteins to the infected site or tissue damage in order to eradicate the causal agent and start the healing process. Although inflammation is important for survival, it can cause immense harm if it is severe, unable to eliminate the causative agent, or directed against the host. The production of free radicals is directly linked to the inflammatory process. When the molecular weight of the chitosan is lowered, chitooligosaccharides show increased activity which appears to be even more amazing.

5 Characterization of Chitosan

Chitosan characterization is critical since its structure determines its qualities, which in turn serves to define a potential application in industry. As listed below, a variety of procedures are employed to determine the degree of deacetylation in chitosan [46].

(i) Elemental analysis: At 600 °C, a known amount of chitosan is heated for 1 h. The residue is then weighed to decide the amount of inorganic substance. Elemental analysis (EA) results are not always precise, especially when pollutants are found. This approach was used to define chitosan as having a nitrogen content of more than 7% and chitin as having a nitrogen concentration of less than 7% [47, 48].

(ii) Titration method

(a) Acid-base titration: After dissolving a known amount of chitosan in 0.1 N HCl overnight, it is titrated with 0.1 N NaOH using D-glucosamine and *N*-acetyl-D-glucosamine as 100% and 0% deacetylated controls, respectively. The DA is determined by the first and second inflection points of the titration curve [49].

(b) Potentiometric titration: A known volume of HCl, with dissolved chitosan in it, is titrated against NaOH. The base graph's fluctuation in pH versus volume yields two inflection points, one for HCl neutralization and the other for ammonium ion neutralization. The difference between these two gives the degree of deacetylation [50, 51].

$$\%DA = 100 - \%DD$$

(c) Colloid titration: The main premise behind this sort of titration is a stoichiometric mixture of positive and negative ions. A sample of chitosan is dissolved in aqueous acetic acid and diluted with deionized water before being titrated with potassium polyphosphate (vinyl sulfate). The protonated chitosan's positive ammonium groups were directly linked to the polysulfate's negatively charged sulfate groups (vinyl sulfate). Excess poly(vinyl sulfate) binds to toluidine blue (indicator) a minute after the equivalent point is reached, and the color shifts from blue to red [52].

(d) Conductometric titration: The measurement of the conductance of a solution is called conductimetry. After dissolving chitosan samples in HCl, it is titrated with portions of NaOH in 20-s. interval. To identify the linear fluctuation before and after the equivalence point, a graph is plotted for the conductance values with the matching titrant volumes [53].

(iii) Hydrolytic methods

(a)   Acid hydrolysis, distillation, and titration

To convert chitosan into acetic acid, a known amount of the salt is first hydrolyzed with sodium hydroxide and then acidified with phosphoric acid. When the distillation flask begins to dry after the aqueous acetic acid distillation, 15 mL of hot distilled water is placed into the flask. Aliquots are titrated with 0.01 N sodium hydroxide, with phenolphthalein as an indicator. The total volume of the distillate is calculated by multiplying the volume of the base by 10 [47].

(b)   Acid hydrolysis-HPLC

A Varian Model is used for the HPLC analyses. At 155 °C for 1 h, a fine powder of the polymer is hydrolyzed with aqueous sulfuric, propionic, and oxalic acids. After cooling for 2 h, the solution is filtered, and 10 l is poured into the HPLC. A UV detector set at 210 nm is then used to identify carboxylic acid. The calibration curve is created using various acetic acid dilutions [54].

(iv)   Spectrometric methods

(a)   HPLC-ultraviolet spectroscopy: The DA of unidentified polymers is determined by comparing them to described chitosan samples used as standards. The hydrolysis of sample C-1 yields the chitosan standards (C-2, C-3, C-4, C-5, and C-6). The DA could be accurately computed because the acetamide groups' UV absorption peak areas were related to the concentration of the chitosan samples.

(b)   Infrared spectroscopy: Structure determination and quantitative analysis can both benefit from infrared absorption spectroscopy. The idea that few groups of atoms emit bands at or near the same frequency, as well as the distinctive IR fingerprint of molecules, allows scientists to analyze the structure of a complex using other techniques [55].

(c)   Liquid-state ^{1}H nuclear magnetic resonance (NMR): Chitosan is dissolved in DCl and swirled at room temperature for 24 h. The 1H NMR spectra were acquired with 16 transients, a 3.642 s acquisition time, and a 1.500 s delay. To improve the solubility of chitosan, the temperature was kept at 70 °C [53].

(d)   Solid-state ^{13}C nuclear magnetic resonance: The majority of structural information is hidden behind a large hump in a solid-state NMR spectrum. It is not always possible to find a suitable sample dissolution for NMR analysis. The degree of deacetylation of chitosan samples is determined by comparing the overall area of the CH_3 peak to the area of the glucoside carbons in solid-state ^{13}C NMR [56].

(e)   Solid-state ^{15}N nuclear magnetic resonance: Due to the difficulties in producing high-resolution solid-state 1H NMR spectra, ^{13}C, ^{31}P, and ^{15}N are the nuclei most commonly examined by solid-state NMR. Low spin polarization, low isotope abundances, and low signal strength are some of the drawbacks of properly sensing ^{13}C and ^{15}N [57].

6 Need for Nanotechnology

Chitosan in the form of pure matrix material may display poor mechanical and thermal properties [6]. Due to this reason, chitosan as a nanocomposite base matrix seems like a rather preferable option because of its high availability in nature, low cytotoxicity, low cost, and high versatility, as well as biocompatibility. The amino and hydroxyl groups present in its glycosidic residue make it a preferable carrier matrix for the synthesis of chitosan nanocomposites. In addition to this, the solubility of chitosan in an acidic medium makes it possible to prepare composites without the use of hazardous organic solvents. Chitosan can behave as a structure-directing agent, and its pH responsiveness and ability to chelate with metals prove to be an added advantage [58].

7 Chitosan Nanocomposites: Structural Analysis

7.1 Structure of Chitosan Nanocomposites: Nanofillers and Nanocoatings

Chitosan nanocomposites are composed of two major components: nanostructures and chitosan. The coupling of chitosan with different nanostructures may be completed in the following ways (Fig. 8):

(i) By embedding the nanostructures into the bulk of the material as a filler (nanofillers)
(ii) By depositing on the surface of chitosan as a coating (nanocoatings or nanofilms).

7.1.1 Nanofillers

This kind of nanocomposite structure has the nanostructures dispersed inside the chitosan matrix. The steps involved in its production are as follows [1]:

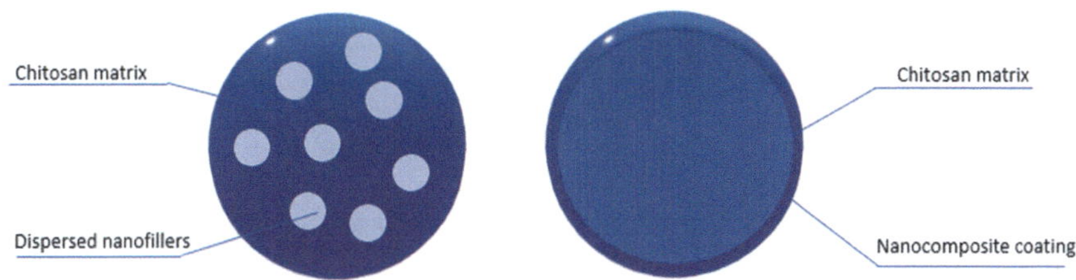

Fig. 8 Schematic showing **a** nanofillers and **b** nanocoatings

(a)   Blending of nanofillers in a solution containing chitosan
(b)   Casting the slurry
(c)   Drying at room temperature.

For the development of nanocomposites consisting of nanofillers embedded inside chitosan matrix, the following essential key parameters are extremely important:

(a)   Surface to volume ratio: When nanofillers are dispersed into chitosan, the area in contact with the matrix is a direct result of the surface area of the fillers, implying that a larger total surface area will result in a larger filler effect on chitosan. A large surface-to-volume ratio of the nanoparticles is essential in order to achieve a strong and extensive interfacial binding between the nanofiller and chitosan base and to pass on the characteristics of NP into the nanocomposite.

(b)   Aspect ratio: The aspect ratio is nothing but the ratio of the longest to the shortest dimension of a nanofiller. Nanofillers can have numerous possible shapes (Fig. 9) and may be 1-dimensional (one of the dimensions being less than 100 nm), 2-dimensional (having two dimensions less than 100 nm), or three-dimensional (having three dimensions) (Fig. 10). Measurement for all the possible filler shapes may be reduced to their aspect ratio, which acts as a

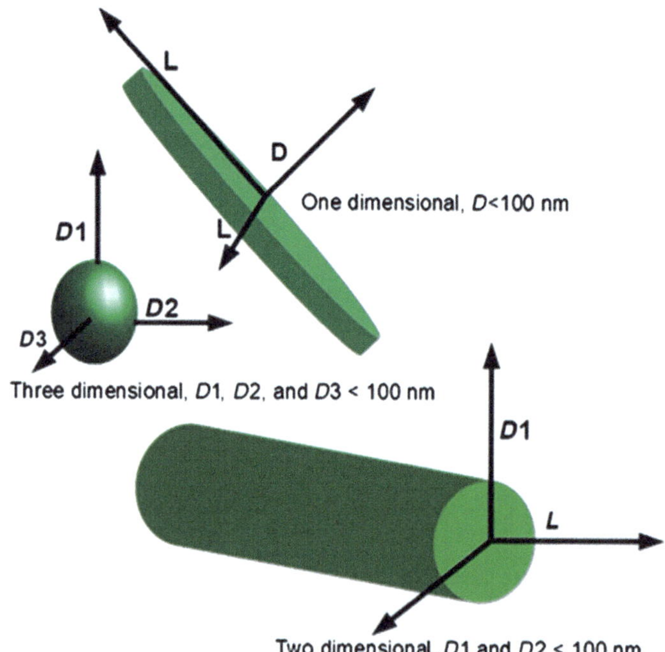

Fig. 9 Classification of nanofillers according to their dimensions: one-dimensional, two-dimensional, and three-dimensional. Reprinted with permission from [59]. Copyright 2019 Elsevier

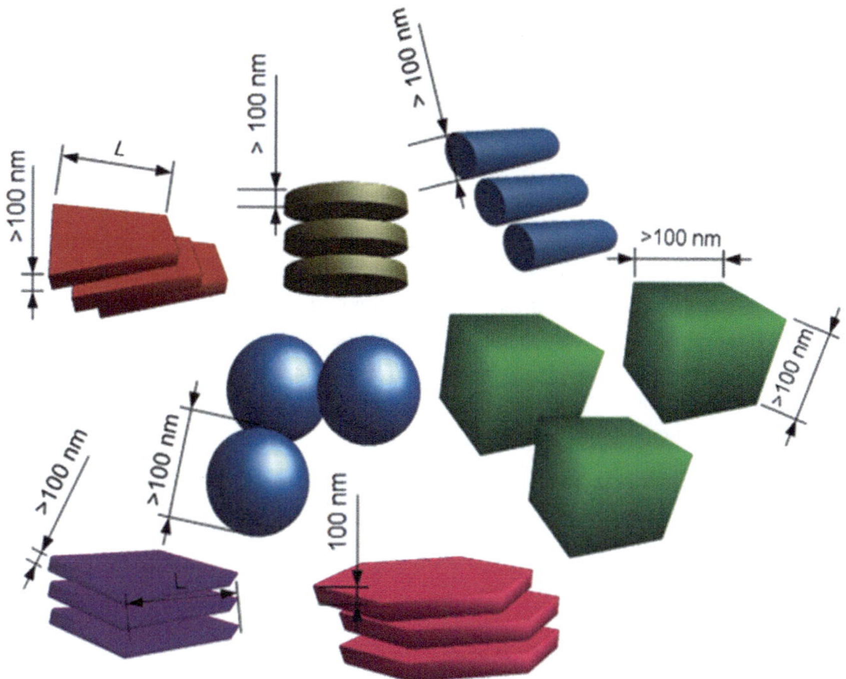

Fig. 10 Various shapes of nanofillers. Reprinted with permission from [59]. Copyright 2019 Elsevier

major factor in the determination of several physical and mechanical properties of chitosan nanocomposites.

(c)   Interface: Mechanical, electrical, and thermal performances of chitosan nanocomposites depend upon the interaction between the filler and chitosan matrix. Strong bonding between the matrix and filler allows an efficient and controlled stress transfer across the interface. Factors like modulus as well as strength of the NCs are strongly dependent on the efficacy of the stress transfer. Nanofillers, after dispersion in the polymer matrix, create an "interaction zone," which leads to certain alterations in the behavior of the polymer matrix and the morphology. Thus, this implies that the interaction between filler and chitosan shall determine the area of the interaction which in turn will determine the magnitude of other effects generated by the presence of nanofillers.

(d)   Orientation: Controlling the orientation of nanofillers in the chitosan matrix can prompt certain nanoeffects. For instance, the effective orientation of carbon nanotubes (CNTs) in the matrix of polymer can either annihilate or trigger a photoelastic response or van der Waals.

(e)   Nanofiller distribution: A good and uniform nanofiller distribution within the chitosan matrix is equally important in order to overcome the aggregation of nanoparticles and enhance the mechanical, electrical, and optical properties of the nanocomposites.

Commonly used nanofillers in the development of chitosan nanocomposites are as follows [60]:

(a) Layered silicates-clays

Polymer-clay nanocomposites are hydrous silicates having a layer-structured nanofiller containing silicon (Si), aluminum (Al), magnesium (Mg), oxygen, and hydroxyl groups with numerous associated cations. In order to build the layers, octahedrally bonded Al or Mg atoms surrounded by eight oxygen atoms or tetrahedrally bonded Si atoms surrounded by four oxygen atoms may be utilized. Layered silicate nanostructures can be homogeneously dispersed into the chitosan matrix, providing three different types of conformations (Fig. 11).

(i)   tactoid structures: The type of structure in which intercalation of chitosan polymer does not take place as there is no expansion of the interlayer space of clay

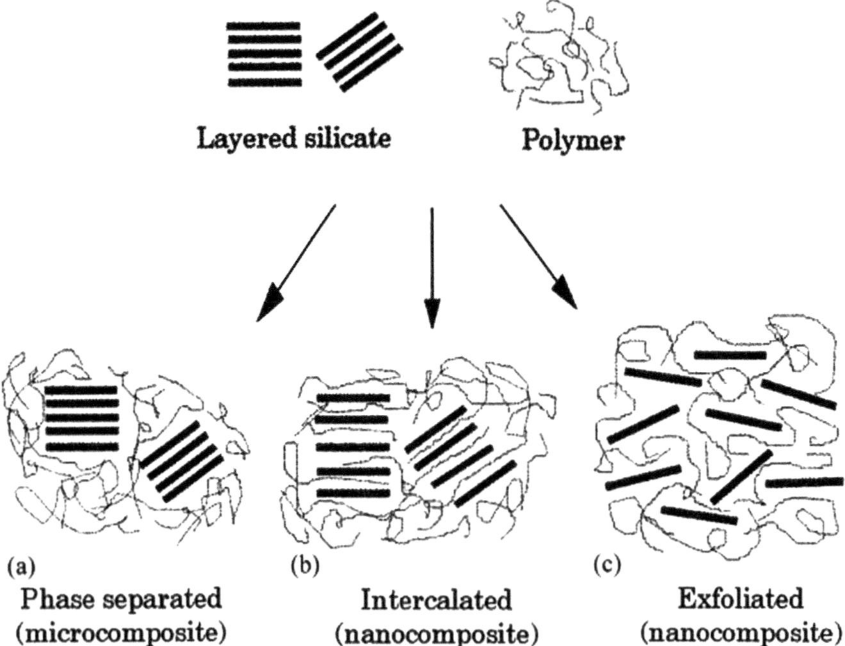

Fig. 11 Schematic depicting the different composites formed by the interaction of polymers and the layered silicates: tactoid, intercalated, and exfoliated nanocomposite. Reprinted with permission from [61]. Copyright 2000 Elsevier

(ii)   intercalated structures: The type of structure in which the penetration of chitosan polymer is possible as the interlayer distance increases between the layers of clay, while also maintaining the layered structure

(iii) exfoliated structures: The type of structure in which dispersion of clay and mixing with the polymer phase takes place as the clay layers are well separated from each other.

(b) **Ceramic/metallic nanoparticles**

The three main groups of particles are as follows:

I.   Bioactive glass: Bioactive glass, produced by melt or sol–gel methods, is made of silicates composed of sodium, phosphorus, or cadmium. They present a high specific surface area and act as a good substrate for the adsorption of proteins.

II.   Ceramic nanoparticles: Ceramic NPs form NCs with improved mechanical properties and enhanced interaction with neighboring tissues.

III. Metal nanoparticles: Several elements, viz. silver (Ag), gold (Au), zinc oxide (ZnO), etc., have been reportedly used for the generation of metal NPs, that can be combined with chitosan to provide biosensing and antibacterial properties. This portion is discussed in detail in subsequent sections.

(c) **Carbon nanotubes (CNTs)**

Carbon nanotubes, made up of hexagonal sp^2 hybridized carbon rolled up into a tube, are of two types: (i) single-walled CNTs, consisting of a single graphene sheet (with a diameter of 1 nm), and (ii) multi-walled CNTs, composed of multiple concentric graphene cylinders held by van der Waals forces (with a diameter ranging from 5 to 20 nm). Excellent properties of CNTs such as high elastic modulus, high thermal conductivity, and good tensile strength make them highly preferable for nanocomposite applications. However, for biomedical applications, it is extremely important that the metallic impurities are eliminated. Functionalization of CNTs with different biomolecules, polymers, surfactants, etc., may be performed to make them less cytotoxic and apply them to different biomedical applications [62] (Fig. 12). Furthermore, due to the strong electrostatic attraction, CNTs show very poor interfacial interaction and are not easily dispersed in the chitosan matrix [63].

(d) **Graphene-based materials**

Over the years, graphene has become increasingly popular due to its excellent properties like high elastic modulus, intrinsic electrical as well as thermal conductivity, and the possibility to attain thermally and electrically conductive graphene NCs. However, due to its surface chemical inertia, its interfacial bonding with the polymer matrix still poses problems [64, 65]. Graphene oxide (GO) can strongly bind with many polymers to give nacre-like structures, because of the oxygen-containing functional groups present on its surface. One major drawback of loss of electrical conductivity that results from oxidation of graphene may be overcome by partial restoration of the lost conductivity through chemical reduction of GO.

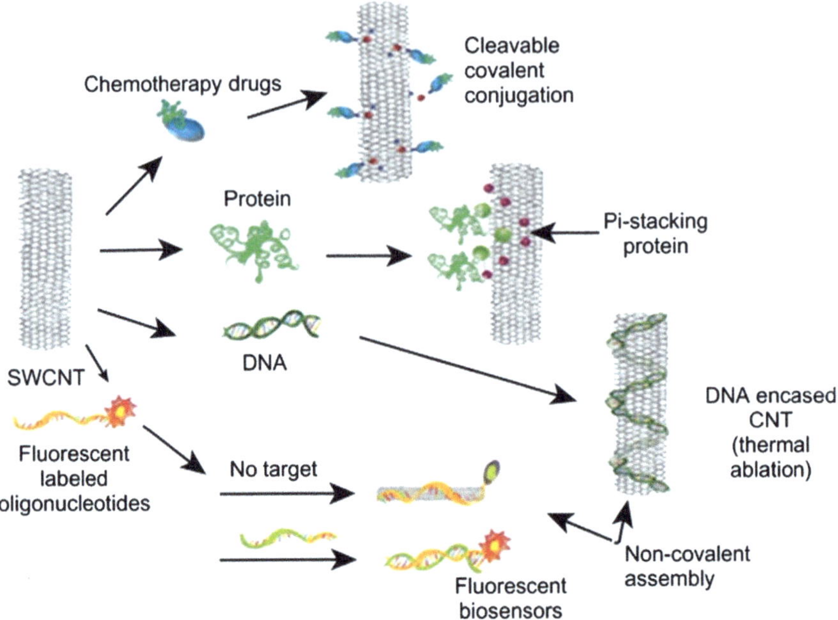

Fig. 12 Functionalization of carbon nanotubes and their biomedical applications. Reprinted from [62] under the Creative Commons Attribution License (CC by 4.0)

7.1.2 Nanocoatings

Nanocoating refers to the kind of NC structure having the nanostructures only at the surface of the chitosan base matrix. Layer-by-layer approach is used wherein positively or negatively charged nanostructures in solution are electrostatically bound or assembled with chitosan. Kumar et al. [66] exploited the high affinity of amine and hydroxyl groups available on the chitosan structure toward Ag^+ ions and introduced Ag nanoparticles in the chitosan matrix. Chemical covalent modification of the chitosan surface may be performed by introducing the functional groups to facilitate its reaction with the nanostructures that need to be attached. The amine and hydroxyl groups present in the chitosan framework encourage this attachment by enhancing covalent or protonation reactions.

7.2 Functionalization of Nanofillers in the Synthesis of Chitosan Nanocomposites

Functionalization of nanofillers is usually accomplished to promote good interfacial adhesion between the chitosan matrix and the nanofiller. It can be of two types: covalent (chemical) or noncovalent (physical) [59].

(i)   Covalent (chemical) functionalization: It involves linking functional groups to the filler surface by either direct functionalization to the sidewalls of fillers through bonding, or by indirect functionalization, i.e., by attaching functional groups to the defects on the filler surface. Deliberate generation of defects may be achieved by carrying out oxidative processes with oxidants. However, the formation of a large number of defects or the breaking up of fillers into small pieces may prove to be major drawbacks of covalent functionalization.

(ii)   Noncovalent (physical) functionalization: It entails the functionalization of filler surface by the use of active species like polymers and surfactants as assembly mediators, through hydrophobic, electrostatic, pi-pi stacking, and polymer wrapping interactions.

8 Synthesis Scheme for Chitosan Nanocomposites

8.1 Synthesis Strategies to Produce Chitosan Nanocomposites

First characterized in 1994 by Ohya et al. [8], chitosan nanocomposites have been studied extensively over the past years with researchers coming up with several synthesis methods, taking into consideration numerous factors like size, stability, retention time, and drug loading capacity [11]. Some of the commonly deployed manufacturing methodologies are as follows [1, 60, 67]:

(i)   Solvent casting:

Used for the synthesis of chitosan nanocomposite films and membranes, the solvent casting method requires the polymer to be dissolved in a solvent and cast onto a surface, followed by evaporation of the solvent. The membranes/films are then detached from the cast form. This cost-effective manufacturing technique that does not require high temperatures makes it possible to incorporate drugs or chemicals within the structures. However, one major limitation is the retention of solvents as residue. Furthermore, this method cannot be used for manufacturing complex shapes. The low interconnectivity present in the structures is unfavorable for tissue engineering. Figure 13 is a schematic showing the solvent casting method.

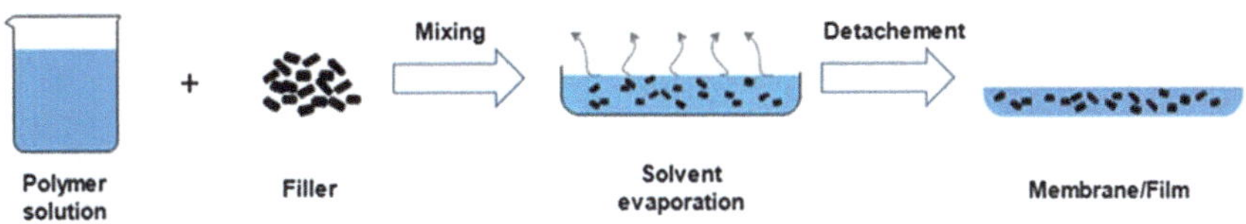

Fig. 13 Schematic showing the solvent casting methods for obtaining chitosan nanocomposite films. Reprinted from [60] under the Creative Commons CC by license

(ii)   Freeze-drying:

This method, used for the synthesis of highly porous scaffolds, requires the solution temperature to be reduced until solid–liquid phase separation takes place, leading to the formation of the polymer phase and frozen solvent. The frozen solvent then disperses through sublimation, via pores formed in the polymeric structure. Anisha et al. [68] used this method to make sponges consisting of chitosan, silver nanoparticles, and hyaluronic acid, showing antimicrobial activity, which could be used as a wound dressing for drug-resistant bacteria. Another group, Mohandes and Salavati-Niasari [69], produced a composite of chitosan, graphene oxide, and hydroxyapatite NPs using the freeze-drying methodology and studied its bioactivity.

(iii)   Layer-by-layer deposition (LbL)

This type of assembly, proposed in 1996 by Iler [70], is based on the adsorption of different components, attracted to each other by van der Waals, H-bonding, electrostatic interactions, etc. Using this method, highly ordered polymeric films and NPs can be fabricated over different kinds of substrates. One major advantage offered by this methodology is the possibility to fabricate multi-layered devices of any nature, composition, shape, and size; thus, nanostructures with desired functionalities can be developed. Further tuning of the properties of multi-layered devices may be accomplished through solution pH, ionic strength, and temperature. This technology has been put to use by various research groups over the years to produce scaffolds of different compositions and types. Different LbL approaches have been used to make a multilayer film including dip coating, spin coating, and spray coating (Fig. 14).

(iv)   Electrospinning

This type of manufacturing process uses an electric field created between the collector and polymer solution, thus producing internal repulsion in the solution, thereby leading to the expulsion of the polymer solution in the shape of fibers toward the collector at a critical point. The three known types are (i) wet-dry electrospinning, which uses a volatile solvent that evaporates when fibers are spun through the collector, (ii) wet-wet method, wherein a non-volatile solvent is spun to a collector with a second solvent, and (iii) co-axial electrospinning, that spins two different components at the same time, thus leading to the possibility of obtaining fibers with a core-sheath structure (Fig. 15).

   This is a low-cost method and is widely popular for the fabrication of NCs for wound dressing applications, implants, and scaffolds.

(v)   Three-dimensional (3D) printing

3-dimensional (3D) printing (or additive manufacturing) is based on the use of a 3-dimensional model to produce a physical object through a compilation of techniques comprising biological and non-biological approaches. The main advantages of this methodology are its cost-effectiveness and control over the geometries of synthesized biological structures. There is no requirement for organic solvents or high processing

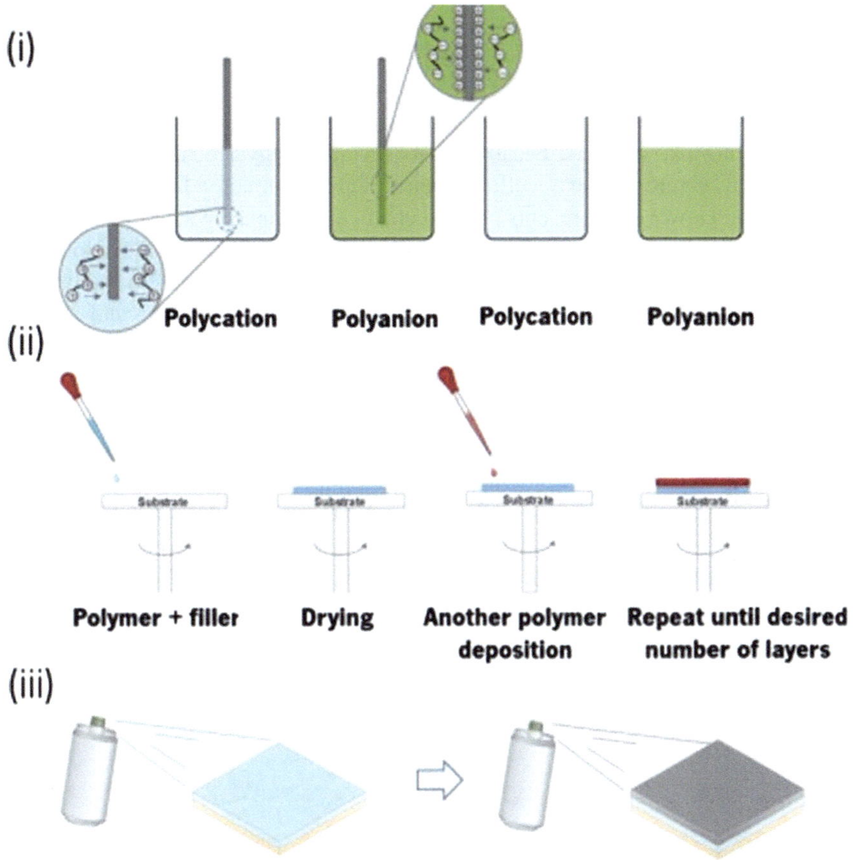

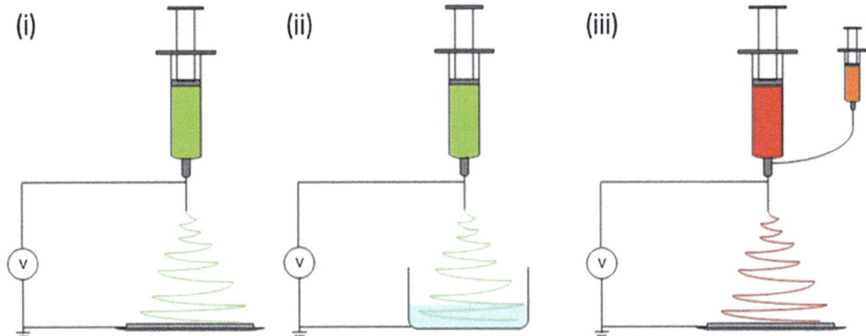

Fig. 14 Schematic showing the three layer-by-layer deposition approaches: (i) dry coating, (ii) spin coating, and (iii) spray coating. Reprinted from [60] under the Creative Commons CC by license

Fig. 15 Schematic showing the different electrospinning approaches: **i** wet-dry, **ii** wet-wet, and **iii** co-axial spinning. Reprinted from [60] under the Creative Commons CC by license

temperatures or difficulties associated with dust removal. Bio-inks are common raw materials that are employed in 3-dimensional printers for biological applications. They are selected keeping in mind various parameters like printability, viscoelasticity, shelf life, fidelity, yield stress, shear-thinning, cost, etc. Chitosan proves to be a good bio-ink candidate because of its favorable physicochemical properties and essential features for extracellular matrix deposition, cell adhesion, and tissue regeneration. However, it presents poor mechanical resistance which turns out to be a major limitation. Blending or coating nanomaterials helps overcome this drawback [71]. Sommer et al. [72] developed a bio-ink containing chitosan with modified Si NPs, that present high yield stress, elastic recovery, and storage modulus and are ideal for 3D printing. In recent years, there has been an increasing interest in 3D printed chitosan nanocomposites for potential biomedical applications [73] (Fig. 16).

(vi)   Reverse micellar method

This method requires the deployment of a lipophilic surfactant (usually sodium bis-(2-ethylhexyl)-sulfosuccinate or cetyltrimethylammonium bromide (CTAB)) for solubilization in an organic solvent (preferably, *n*-hexane) to develop an organic phase. Chitosan solution, glutaraldehyde, and the drug are then added to it with constant stirring, to prevent turbidity. The last step involves the extraction of chitosan nanoparticles (Fig. 17).

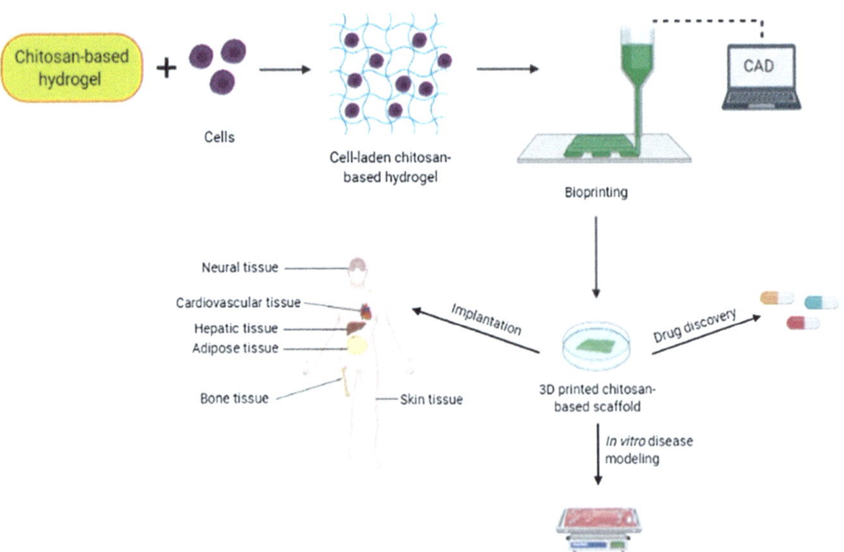

Fig. 16 Representation of chitosan hydrogels in 3D printing for biomedical applications. Reprinted with permission from [74]. Copyright 2021 Elsevier

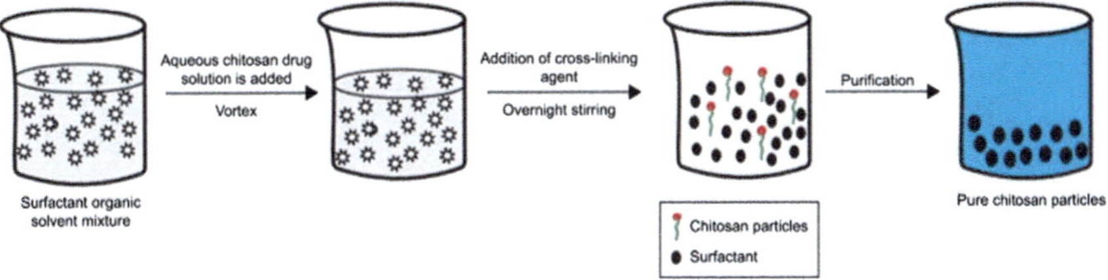

Fig. 17 Schematic representing the reverse micellar method. Reprinted from [75] under the Creative Commons Attribution—Non-Commercial 3.0 unported (CC BY-NC 3.0)

(vii) Ionic gelation

It involves the addition of an anionic form of tripolyphosphate (TPP) solution to the cation of chitosan solution, formed by the addition of chitosan in an aqueous acidic solution. Electrostatic forces between the anionic TPP and chitosan cation result in a complex, that leads to ionic gelation of chitosan. Following this, spherical nanoparticles are created after precipitation (Fig. 18).

(viii) Ionic gelation with radical polymerization

It entails the addition of an aqueous solution of acid monomer to chitosan solution at room temperature. Interaction between the anionic acrylic monomer and chitosan cation results in the ionic gelation of chitosan, and radial polymerization between acrylic acid monomer and potassium persulfate is initiated under stream nitrogen at

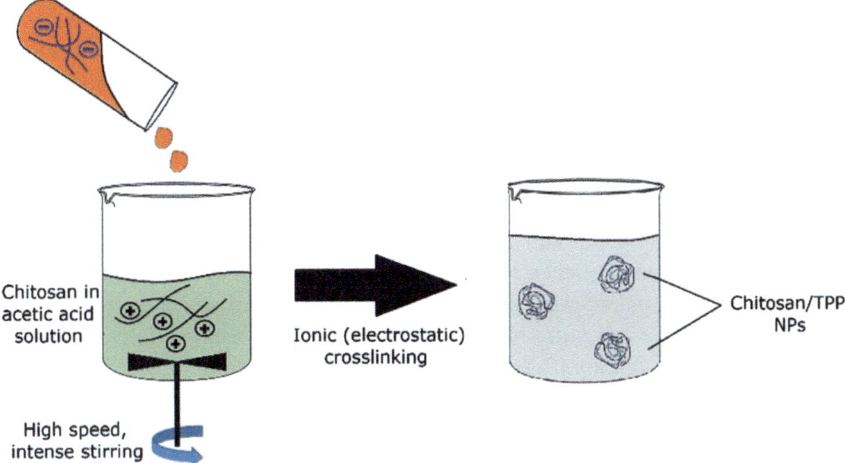

Fig. 18 Schematic illustration of the ionic gelation method. Reprinted from [11] under the Creative Commons license (CC by 4.0)

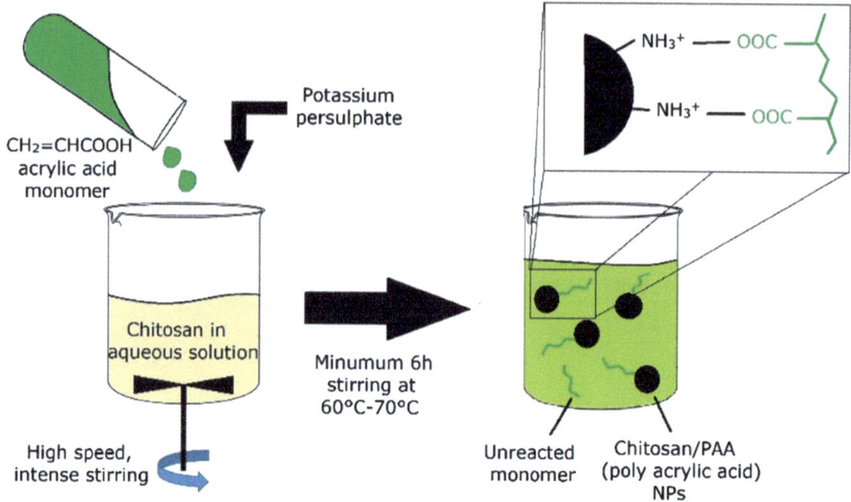

Fig. 19 Schematic illustration of the ionic gelation method with radical polymerization. Reprinted from [11] under the Creative Commons license (CC by 4.0)

temperatures of 60–70 °C. Following this, suspended nanoparticles are allowed to settle overnight, while the unreactive monomer is removed by dialysis (Fig. 19).

(ix)   Emulsification

(a)   Emulsion droplet coalescence

The water-in-oil emulsion is developed by the addition of chitosan solution, together with the drug, to the liquid paraffin-containing stabilizer (such as span 83), under high-speed homogenization. Another emulsion is made in a similar way but with sodium hydroxide. As the two emulsions are mixed, droplets of both emulsions randomly collide and coalesce, precipitating chitosan droplets that form nanoparticles (Fig. 20).

(b)   Emulsion solvent diffusion

In this method, the water-in-oil emulsion is developed by the addition of an organic phase (such as methylene chloride/acetone) containing a hydrophilic drug to a chitosan solution containing a stabilizer (such as poloxamer/lecithin), under continuous stirring. The organic phase is made to evaporate under high-pressure homogenization. Precipitation of chitosan leads to the formation of nanoparticles, which are separated by centrifugation (Fig. 21).

(c)   Emulsification cross-linking

The water-in-oil emulsion is made by adding chitosan solution in an oil phase followed by the stabilization of aqueous droplets by the use of a surfactant (like span 80). Glutaraldehyde is added to facilitate the cross-linking of chitosan and glutaraldehyde, thus leading to the formation of chitosan NPs (Fig. 22).

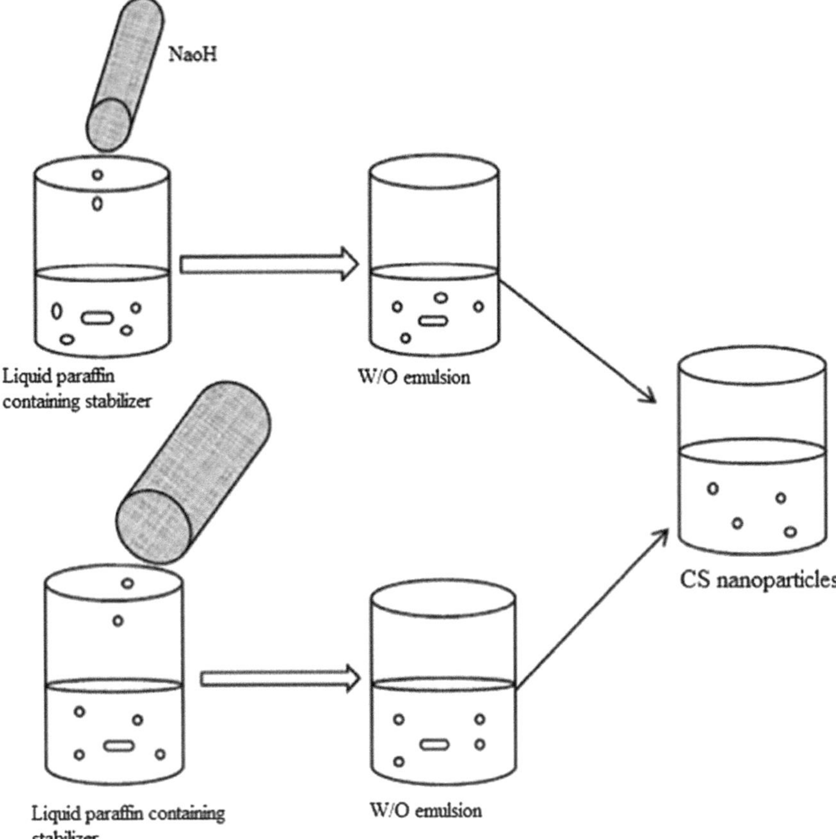

Fig. 20 Schematic representing emulsion droplet coalescence method of synthesis. Reprinted with permission from [67]. Copyright 2018 Elsevier

(x)   Desolvation

Precipitating agent (such as sodium sulfate) is added to an aqueous chitosan solution containing a stabilizer (like Tween 80). Because of the salty surroundings of the aqueous chitosan solution, elimination of salvation water is observed. Precipitation of chitosan is followed by the addition of glutaraldehyde to harden the formed nanoparticles.

(xi)   Nanoprecipitation

A diffusing phase, formed by dissolving chitosan in an appropriate solvent, is added to a dispersing phase like methanol under magnetic stirring by means of a needle present above the surface with a peristaltic pump. The stabilizer is then added to the dispersing phase, leading to the formation of nanoparticles.

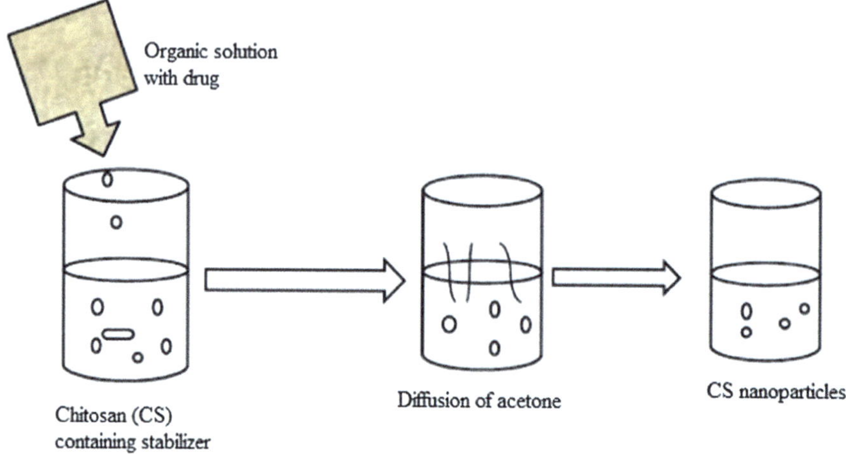

Fig. 21 Schematic representing emulsion solvent diffusion method of synthesis. Reprinted with permission from [67]. Copyright 2018 Elsevier

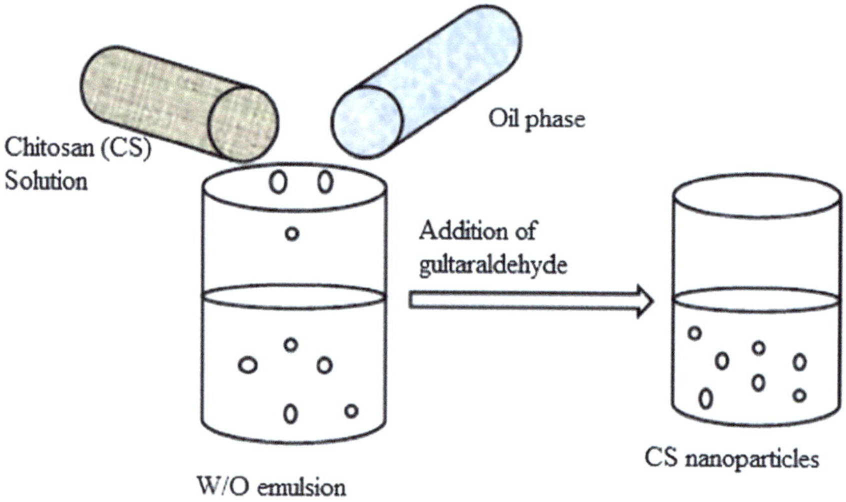

Fig. 22 Schematic representing emulsification cross-linking method of synthesis. Reprinted with permission from [67]. Copyright 2018 Elsevier

(xii)   Spray drying

Chitosan solution, prepared by dissolving chitosan directly in water with glacial acetic acid, is stored overnight. Small droplets are made through atomization of the chitosan solution, which is then mixed with a drying gas to evaporate the liquid, thus forming the nanoparticles (Fig. 23).

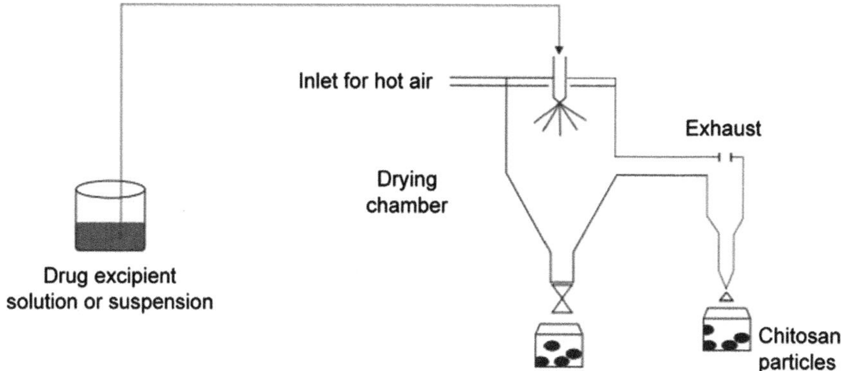

Fig. 23 Schematic representing the spray drying method. Reprinted from [75] under the Creative Commons Attribution—Non-Commercial 3.0 unported (CC BY-NC 3.0)

8.2 Cross-Linking of Chitosan Nanocomposites

Cross-linked chitosan NCs have been of great therapeutic use for controlled drug release. Cross-linking of chitosan is achieved to control swelling, rates of degradation, and drug release profile. Chitosan can be cross-linked in two ways:

(i) Covalent cross-linking: Covalent cross-linking of chitosan is the preferred type because it provides stability under acidic conditions, unlike ionic cross-linking. Herein, covalent bonds form between the cross-linking agent and the polysaccharide chains. Common covalent cross-linkers/cross-linking agents are glutaraldehyde and epichlorohydrin, viz. compounds having at least two functional groups, which can undergo condensation reactions.

(ii) Ionic cross-linking: In this type, anionic cross-linkers with multivalent ions, like TPP, cross-link with chitosan via electrostatic interactions.

Due to the presence of amine groups in the chitosan framework, it can be easily cross-linked with numerous cross-linking agents.

9 Chitosan Metal Nanocomposites

A specific type of chitosan nanocomposite, chitosan metal nanocomposites, is increasingly being used in numerous biomedical applications which are specifically discussed in terms of their characteristics. Several metals like copper (Cu), zinc (Zn), silver (Ag), iron (Fe), gold (Au), etc., can be integrated into several forms of nanoparticles, to synthesize chitosan metal NCs with improved physicochemical and biological characteristics. The strong metal affinity of chitosan is a result of the presence of free amine groups in the polymeric framework of chitosan. In the case of metal cations, the adsorption occurs through a chelation mechanism by the amine groups in a neutral medium, while the adsorption of metal anions occurs in acidic solutions via electrostatic attraction forces between the protonated amine groups of chitosan. Chitosan metallic nanoparticles can be synthesized in two steps [76]:

(i)   The first step is the reduction which requires the use of a reducing agent
(ii)   Secondly, the nanoparticles formed are stabilized using a stabilizing agent.

Toxicity levels of the formed nanoparticles depend upon the properties of the stabilizing agent; hence, it is very essential that the stabilizing agent used is non-toxic in nature. Chitosan is less toxic in nature and highly permeable across the cell membrane and thus proves to be a highly preferable candidate to be used as a stabilizing agent. When used as a stabilizer, it enhances the stability of nanoparticles by displaying improved surface absorption, specific recognition, and electrostatic interactions.

Chitosan metal NCs show a wide spectrum of activity against fungi and Gram-positive and Gram-negative bacteria. However, at the same time, there are also certain limitations with respect to their use. For instance, Cu NPs oxidize rapidly upon air exposure. Chitosan when used as a stabilizer helps overcome this drawback. Also, the chitosan coating of such NPs facilitates the improvement of their antimicrobial activity. Some of the commonly known chitosan metallic nanocomposites are discussed below [75, 77]:

9.1 Chitosan-Gold Nanocomposites

Gold (Au) nanoparticles are known to be useful in numerous biomedical fields, like drug delivery and diagnosis. However, Au nanoparticles produced by conventional methodologies (Turkevich's method) employ the use of toxic reducing agents to reduce Au(III) and facilitate stabilization of Au(0) on the nanoparticle surfaces. In this case, the replacement of these toxic reducing agents with non-toxic chitosan can greatly enhance the biocompatibility of the ensued NCs. Chitosan is used as the reducing agent in the reduction step of the synthesis of chitosan-gold NC. In addition to this, chitosan also enhances the penetration and uptake of therapeutic agents across the mucous membrane. Bhumkar et al. [78] synthesized chitosan-Au NPs loaded with insulin that did not aggregate for around six months and successfully reduced glucose levels in the blood of diabetic rats. Another group, Salehizadeh et al. [79] prepared Fe_3O_4 gold chitosan nanocomposites and established their usefulness in various biomedical applications.

9.2 Chitosan-Nickel Nanocomposites

Owing to the highly exceptional properties of nickel oxide (NiO) like electrocatalytic activity, high isoelectric point, and the possibility of altering surface properties in accordance with the size, it has been highly preferred for the immobilization of different enzymes and other biomolecules. The major drawback of NiO nanoparticles getting agglomerated during their use is overcome with the help of chitosan which is employed as a stabilizer. Solanki et al. [80] synthesized chitosan-nickel NCs for biosensing applications. The preparation method involved the addition of pre-formed nickel oxide nanoparticles in a 0.5% solution of chitosan, made in acetate buffer. On subjected to mechanical stirring at room temperature, the desired nanoparticles were obtained in a viscous chitosan solution. The NCs were formed through electrostatic interactions between the nickel oxide nanoparticles and the hydroxyl group of chitosan. The formed NCs were further used to fabricate a highly sensitive and selective immunesensor, and it was concluded that chitosan played a significant role in the dispersion of the nickel oxide NPs and enhanced the loading capacity of antibody molecules.

9.3   Chitosan-Iron Nanocomposites

Fe_2O_3 nanoparticles display numerous favorable properties like biocompatibility, non-toxic nature, high solubility in water, cost-effectiveness, crystallinity, super-paramagnetic behavior, etc. It acts as a strong antimicrobial agent due to its highly crystalline structure that renders them with many corners and edges that can act as potentially reactive sites. These nanoparticles are coated with chitosan in order to alter the functional groups on the NPs and change the surface potential. Chitosan can strongly bind with Fe_2O_3 NPs through H-bonding because of the availability of hydroxyl groups in the chitosan framework. It has even been proved that chitosan-coated iron oxide NPs act as a better antimicrobial agent when compared with bare iron oxide NPs. The inhibition of the growth of microbes by chitosan is a direct result of the chelation between its positively charged amino group and the negatively charged components of the microbial cell membrane. Furthermore, the coating of chitosan protects the iron oxide nanoparticles from getting aggregated in an aqueous medium, consequently resulting in their enhanced antimicrobial activity. Unsoy et al. [81] synthesized chitosan-coated Fe_2O_3 nanocomposites by the cross-linking method. Cationic chitosan was adsorbed on the surface of anionic Fe_3O_4 nanoparticles via electrostatic interactions, with TPP used as an ionic cross-linker. The prepared NCs were found to be non-cytotoxic on cancer cells, and their potential applications in biomedicine were reported. Figure 24 is a schematic depicting the structure of chitosan-iron nanocomposites.

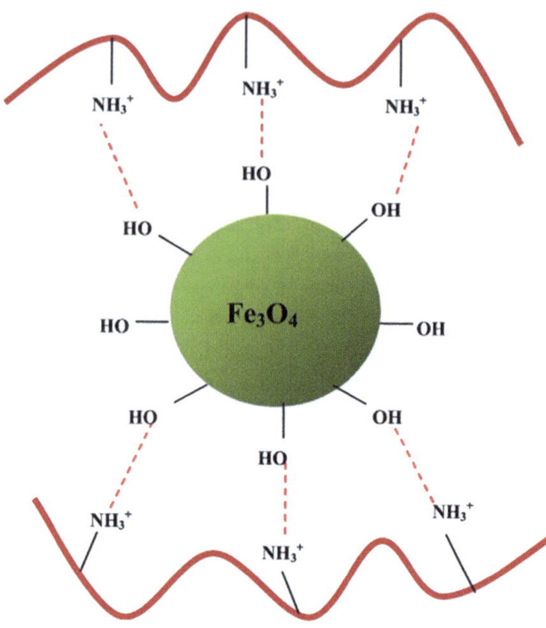

Fig. 24 Representation of the structure of chitosan-iron nanocomposite. Reprinted with permission from [77]. Copyright 2021 Elsevier

9.4 Chitosan-Copper Nanocomposites

Copper nanoparticles have been receiving major attention because of their low cost in contrast to Ag and Au nanoparticles. However, one major limitation is the oxidation of copper to copper (I/II) oxide during their preparation. To provide stability to the copper nanoparticles, chitosan is used as a stabilizer. Due to its ability to chelate metals, it ensures that the nanoparticles formed are of appropriate shape and size. Qi et al. [82] prepared copper-loaded chitosan NCs by ionic gelation between TPP and chitosan that displayed a significant growth inhibition of a vast range of microorganisms. Figure 25 is a schematic showing the structure of chitosan-copper nanocomposite.

9.5 Chitosan-Silver Nanocomposites

The bactericidal effects of silver (Ag) have been realized since ancient times. Moreover, Ag nanoparticles are known for their antibacterial, antifungal, and anti-inflammatory action. Silver nanoparticles are believed to alter the structure and permeability of the cell membrane of bacteria by binding the positively charged silver with the negatively charged bacterial cell wall. The incorporation of chitosan as a stabilizer or reducing agent further provides stability (steric/electrostatic) to the NPs, through an interplay between the amino group of chitosan and Ag^+. Sanpui et al. [83]

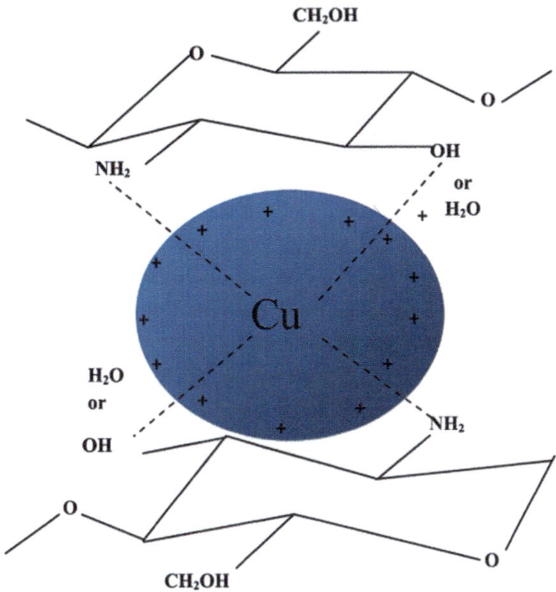

Fig. 25 Representation of the structure of chitosan-copper nanocomposite. Reprinted with permission from [77]. Copyright 2021 Elsevier

followed a chemical reduction method to synthesize Ag NPs coated with chitosan that displayed excellent antimicrobial activity, biodegradability, and prolonged action of Ag on the affected cells and thus were concluded to have potential applications in biosensing and cancer therapy. Figure 26 is a schematic showing the structure of chitosan-silver nanocomposites.

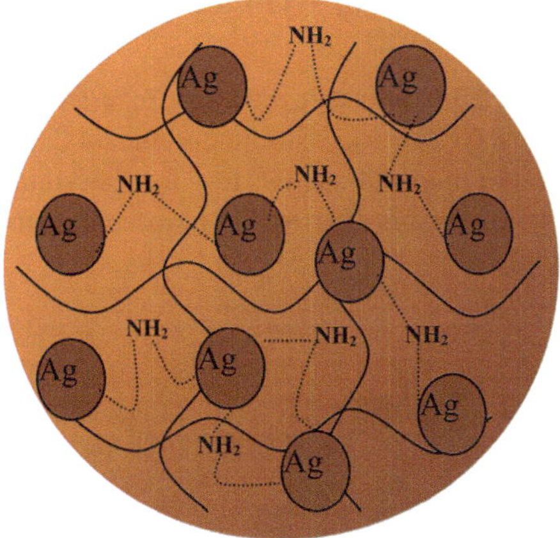

Fig. 26 Representation of the structure of chitosan-silver nanocomposite. Reprinted with permission from [77]. Copyright 2021 Elsevier

10 Properties of Chitosan Nanocomposites

10.1 Loading and Release of Drugs

Generally, two methods for putting medicines into chitosan nanoparticles comprise absorption and integration. On the other hand, diffusion, swelling, and erosion are the three methods for drug release from chitosan nanoparticles. In the context of protein, meanwhile, the drug-releasing actions are diffusion, desorption, and release. Dissolution rate, diffusion, and nanoparticle size are all important parameters that influence the drug release rate from chitosan nanocomposites.

10.2 Particle Size and Zeta Potential

The zeta potential of nanoparticles can be determined using electrical potential and the movement of charged particles. The value of zeta potential can be positive, negative, or neutral depending on how the polymer's surface is modified. The average particle size can be determined using photon correlation spectroscopy, which is focused on light scattering generated by Brownian motion of the particles. Scanning or transmission electron microscopy or atomic force microscopy can be used to determine particle size and shape. The dimension of chitosan nanoparticles has thus been reported to be between 100 and 400 nm.

10.3 Study of Cytotoxicity and Cellular Uptake

Cell viability must be measured in order to study the cytotoxicity of chitosan nanoparticles. MTT (3-(4, 5-dimethylthiazol-2-yl)-2, 5-diphenyltetrazolium bromide) is a simple, non-radioactive test for determining cell viability. Confocal microscopic examination is also used to ensure efficient cellular uptake of chitosan nanoparticles (CMA).

10.4 Stability

The therapeutic efficacy of pharmaceutical products depends upon a very important factor which we refer to as the stability of nanoparticles. Agglomerations of particles, coagulation, and bridging flocculation affect the physical stability of nanoparticles. The chemical stability of nanoparticles, on the other hand, is impacted by elements such as formulation composition, temperature, medium pH, molecular weight, and polymer type.

10.5 Antimicrobial Properties of Chitosan Nanocomposites

Chitosan's antibacterial capabilities are due to its positively charged characteristics. Studies on the antibacterial characteristics of chitosan nanocomposites are being conducted in a variety of fields, including water management and treatment systems. The killing mechanism of chitosan is the disruption of germs' natural structure caused by interactions between both the positive charge of chitosan and the negatively charged bacterial walls of bacteria and/or protein within. Yet, due to the very weak positive charge center of amino groups on the chitosan backbone, this activity is limited. As a result, the positive charge within chitosan must be strengthened. Chemical modification of chitosan is one method that can be employed to accomplish this. Using the agar disk diffusion experiment, Hajji et al. [84] demonstrated the antibacterial capabilities of a nanocomposite containing chitosan-poly(vinyl alcohol)-Ag nanoparticles against Gram-positive and Gram-negative bacteria. They claimed that the nanocomposite films slowed the growth of these bacteria, and they offered mechanisms for their success. The antimicrobial properties of chitosan and its nanocomposites are being investigated further for the creation of antifouling and anti-biofouling nanocomposites. Fouling is among the major challenges that membrane-based wastewater treatment technology faces, particularly in a classic membrane reactor. Membrane fouling can lower permeate flux and efficiency, boost energy usage and cleanup frequency, shorten membrane longevity, and thereby raise the cost of maintenance. Natarajan et al. [85] found that combining TiO_2 and Ag nanoparticles with chitosan significantly improved the antifouling mechanism toward *Scenedesmus* sp. and *Chlorella* sp. algae. With the photocatalytic activity of TiO_2 and dissolution of Ag ions from the films, UV light irradiation raised the hydrophilicity of the nanocomposite films while also increasing their toxic effects on both algae species, limiting the development of slime that causes fouling of the films. Figure 27 summarizes the different physical and chemical properties of chitosan nanocomposites.

11 Characterization of Chitosan Nanocomposites

Numerous characterization techniques can be employed for the investigation of chitosan nanocomposites, either individually or as a combination. Chitosan NCs may be characterized for various properties, such as size and morphology, structure, and optical properties. The various popular techniques used for the characterization of chitosan nanocomposites are as follows [75, 86]:

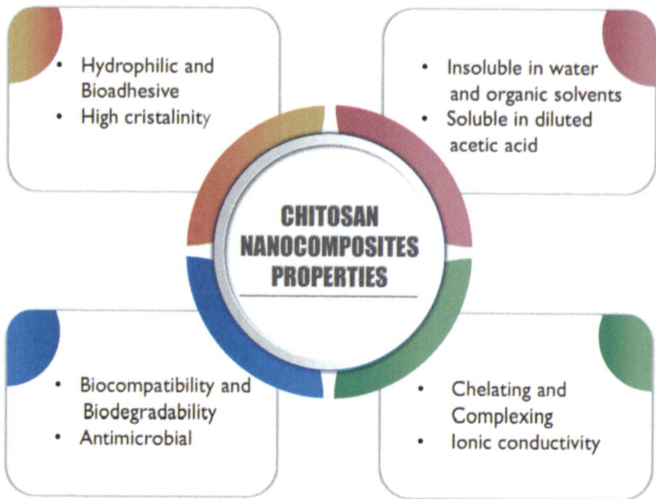

Fig. 27 Different physical and chemical properties of chitosan nanocomposites. Reprinted with permission from [1]. Copyright 2021 Elsevier

11.1 Measurement of Size

One of the main parameters to be assessed in the characterization of chitosan nanocomposites is their size, which greatly influences their physical and biological properties. Some of the commonly used characterization techniques for the assessment of their size are as follows:

(i) Dynamic light scattering (DLS): It determines the size of nanoparticles dispersed in colloidal suspensions, by measuring their light scattering ability and Brownian motion. Due to the involvement of low nanoparticle concentrations, the multiple scattering effects are avoided. Size values determined by DLS are governed by certain factors like particle shape and concentration, NP surface coating, and colloidal stability.

(ii) Nanoparticle tracking analysis (NTA): A recently developed technique for determining the dimensions of NPs that exploits the light scattering ability and Brownian motion of NPs dispersed in a liquid to determine their size distribution. It proves to be a better method than DLS as the average size values are more precise as compared to the ones obtained in the case of DLS. NTA may also be used to identify various concentrations in a polydisperse sample making it possible to track single particles, unlike DLS.

(iii) Differential centrifugal sedimentation (DCS): Measurement of particle size by DCS is done based on their sedimentation rates. It provides high sensitivity, ultra-high resolution, accuracy, and reproducibility in results.

(iv) Small-angle X-ray scattering (SAXS) method is a scattering method that investigates the interaction between the sample and an X-ray beam of high energy.

Scattered rays are detected using a two-dimensional detector. The refractive index of particles is not any determining factor in this technique.

11.2 Measurement of Morphology

Morphology of chitosan nanoparticles (dimensions, shape, and tendency to aggregate) is assessed using characterization techniques like:

(i)   Scanning electron microscopy (SEM): This tool is used to study the purity, homogeneity, and dispersion degree of nanoparticles. It also measures their tendency to produce aggregates along with the degree of aggregation. It is carried out inside the sample chamber at the relative temperature and humidity.

(ii) Transmission electron microscope (TEM): TEM may prove to be a more preferable option when it comes to distinguishing individual particles in systems that tend to agglomerate, where other techniques like SEM fail. This is due to the high resolution of sample images and precise measurement of nanoparticle dimensions and homogeneity. However, one major limitation is the incapability to characterize a large number of particles and thus needs to be combined with other microscopic techniques.

(iii) Confocal laser electron microscopy (CLSM): It assists in obtaining 3-dimensional images of chitosan nanoparticle distribution at a resolution of micrometers, through optical sectioning. Images are obtained at higher resolutions, and image quality is enhanced due to the laser sectioning method. In order to visualize cells and tissues, long exposure of samples to the laser is possible. CLSM is an efficient tool that provides information regarding the morphology of nanoparticles, particularly after internalization within cells and tissues.

(iv) Atomic force microscopy (AFM): Owing to its high spatial resolution, versatility (ability to work in different environments), and the possibility to study any kind of material or biological sample, AFM can provide resolutions with a minimum amount of sample and thus is commonly used for analyzing the size and three-dimensional topography of chitosan nanoparticles.

(v)   Scanning probe microscopy (SPM) refers to the AFM technique with modified tips and software that is deployed for almost all types of measurable attraction forces, like van der Waals, thermal, electric, and magnetic interactions. In the case of polydisperse samples, AFM provides higher accuracy in contrast to DLS.

11.3 Measurement of Specific Surface Area

Brunauer Emmett Teller (BET): It determines the specific surface area (SSA) of particles based on the adsorption of a gas on a sample surface. It provides an edge over

DLS and DCS methods as even smaller SSA values can easily be measured; however, the aggregation tendency of nanoparticles might lead to errors in the measurement. SSA on chitosan nanoparticles is governed by the synthesis process. For instance, Villegas-Peralta et al. [87] synthesized chitosan nanocomposites through two ionic gelation methods by varying the stirring time and amount of TPP, concluding that the highest TPP amount showed higher SSA in particles, because of the different size, morphology, and porous surface.

11.4 Measurement of Chemical Composition

(i) Fourier transform infrared spectroscopy (FTIR): It helps to identify typical functional groups and the interactions between them by detecting electromagnetic radiation (EMR) absorption within the infrared region. The data obtained is related to molecular structures and interactions.

(ii) X-ray photoelectron spectroscopy (XPS): Based on the photoelectric effect working under ultra-high vacuum surroundings, this technique allows quantitative chemical investigation of the surface of samples. It is employed to obtain data about the composition, electronic structure, ligand exchange interactions, oxidation state of an element, and surface functionalization of the nanoparticles.

(iii) Nuclear magnetic resonance spectroscopy (NMR): This technique is used to examine the interactions between the diamagnetic or antiferromagnetic nanoparticle surface and ligand. It identifies specific chemical shifts of the ^{1}H and ^{13}C NMR signals of chitosan and estimates the degree of cross-linking of chitosan to produce the chitosan nanocomposites.

11.5 Measurement of Zeta Potential

Zeta potential is measured using ZetaMeter, which applies the principle of electrophoresis. On the application of an electric field, dispersed particles move toward an oppositely charged electrode carrying a velocity that is proportional to the zeta potential. This velocity is determined using Laser Doppler Anemometer. This technique was used by Rodrigues et al. [88] for measuring the zeta potential for chitosan/carrageenan NPs.

11.6 Measurement of Antimicrobial Properties

(i) Plate count method: A cost-effective yet time-consuming methodology that requires counting the colony-forming units of bacteria on an agar plate.

(ii)   Microtiter plate assay: In this method, dilutions of nanoparticles and mycobacteria are kept in an incubation period of 7 days at 37 °C, followed by the addition of resazurin at the end of the incubation period and further incubation. A color change (blue to pink or colorless) indicates the bacterial growth, and the lowest concentration of nanoparticle that is able to prevent the growth of bacteria is called the minimum inhibitory concentration (MIC).

11.7   Drug Content, Loading Efficiency, and Drug Release Measurements

The nanoparticles system is subjected to centrifugation at high speed, and the amount of drug in the supernatant is examined. Following this, the drug content is measured using high-performance liquid chromatography (HPLC) or spectrophotometry. Loading efficiency is then determined using the formula:

$$\text{Loading efficiency} = \frac{\text{Amount of drug in a definite mass of the particles}}{\text{Total mass of the particles}}$$

The drug release profile from the synthesized chitosan nanoparticles is determined using the in vitro release test.

12   Biomedical Applications of Chitosan Nanocomposites

Because of their non-toxic character, chitosan nanoparticles have become quite important in biomedical research. The following is an overview of the different biomedical applications of chitosan-based nanoparticles.

12.1   Chitosan-Based Nanoparticles for the Treatment of Cancer

The anticancer efficacy of chitosan nanoparticles is due to their small size. The tiny particle size increases the surface-to-volume ratio and specific surface area, which aids in the solubility of chitosan and hence its bioavailability. Mathew et al. [89] created a novel folic acid-linked carboxymethyl chitosan quantum dot nanoparticle that was connected to manganese doped zinc sulfide quantum dots. This strategy can be used to identify cancer cells, manage drug distribution, and image them. 5-Fluorouracil, an anticancer medication that can be used to treat breast cancer, was chosen for this investigation.

12.2   Chitosan Nanoparticles for Tissue Engineering

Chitosan and its derivatives are potential biomaterials for tissue engineering scaffold applications because of their biocompatibility, biodegradability, and non-toxicity. The mechanical characteristics of a chitosan-based scaffold are affected by the pore shape and orientations, according to the findings. Furthermore, chitosan's cationic properties allow for pH-dependent electrostatic interactions with anionic species.

12.3   Chitosan Nanoparticles for Wound Dressing

Nanocomposites' wound-healing qualities are one of the most studied features for biomedical applications. Because of the potential applications for chitosan-based wound dressings, many chitosan nanocomposites have been proposed and developed with the goal of creating an optimal wound treatment that provides a superior physical shield, accelerates the healing of wounds, and boosts antibacterial activity. The application of chitosan nanocomposite films as a wound dressing is depicted in Fig. 28.

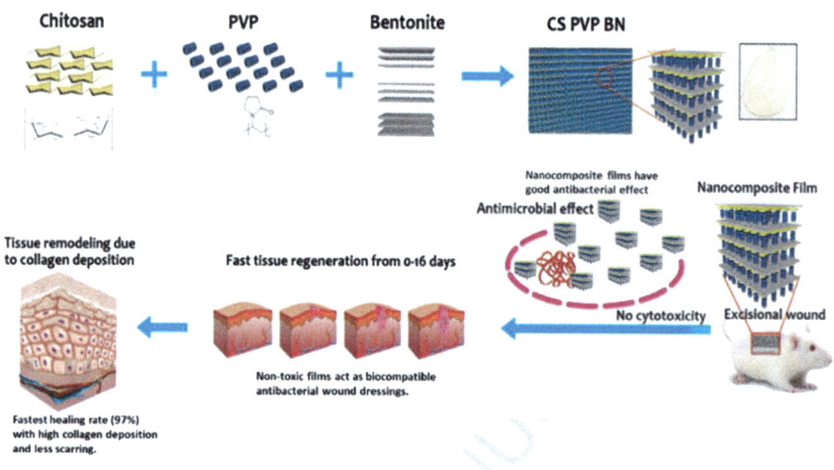

Fig. 28 Schematic illustrating would healing using chitosan nanocomposite films. Reprinted with permission from [90]. Copyright 2018 Elsevier

12.4   Chitosan Nanoparticles for Encapsulation of biologically Active Compounds

In the food and biochemical sector, chitosan-based products have an ever-increasing type of use. Encapsulating any component, whether hydrophobic, hydrophilic, or bacterial, is possible where chitosan does not lose the biocompatibility of macromolecules like proteins and DNA. Chitosan's positive charge interacts strongly with negatively charged molecules without affecting its function. Jang and Lee [91] investigated the properties and longevity of vitamin C-loaded chitosan nanoparticles made by ionic gelation of chitosan with TPP anions in an aqueous medium during thermal treatment. Chitosan nanoparticles were discovered to be stable at high temperatures, and the constant generation of vitamin C from chitosan nanoparticles was observed in this study.

12.5   Chitosan Nanoparticles for Drug Delivery

Chitosan and its derivative products can be used to improve peptide- and protein-based medication delivery systems. Chitosan has also been extensively examined in the delivery of drugs to the brain. Because it affects the tight junction, it can increase drug permeability across the blood-brain barrier. The chitosan polymers have also stabilized insulin's natural structure. Because of the positive charge on the surface, chitosan nanoparticles can be absorbed on the negatively charged cell membrane, improving the residence period on the nasal mucosa. As a result, medication transport from the nasal passages to the brain is optimized. The ability to target a brain tumor is improved even more by surface modification of chitosan nanoparticles with tumor-targeting peptides such as chlorotoxin and transferrin [92]. Figure 29 is a schematic drawing of a multifunctional polymeric nanocomposite used as a drug carrier.

12.6   Chitosan Nanoparticles for Bone Engineering

In bone engineering, chitosan nanocomposites with nanofillers such as hydroxyapatite, bioactive glass, copper nanoparticles, zeolite, and carbon filler are commonly employed. Sun et al. [94] investigated the mechanical properties of chitosan/nanodiamond nanocomposites, such as tensile modulus and indentation hardness. When 5 wt% nanodiamond filler was added to chitosan nanocomposites, the indentation hardness and tensile modulus increased by 127% and 343%, respectively.

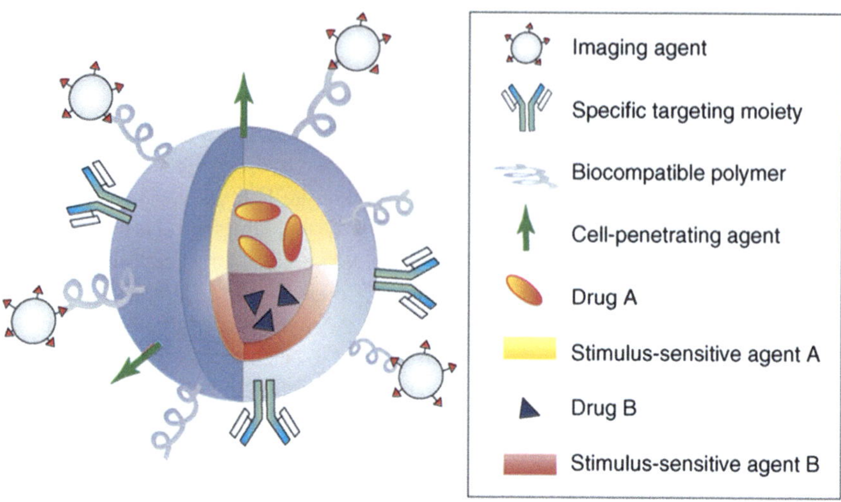

Fig. 29 Schematic representation of a multifunctional polymeric nanocomposite used for drug delivery. Reprinted with permission from [93]. Copyright 2008 Elsevier

12.7 Chitosan Nanoparticles for Antioxidant Activity

Chitosan is perceived to be an antioxidant. It can donate hydrogen or a lone pair of electrons to hunt for free radicals and bind metal ions. Chitosan's hydroxyl and amino functional groups react with metallic ions, causing a variety of reactions such as adsorption, chelation, and ion exchange. Chitosan's strong hydrogen bonds and semi-crystalline structure ensure that it cannot be separated from the metal ions. Hence, chitosan nanocomposites show significant antioxidant activity.

12.8 Chitosan Nanoparticles for Biosensor Applications

In contrast to pure chitosan, chitosan nanocomposites-based biosensors have been found to have improved sensitivity, efficiency, and durability. Many nanostructured inorganic materials, such as cuprous oxide nanoparticles, $NiFe_2O_4$ nanoparticles, Fe_3O_4 nanoparticles, TiO_2 nanoparticles, and cerium oxide nanoparticles, are frequently utilized in nanocomposites to boost the electronic characteristics and conductivity of chitosan-based materials. Biosensors to measure glucose in human serum, cholesterol in human blood serum, and immunesensors for ochratoxin-A are typical examples of applications. A chitosan-based biosensor is depicted schematically in Fig. 30.

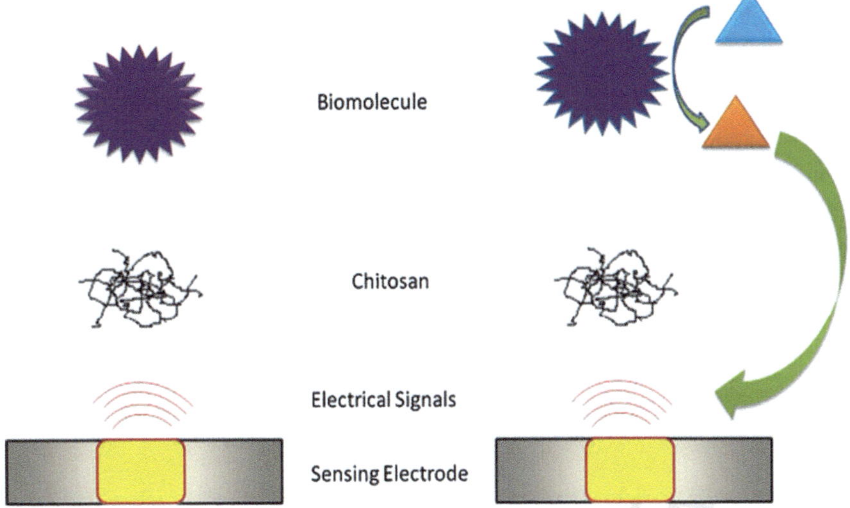

Fig. 30 Schematic representing a chitosan-based biosensor. Reprinted with permission from [95]. Copyright 2013 Elsevier

12.9   Chitosan Nanoparticles for Enzyme Immobilization Support

Chitosan is a wonderful ingredient for enzyme immobilization because of its many features, including increased chemical resistance and the ability to keep metal ions from disrupting an enzyme. Chitosan is considered appropriate for enzyme immobilization because it has an amino functional group. Using glutaraldehyde (GA) as a cross-linker, trypsin immobilized on linolenic acid-modified chitosan nanoparticles has been examined, and it is discovered that the thermal stability and optimal temperature of immobilized trypsin are enhanced [96]; lipase enzymes can be immobilized using chitosan magnetic core–shell nanoparticles. Because of the tight connection between chitosan and lipase, enzyme loading and adsorption are improved. D-carbamoylase and D-hydantoinase are incorporated in chitosan-based materials for the manufacture of D-p-hydroxyphenylglycine, as shown in Fig. 31.

12.10   Chitosan Nanoparticles as Antimicrobial Agents

Chitosan nanoparticles have been used as antibacterial agents to prevent the usage of synthetic materials. The positively charged chitosan combines with the negatively charged phospholipid of the cell membrane, causing the cell's porosity to change, causing cell death and outflow of cell components. Chitosan's ability to chelate

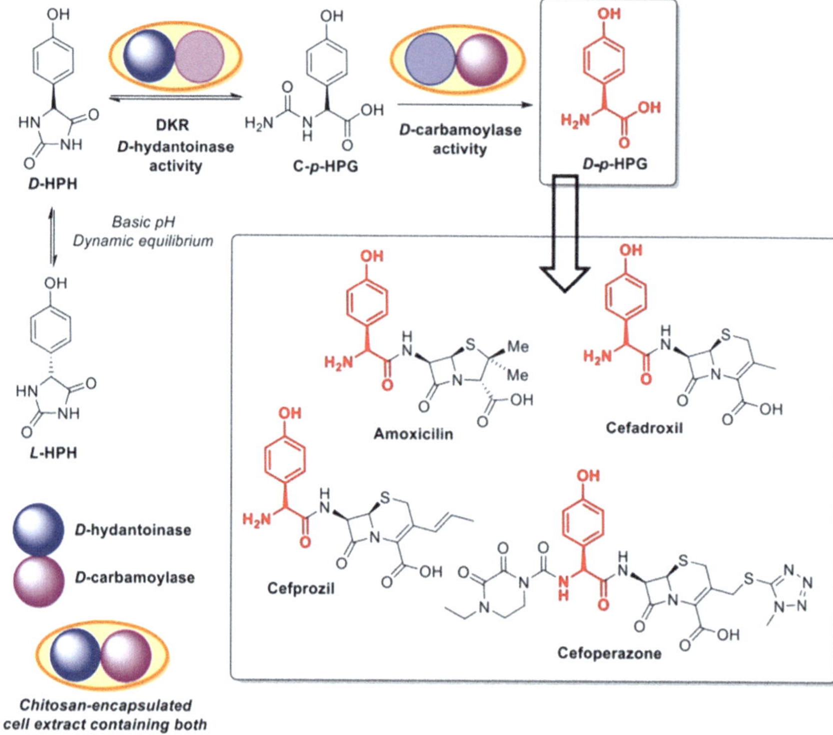

Fig. 31 Schematic showing encapsulation of D-carbamoylase and D-hydantoinase encapsulated in chitosan-based materials for production of D-*p*-hydroxyphenylglycine. Reprinted from [97] under the Creative Commons (CC BY) license

metal ions is one of the reasons for its antibacterial properties. The mechanism of chitosan's antibacterial effect is depicted in Fig. 32. Chitosan can also get through the cell wall and attach to DNA, preventing the production of mRNA [98, 99].

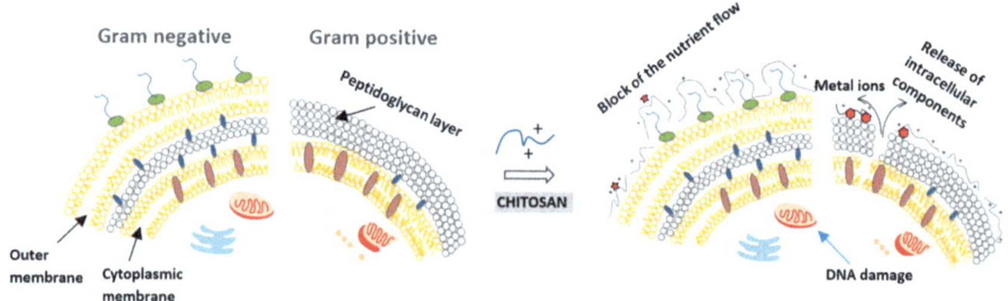

Fig. 32 Schematic showing the mechanism of antimicrobial action of chitosan. Reprinted from [101] under the Creative Commons BY license

Chitosan Ag nanoparticle composites were found to be effective against the disease *Colletotrichum gloeosporioides*, which causes mango anthracnose [100].

12.11 Chitosan-Based Nanoparticles for Gene Therapy

Chitosan and its variants have recently been investigated as a nonviral gene therapy vector for the retina. Pure oligo-chitosan polyplexes are created, described, and proved to be efficient in protecting the plasmid from enzymatic digestion and improving transfection [102]. An oligomer-based example is a chitosan–DNA nanoparticles, which have recently been sold and utilized as transporters for nonviral gene therapy. These freshly formed nanoparticles, ultrapure chitosan–DNA nanoparticles, could be examined for gene therapy applications and the cure of eye illnesses outside the cornea [103]. Furthermore, chitosan–TPP nanoparticles show possibilities as viable vector options for siRNA delivery that is both safe and cost-effective.

13 Limitations and Challenges of the Use of Chitosan Nanocomposites in Biomedical Research

One major drawback that often limits the use of chitosan nanocomposites for biomedical applications (and related applications) is their lack of stability. However, there is enough scope for overcoming this limitation, and that is by controlling the environmental factors, introducing a proper stabilizer, and maintaining the temperature. The chitosan may be blended with other polymers to enhance its stability features, or its structure may be modified by introducing certain ionic or chemical agents. Another major inadequacy may be their weak solubility, which limits the possibility of encapsulation to only hydrophilic drugs. Modification of chitosan nanocomposites is necessary for the encapsulation of hydrophobic drugs. In today's time, when researchers from around the world are emphasizing green chemistry and applying its principles to actual practical use, the need of the hour is to shift the production of chitosan and chitosan NCs from laboratory scale to industrial scale via greener and sustainable pathways. To put chitosan NCs to commercial use, proper regulations of use and toxicology studies are needed to affirm its biocompatibility in humans [67]. Furthermore, there is still a need for more investigation on economic sources from which chitosan can be isolated. The current preparation methods are expensive, even though the marine sources are abundant and renewable in nature. Thus, more research is warranted on developing more economical production protocols for chitosan and its derived products for upscaling to industrial use. Despite the fact that chitosan and its derivatives do not have much hazardous environmental impact, there are still doubts about the harmful chemical solvents deployed in their processing.

14 Conclusion and Future Prospects

Owing to its biological properties like biocompatibility, non-toxic nature, and good antimicrobial action, among other attributes, chitosan, a natural biopolymer extracted from chitin, is widely explored in the field of biomedicine. Having said that, certain drawbacks with respect to its biological nature yet exist. Poor solubility, low thermal stability, weak mechanical properties, and expensive production methodology still restrict its mass commercial use. However, its superb chelating ability, rendered by the free amino and hydroxyl groups in its carbon framework, is exploited by researchers to produce functionalized chitosan derivatives with improved properties compared to their bulk counterparts. Since the advent of nanotechnology, chitosan nanocomposites have become increasingly popular in the biomedical research field including cancer therapy, drug delivery, wound healing, tissue engineering, bioimaging, and dentistry, among others; they are seen as excellent antibacterial, antifungal, and antiviral agents. Chitosan nanocomposites are biocompatible, biodegradable, non-toxic, renewable, abundant, and versatile. There is also future scope for improvement in their potential commercial expansion if more research is carried out and further investigations are performed to render them industrially feasible. Additional research on making the extraction process of chitosan more cost-effective, finding alternatives to harmful chemical solvents, and further optimization of the synthesis steps shall open new windows of opportunities with respect to their potential applications in the near future. Furthermore, additional explorations probing the in vivo interactions of nanocomposites and the degradation of nanocomposites would be highly meaningful for their market expansion.

References

1. Silva AO, Cunha RS, Hotza D, Machado RAF (2021) Chitosan as a matrix of nanocomposites: a review on nanostructures, processes, properties, and applications. Carbohydr Polym 272:118472
2. Kaya M, Baran T, Erdoğan S, Menteş A, Aşan Özüsağlam M, Çakmak YS (2014) Physico-chemical comparison of chitin and chitosan obtained from larvae and adult Colorado potato beetle (*Leptinotarsa decemlineata*). Mater Sci Eng C 45:72–81
3. Crini G (2019) Historical review on chitin and chitosan biopolymers. Environ Chem Lett 17:1623–1643
4. Periayah MH, Halim AS, Saad AZM (2016) Chitosan: a promising marine polysaccharide for biomedical research. Pharmacognosy Rev 10:39–42
5. Aizat MA, Aziz F (2018) 12—Chitosan nanocomposite application in wastewater treatments. Elsevier Inc., Amsterdam
6. Bakshi PS, Selvakumar D, Kadirvelu K, Kumar NS (2020) Chitosan as an environment friendly biomaterial—a review on recent modifications and applications. Int J Biol Macromol 150:1072–1083
7. Biao L, Tan S, Wang Y, Guo X, Fu Y, Xu F, Zu Y, Liu Z (2017) Synthesis, characterization and antibacterial study on the chitosan-functionalized Ag nanoparticles. Mater Sci Eng C 76:73–80

8. Ohya Y, Shiratani M, Kobayashi H, Ouchi T (1994) Release behavior of 5-fluorouracil from chitosan-gel nanospheres immobilizing 5-fluorouracil coated with polysaccharides and their cell specific cytotoxicity. J Macromol Sci Part A 31:629–642

9. Pinto Reis C, Neufeld RJ, Ribeiro AJ, Veiga F (2006) Nanoencapsulation I. Methods for preparation of drug-loaded polymeric nanoparticles. Nanomed Nanotechnol Biol Med 2:8–21

10. Chevalier MT, Gonzalez J, Alvarez V (2015) Biodegradable polymeric microparticles as drug delivery devices. In: IFMBE Proceedings, vol 49, pp 187–190

11. Yanat M, Schroën K (2021) Preparation methods and applications of chitosan nanoparticles; with an outlook toward reinforcement of biodegradable packaging React Funct Polym 161

12. Huang X, Du YZ, Yuan H, Hu FQ (2009) Preparation and pharmacodynamics of low-molecular-weight chitosan nanoparticles containing insulin. Carbohydr Polym 76:368–373

13. Rajitha P, Gopinath D, Biswas R, Sabitha M, Jayakumar R (2016) Chitosan nanoparticles in drug therapy of infectious and inflammatory diseases. Expert Opin Drug Deliv 13:1177–1194

14. des Rieux A, Fievez V, Garinot M, Schneider YJ, Préat V (2006) Nanoparticles as potential oral delivery systems of proteins and vaccines: a mechanistic approach. J Controlled Release 116:1–27

15. Shariatinia Z (2019) Pharmaceutical applications of chitosan. Adv Colloid Interface Sci 263:131–194

16. Ding L, Shan X, Zhao X, Zha H, Chen X, Wang J, Cai C, Wang X, Li G, Hao J, Yu G (2017) Spongy bilayer dressing composed of chitosan–Ag nanoparticles and chitosan–*Bletilla striata* polysaccharide for wound healing applications. Carbohydr Polym 157:1538–1547

17. Kankala RK, Liu CG, Yang DY, Bin WS, Chen AZ (2020) Ultrasmall platinum nanoparticles enable deep tumor penetration and synergistic therapeutic abilities through free radical species-assisted catalysis to combat cancer multidrug resistance. Chem Eng J 383:123138

18. Nasrollahzadeh M, Sajjadi M, Iravani S, Varma RS (2021) Starch, cellulose, pectin, gum, alginate, chitin and chitosan derived (nano)materials for sustainable water treatment: a review. Carbohydr Polym 251:116986

19. Motahharifar N, Nasrollahzadeh M, Taheri-Kafrani A, Varma RS, Shokouhimehr M (2020) Magnetic chitosan-copper nanocomposite: a plant assembled catalyst for the synthesis of amino- and N-sulfonyl tetrazoles in eco-friendly media. Carbohydr Polym 232:115819

20. Ashrafizadeh M, Delfi M, Hashemi F, Zabolian A, Saleki H, Bagherian M, Azami N, Farahani MV, Sharifzadeh SO, Hamzehlou S, Hushmandi K, Makvandi P, Zarrabi A, Hamblin MR, Varma RS (2021) Biomedical application of chitosan-based nanoscale delivery systems: potential usefulness in siRNA delivery for cancer therapy. Carbohydr Polym 260:117809

21. Khorsandi Z, Borjian-Boroujeni M, Yekani R, Varma RS (2021) Carbon nanomaterials with chitosan: a winning combination for drug delivery systems, vol 66. Elsevier B.V., Amsterdam

22. Silvestri D, Wacławek S, Sobel B, Torres-Mendieta R, Novotný V, Nguyen NHA, Ševců A, Padil VVT, Müllerová J, Stuchlík M, Papini MP, Černík M, Varma RS (2018) A poly(3-hydroxybutyrate)-chitosan polymer conjugate for the synthesis of safer gold nanoparticles and their applications. Green Chem. 20:4975–4982

23. Kiani M, Bagherzadeh M, Kaveh R, Rabiee N, Fatahi Y, Dinarvand R, Jang HW, Shokouhimehr M, Varma RS (2020) Novel Pt-Ag$_3$PO$_4$/CdS/chitosan nanocomposite with enhanced photocatalytic and biological activities. Nanomaterials 10:1–21

24. Gulati S, Singh P, Diwan A, Mongia A, Kumar S (2020) Functionalized gold nanoparticles: promising and efficient diagnostic and therapeutic tools for HIV/AIDS. RSC Med Chem 11:1252–1266

25. Gulati S, Lingam BHN, Kumar S, Goyal K, Arora A, Varma RS (2022) Improving the air quality with functionalized carbon nanotubes: sensing and remediation applications in the real world. Chemosphere 299:134468

26. Kumar S, Diwan A, Singh P, Gulati S, Choudhary D, Mongia A, Shukla S, Gupta A (2019) Functionalized gold nanostructures: promising gene delivery vehicles in cancer treatment. RSC Adv 9:23894–23907

27. Sharma RK, Gulati S, Mehta S (2012) Preparation of gold nanoparticles using tea: a green chemistry experiment. J Chem Educ 89:1316–1318

28. Muanprasat C, Chatsudthipong V (2017) Chitosan oligosaccharide: biological activities and potential therapeutic applications. Pharmacol Ther 170:80–97
29. Arbia W, Arbia L, Adour L, Amrane A (2013) Chitin extraction from crustacean shells using biological methods—a review. Food Technol Biotechnol 51:12–25
30. Beaney P, Lizardi-Mendoza J, Healy M (2005) Comparison of chitins produced by chemical and bioprocessing methods. J Chem Technol Biotechnol 80:145–150
31. Hamed I, Özogul F, Regenstein JM (2016) Industrial applications of crustacean by-products (chitin, chitosan, and chitooligosaccharides): a review. Trends Food Sci Technol 48:40–50
32. Riaz Rajoka MS, Zhao L, Mehwish HM, Wu Y, Mahmood S (2019) Chitosan and its derivatives: synthesis, biotechnological applications, and future challenges. Appl Microbiol Biotechnol 103:1557–1571
33. Kumar MNVR (2015) A review of chitin and chitosan applications. React Funct Polym 46:1–27
34. El-Sherbiny IM, El-Baz NM (2015) A review on bionanocomposites based on chitosan and its derivatives for biomedical applications, vol 74
35. Ji J, Wang L, Yu H, Chen Y, Zhao Y, Zhang H, Amer WA, Sun Y, Huang L, Saleem M (2014) Chemical modifications of chitosan and its applications. Polym Plast Technol Eng 53:1494–1505
36. Payne GF, Chaubal MV, Barbari TA (1996) Enzyme-catalysed polymer modification: reaction of phenolic compounds with chitosan films. Polymer (Guildford) 37:4643–4648
37. Lee CC, MacKay JA, Fréchet JMJ, Szoka FC (2005) Designing dendrimers for biological applications. Nat Biotechnol 23:1517–1526
38. Jayakumar R, Prabaharan M, Nair SV, Tokura S, Tamura H, Selvamurugan N (2010) Novel carboxymethyl derivatives of chitin and chitosan materials and their biomedical applications. Prog Mater Sci 55:675–709
39. Hall D, Yalpani M (1980) Formation of branched-chain, soluble polysaccharides from chitosan. J Chem Soc Chem Commun 1153–1154
40. Chen RH, Hwa HD (1996) Effect of molecular weight of chitosan with the same degree of deacetylation on the thermal, mechanical, and permeability properties of the prepared membrane. Carbohydr Polym 29:353–358
41. Vårum KM, Ottøy MH, Smidsrød O (1994) Water-solubility of partially N-acetylated chitosans as a function of pH: effect of chemical composition and depolymerisation. Carbohydr Polym 25:65–70
42. Kasaai MR, Joseph Arul GC (2000) Intrinsic viscosity–molecular weight relationship for chitosan. J Polym Sci Part B Polym Phys 38:2591–2598
43. Kogan G, Skorik YA, Žitňanová I, Križková L, Ďuračková Z, Gomes CAR, Yatluk YG, Krajčovič J (2004) Antioxidant and antimutagenic activity of N-(2-carboxyethyl)chitosan. Toxicol Appl Pharmacol 201:303–310
44. Xing R, Yu H, Liu S, Zhang W, Zhang Q, Li Z, Li P (2005) Antioxidant activity of differently regionselective chitosan sulfates in vitro. Bioorg Med Chem 13:1387–1392
45. Zhou J, Wen B, Xie H, Zhang C, Bai Y, Cao H, Che Q, Guo J, Su Z (2021) Advances in the preparation and assessment of the biological activities of chitosan oligosaccharides with different structural characteristics. Food Funct 12:926–951
46. de Alvarenga ES (2011) Characterization and properties of chitosan
47. Davies DH, Hayes ER (1988) Determination of the degree of acetylation of chitin and chitosan. Methods Enzymol 161:442–446
48. dos Santos ZM, Caroni ALPF, Pereira MR, da Silva DR, Fonseca JLC (2009) Determination of deacetylation degree of chitosan: a comparison between conductometric titration and CHN elemental analysis. Carbohydr Res 344:2591–2595
49. Arcidiacono S, Kaplan DL (1992) Molecular weight distribution of chitosan isolated from *Mucor rouxii* under different culture and processing conditions. Biotechnol Bioeng 39:281–286
50. Balázs N, Sipos P (2007) Limitations of pH-potentiometric titration for the determination of the degree of deacetylation of chitosan. Carbohydr Res 342:124–130

51. Zhang Y, Zhang X, Ding R, Zhang J, Liu J (2011) Determination of the degree of deacetylation of chitosan by potentiometric titration preceded by enzymatic pretreatment. Carbohydr Polym 83:813–817

52. Tôei K, Kohara T (1976) A conductometric method for colloid titrations. Anal Chim Acta 83:59–65

53. de Alvarenga ES, Pereira de Oliveira C, Roberto Bellato C (2010) An approach to understanding the deacetylation degree of chitosan. Carbohydr Polym 80:1155–1160

54. Niola F, Basora N, Chornet E, Vidal PF (1993) A rapid method for the determination of the degree of N-acetylation of chitin-chitosan samples by acid hydrolysis and HPLC. Carbohydr Res 238:1–9

55. Baxter A, Dillon M, Anthony Taylor KD, Roberts GAF (1992) Improved method for i.r. determination of the degree of N-acetylation of chitosan. Int J Biol Macromol 14:166–169

56. Raymond L, Morin FG, Marchessault RH (1993) Degree of deacetylation of chitosan using conductometric titration and solid-state NMR. Carbohydr Res 246:331–336

57. Kasaai MR (2009) Various methods for determination of the degree of N-acetylation of chitin and chitosan: a review. J Agric Food Chem 57:1667–1676

58. Reddy DHK, Lee SM (2013) Application of magnetic chitosan composites for the removal of toxic metal and dyes from aqueous solutions. Adv Colloid Interface Sci 201–202:68–93

59. Akpan EI, Shen X, Wetzel B, Friedrich K (2018) Design and synthesis of polymer nanocomposites. Elsevier Inc., Amsterdam

60. Moura D, Mano JF, Paiva MC, Alves NM (2016) Chitosan nanocomposites based on distinct inorganic fillers for biomedical applications. Sci Technol Adv Mater 17:626–643

61. Alexandre M, Dubois P (2000) Polymer-layered silicate nanocomposites: preparation, properties and uses of a new class of materials. Mater Sci Eng R Rep 28:1–63

62. Vardharajula S, Ali SZ, Tiwari PM, Eroğlu E, Vig K, Dennis VA, Singh SR (2012) Functionalized carbon nanotubes: biomedical applications. Int J Nanomed 7:5361–5374

63. Rahmat M, Hubert P (2011) Carbon nanotube-polymer interactions in nanocomposites: a review. Compos Sci Technol 72:72–84

64. Kuilla T, Bhadra S, Yao D, Kim NH, Bose S, Lee JH (2010) Recent advances in graphene based polymer composites. Prog Polym Sci 35:1350–1375

65. Terrones M, Martín O, González M, Pozuelo J, Serrano B, Cabanelas JC, Vega-Díaz SM, Baselga J (2011) Interphases in graphene polymer-based nanocomposites: achievements and challenges. Adv Mater 23:5302–5310

66. Kumar S, Mukherjee A, Dutta J (2020) Chitosan based nanocomposite films and coatings: emerging antimicrobial food packaging alternatives. Trends Food Sci Technol 97:196–209

67. Naskar S, Sharma S, Kuotsu K (2019) Chitosan-based nanoparticles: an overview of biomedical applications and its preparation. J Drug Deliv Sci Technol 49:66–81

68. Anisha BS, Biswas R, Chennazhi KP, Jayakumar R (2013) Chitosan-hyaluronic acid/nano silver composite sponges for drug resistant bacteria infected diabetic wounds. Int J Biol Macromol 62:310–320

69. Mohandes F, Salavati-Niasari M (2014) Freeze-drying synthesis, characterization and in vitro bioactivity of chitosan/graphene oxide/hydroxyapatite nanocomposite. RSC Adv. 4:25993–6001

70. Iler RK (1966) Multilayers of colloidal particles. J Colloid Interface Sci 21:569–594

71. Semba JA, Mieloch AA, Rybka JD (2020) Introduction to the state-of-the-art 3D bioprinting methods, design, and applications in orthopedics. Bioprinting 18:e00070

72. Sommer MR, Alison L, Minas C, Tervoort E, Rühs PA, Studart AR (2017) 3D printing of concentrated emulsions into multiphase biocompatible soft materials. Soft Matter 13:1794–1803

73. Ahmed J, Mulla M, Maniruzzaman M (2020) Rheological and dielectric behavior of 3D-printable chitosan/graphene oxide hydrogels. ACS Biomater Sci Eng 6:88–99

74. Rajabi M, McConnell M, Cabral J, Ali MA (2021) Chitosan hydrogels in 3D printing for biomedical applications. Carbohydr Polym 260:117768

75. Ahmed TA, Aljaeid BM (2016) Preparation, characterization, and potential application of chitosan, chitosan derivatives, and chitosan metal nanoparticles in pharmaceutical drug delivery. Drug Des Dev Ther 10:483–507
76. Derfus AM, Chan WCW, Bhatia SN (2004) Probing the cytotoxicity of semiconductor quantum dots. Nano Lett 4:11–18
77. Chouhan D, Mandal P (2021) Applications of chitosan and chitosan based metallic nanoparticles in agrosciences—a review. Int J Biol Macromol 166:1554–1569
78. Bhumkar DR, Joshi HM, Sastry M, Pokharkar VB (2007) Chitosan reduced gold nanoparticles as novel carriers for transmucosal delivery of insulin. Pharm Res 24:1415–1426
79. Salehizadeh H, Hekmatian E, Sadeghi M, Kennedy K (2012) Synthesis and characterization of core-shell Fe_3O_4-gold-chitosan nanostructure. J Nanobiotechnol 10:3
80. Solanki PR, Patel MK, Ali MA, Malhotra BD (2015) A chitosan modified nickel oxide platform for biosensing applications. J Mater Chem B 3:6698–6708
81. Unsoy G, Yalcin S, Khodadust R, Gunduz G, Gunduz U (2012) Synthesis optimization and characterization of chitosan coated iron oxide nanoparticles produced for biomedical applications J Nanoparticle Res 14
82. Qi L, Xu Z, Jiang X, Hu C, Zou X (2004) Preparation and antibacterial activity of chitosan nanoparticles. Carbohydr Res 339:2693–2700
83. Sanpui P, Murugadoss A, Prasad PVD, Ghosh SS, Chattopadhyay A (2008) The antibacterial properties of a novel chitosan-Ag-nanoparticle composite. Int J Food Microbiol 124:142–146
84. Hajji S, Ben SRBS, Hamdi M, Jellouli K, Ayadi W, Nasri M, Boufi S (2017) Nanocomposite films based on chitosan–poly(vinyl alcohol) and silver nanoparticles with high antibacterial and antioxidant activities. Process Saf Environ Prot 111:112–121
85. Natarajan S, Shanthana Lakshmi D, Bhuvaneshwari M, Iswarya V, Mrudula P, Chandrasekaran N, Mukherjee A (2017) Antifouling activities of pristine and nanocomposite chitosan/TiO_2/Ag films against freshwater algae. RSC Adv 7:27645–27655
86. Garavand F, Cacciotti I, Vahedikia N, Rehman A, Tarhan Ö, Akbari-Alavijeh S, Shaddel R, Rashidinejad A, Nejatian M, Jafarzadeh S, Azizi-Lalabadi M, Khoshnoudi-Nia S, Jafari SM (2022) A comprehensive review on the nanocomposites loaded with chitosan nanoparticles for food packaging. Crit Rev Food Sci Nutr 62:1383–1416
87. Villegas-Peralta Y, Correa-Murrieta MA, Meza-Escalante ER, Flores-Aquino E, Álvarez-Sánchez J, Sánchez-Duarte RG (2019) Effect of the preparation method in the size of chitosan nanoparticles for the removal of allura red dye. Polym Bull 76:4415–4430
88. Rodrigues S, Da CAMR, Grenha A (2012) Chitosan/carrageenan nanoparticles: effect of cross-linking with tripolyphosphate and charge ratios. Carbohydr Polym 89:282–289
89. Mathew ME, Mohan JC, Manzoor K, Nair SV, Tamura H, Jayakumar R (2010) Folate conjugated carboxymethyl chitosan-manganese doped zinc sulphide nanoparticles for targeted drug delivery and imaging of cancer cells. Carbohydr Polym 80:442–448
90. Shanmugapriya K, Kim H, Saravana PS, Chun BS, Kang HW (2018) Fabrication of multifunctional chitosan-based nanocomposite film with rapid healing and antibacterial effect for wound management. Int J Biol Macromol 118:1713–1725
91. Jang KI, Lee HG (2008) Stability of chitosan nanoparticles for L-ascorbic acid during heat treatment in aqueous solution. J Agric Food Chem 56:1936–1941
92. Yu S, Xu X, Feng J, Liu M, Hu K (2019) Chitosan and chitosan coating nanoparticles for the treatment of brain disease. Int J Pharm 560:282–293
93. Sanvicens N, Marco MP (2008) Multifunctional nanoparticles—properties and prospects for their use in human medicine. Trends Biotechnol 26:425–433
94. Sun C, Wen L, Zeng J, Wang Y, Sun Q, Deng L, Zhao C, Li Z (2016) One-pot solventless preparation of PEGylated black phosphorus nanoparticles for photoacoustic imaging and photothermal therapy of cancer. Biomaterials 91:81–89
95. Shukla SK, Mishra AK, Arotiba OA, Mamba BB (2013) Chitosan-based nanomaterials: a state-of-the-art review. Int J Biol Macromol 59:46–58
96. Liu CG, Desai KGH, Chen XG, Park HJ (2005) Preparation and characterization of nanoparticles containing trypsin based on hydrophobically modified chitosan. J Agric Food Chem 53:1728–1733

97. Aranaz I, Alcántara A R, Civera MC, Arias C, Elorza B, Caballero AH, Acosta N (2021) Chitosan: an overview of its properties and applications. Polymers (Basel) 13

98. Sudarshan NR, Hoover DG, Knorr D (1992) Antibacterial action of chitosan. Food Biotechnol 6:257–272

99. Hernández-Lauzardo AN (2011) Current status of action mode and effect of chitosan against phytopathogens fungi. Afr J Microbiol Res 5:4243–4247

100. Chowdappa P, Gowda S, Chethana SC, Madura S (2014) Antifungal activity of chitosan-silver nanoparticle composite against *Colletotrichum gloeosporioides* associated with mango anthracnose. Afr J Microbiol Res 8:1803–1812

101. Kravanja G, Primožič M, Knez Ž, Leitgeb M (2019) Chitosan-based (nano)materials for novel biomedical applications. Liq Extr 683–723

102. Puras G, Zarate J, Aceves M, Murua A, Díaz AR, Avilés-Triguero M, Fernández E, Pedraz JL (2013) Low molecular weight oligochitosans for non-viral retinal gene therapy. Eur J Pharm Biopharm 83:131–140

103. Klausner EA, Zhang Z, Chapman RL, Multack RF, Volin MV (2010) Ultrapure chitosan oligomers as carriers for corneal gene transfer. Biomaterials 31:1814–1820

Chapter 2
Strategies for Synthesis and Chemical Modifications of Chitosan-Based Nanocomposites: A Versatile Material with Extraordinary Potential for Diverse Applications

Mansi, Shikha Gulati, and Anoushka Amar

Abstract Chitosan is a bio-functional polysaccharide that has a great potential for applications in various fields owing to its chemical functional groups which can be easily modified to achieve specific goals. Chitosan-based nanomaterials are gaining immense interest from researchers due to their versatile physicochemical and biological properties. In the present chapter, we give a complete overview of the preparation strategies of chitosan nanoparticles, including both novel and green methods. Moreover, we have systematically summarized the modification strategies of chitosan for improving their water solubility, biocompatibility, mechanical properties, and antimicrobial activity, which will help the researchers pick the most appropriate strategy for its particular application.

Keywords Chitosan · Chemical modifications · Cross-linking · Chitin · Nanoparticles · Synthesis

Abbreviations

BNNT	Boron nitride nanotube
Ch	Chitosan
COS	Chito oligosaccharide
GTA	Glutaraldehyde
HMW	High molecular weight
LCC	N lauryl-carboxymethyl-chitosan
LMW	Low molecular weight
NPs	Nanoparticles
PCDHN	Physically crosslinked hydrogel double-network
PCL	Polycaprolactone

Mansi · S. Gulati (✉) · A. Amar
Department of Chemistry, Sri Venkateswara College, University of Delhi, Delhi 110021, India
e-mail: shikha2gulati@gmail.com

S. Gulati (ed.), *Chitosan-Based Nanocomposite Materials*,
https://doi.org/10.1007/978-981-19-5338-5_2

PEG   Polyethylene glycol
QAS   Quaternized ammonium salt
TEMPO   (2,2,6,6-tetramethylpiperidin-1-yl)oxyl
TPP   Sodium tripolyphosphate

1 Introduction

The research interest in natural polymers has been rising for the last three decades. The worldwide market for chitosan derivatives has grown at a rate of approximately 6.5% approximately over a period of 5 years, and in 2024, it is anticipated to reach 53 million USD, owing to the increasing investments in newer drug developments, applications in the biomedical field, as well as applications like detoxification of water and wastewater. There is also a growing interest in the use of bio-degradable chitosan products such as fertilizers. Key areas of interest include physical moderation of polysaccharides to increase their applications in the mechanisms of biological activity of the enzymatic methods, polymers, chemical, and products of their physical, enzymatic, chemical, or degradation; and the molecular and biochemical characterization of chitosanolytic and chitinolytic enzymes synthesized by numerous organisms. Chitin deacetylases can be used to bio-convert chitin to chitosan. The aforementioned enzymatic reaction has many advantages over the traditional process used in the conversion, namely, the production of chitosan with the desired degree of deacetylation and higher molecular weight. Chitosanases and chitinases are two other enzymes involved in chitosan and chitin conversion. The various functional groups of chitosan can be moderated using a large number of ligands. The amino group functional group within ligands can be used in a variety of chemical reactions, including metal chelation, sulfonation, alkylation, carboxymethylation, grafting acetylation, quaternization, and so on. Hydroxyethylation, carboxymethylation, sulfonation, and phosphorylation can also be used to modify hydroxyl groups. A plethora of effective strategies for chitosan and its derivatives have been created in order to broaden their applications by increasing their antimicrobial activity and water solubility. Betaine, nitric oxide (NO) releasing donors, a quaternary phosphonium salt, and metal ions, for example, have been added to chitosan derivatives to improve their antimicrobial activity. Various solid materials with good adsorptive properties have been used to improve the heat resistance, strength, and adsorptive activity of chitosan. Chitosan in combination with activated carbon performs well in the removal of some inorganic and metallic pollutants from an aqueous medium. This chapter aims to present recent research on a wide range of chitosan modification strategies [30].

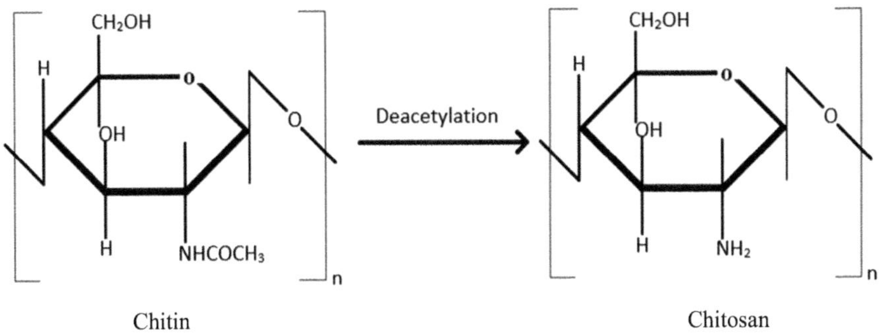

Fig. 1 Structures of chitin and chitosan

2 Structures and Properties of Chitosan

2.1 Structure

Chitosan's structure (Fig. 1) is very similar to that of cellulose, which is made up of hundreds to thousands of –(1-4) linked D-glucose units [57]. The hydroxyl group at the C-2 position of cellulose has been replaced by an acetamide group in the chitosan structure. Chitosan, also known as –(1-4) linked 2-amino-2-deoxy—D-glucopyranose, is an *N*-deacetylated chitin derivative obtained by converting the acetamide groups into primary amino groups [2].

However, chitin deacetylation is never complete, so some chitosan or deacetylated chitin still contains acetamide groups. Chitosan and chitin both contain 5–8% nitrogen, in the form of acetylated amine groups in chitin and primary aliphatic amine groups in chitosan [11] On each repeat unit of chitosan, there are primary and secondary hydroxyl groups, as well as an amine group on each deacetylated unit. These reactive groups can easily change the physical and mechanical properties of chitosan [26].

The presence of amine groups in chitosan is advantageous because it allows for various biological functions as well as the use of modification reactions [3]. Chitosan's excellent properties, such as biocompatibility, biodegradability, bioactivity, non-toxicity, and good absorption, make it an excellent alternative to synthetic polymers [12, 67].

2.2 Physiochemical Properties

Chitosan is well-known for its chelation, solubility, viscosity, mucoadhesive nature, film formation, polyoxysalt formation, and polyelectrolyte behavior. The unbranched and linear forms of chitosan have high viscosity and can be customized by adjusting the deacetylation conditions. Chitosan is distinguished by its low crystalline region content and a high degree of deacetylation. Several factors influence the physicochemical properties of chitosan, including the degree of deacetylation (DD), crystallinity, MW, and degradation methods [19]. Commercial chitosan is classified into two types based on its MW: high MW chitosan and low MW chitosan. High MW chitosan has a molecular weight (MW) of 190–375 kDa and a DD of >75%, whereas low MW chitosan has a MW of 20–190 kDa and a DD of 75%. It was discovered that the rate of chitosan degradation is inversely proportional to DD and is affected by the order and distribution of acetyl groups. Every glycosidic residue contains three reactive positions, one amino group, and two hydroxyl groups. The amino group is very important because it is responsible for the cationic nature of chitosan as well as the regulation of its various physiochemical properties. It is also pH sensitive. Chitosan dissolves and forms soluble cationic polysaccharides due to the pH-responsive amino groups protonated at lower pH. At a pH greater than 6, the amino group is deprotonated, rendering chitosan insoluble. Its solubility is also affected by acetyl group positioning along the chain, deacetylation methods, and ionic strength. Chitosan exhibits chelating properties for several metal ions at an acidic pH, which occurs at a pH = 7 or by electrostatic attraction on protonated ionic groups.

2.3 Biological Properties

According to studies, chitosan is a non-toxic, biocompatible, and biodegradable polymer [4, 36, 62]. Chitosan and its derivatives showcase exceptional biological properties like anti-inflammatory [10, 37], anti-tumor, anti-bacterial [23, 72], antifungal [72], hemostatic [76], and analgesic [5, 76]. Chitosan has a powerful antibacterial effect on a wide variety of pathogenic bacteria due to its cationic nature, which allows negatively charged proteins and lipids in the bacterial cell wall to interact. Chitosan diffusion into the cell membrane causes membrane permeability to expand and disrupt, resulting in cytoplasmic portion leakage and bacterial cell death. Various in-vitro experiments revealed that chitosan's antibacterial function may be due to its DNA binding capacity. When chitosan comes into contact with bacteria nuclei, it binds to DNA and prevents mRNA synthesis [17, 38].

Chitosan's antifungal properties are due to an electrostatic reaction with the negatively charged phospholipids in the cell membrane. Chitosan can penetrate the cell and prevent DNA/RNA synthesis, resulting in cell death when the cell membrane is disrupted. The DD and MW play an important role in regulating chitosan's antifungal properties. Chitosan's antifungal effect is generally enhanced when the DD is higher and the MW is lower [71]. Chitosan has anti-inflammatory properties because it inhibits the release of interleukin-8 (IL-8) and tumor necrosis factor (TNF) from mast cells [53]. Chitosan is superior to NSAIDs (the conventional medications for various inflammatory conditions). It has no gastric side effects because the free amino groups form a defensive shield over the stomach. Chitosan can also be used to heal connective tissues. When chitosan is acid hydrolyzed, glucosamine monosaccharides are formed, which form the proteoglycan structural units of connective tissues and cartilage, assisting in tissue repair. Chitosan derivatives with anti-tumor activity are known as chit oligosaccharides. According to some researchers, the anti-tumor effects were caused by the increased activity of natural killer lymphocytes. Chit oligosaccharides are thought to cause T-cell differentiation and proliferation by stimulating lymphocyte-activating factors. Some research suggests that chit oligosaccharides effectively boost immune responses and modulate the functions of immunocompetent cells [47]. Chitosan (poly-N-acetyl glucosamine) exhibits a hemostatic effect by accelerating erythrocyte accumulation [63]. Chitosan's positive charge promotes erythrocyte adhesion, fibrinogen absorption, and platelet adhesion and activation. Chitosan's hemostatic property is due to its polycationic content and non-specific plasma membrane binding [73]. Several in-vitro studies have revealed that the main analgesic effect of chitosan is achieved by lowering the concentration of inflammatory mediators (bradykinin) at the site of the injury. It also absorbs protons released at the site of inflammation to control pain [50].

3 Some Common Chitosan Derivatives

A. Mono-Carboxymethylated Chitosan (MCC)

MCC is a polyampholyte capable of forming viscous-elastic gels with anionic macromolecules at neutral pH or in aqueous environments. MCC appears to be less potent than the quaternized derivative of chitosan. It improves the permeation and absorption of low molecular-weight heparin (LMWH; an anionic polysaccharide) across intestinal epithelial cells. Because they cause cell membrane damage, chitosan derivatives have no effect on the viability of intestinal epithelial cells. Permeation increases due to an increase in the number of pores formed in the cell membrane, and there is no need to alter epithelial cell viability [29, 69].

B. N-Succinyl Chitosan

It is made by inserting succinyl groups into the chitosan terminals of glucosamine units. In a succinyl chitosan molecule, $-NH_3^+$ and $-COO^-$ form polyionic complexes. At various pH levels, this derivative is water-soluble [60]. N-succinyl-chitosan exhibits distinct properties in vivo and in vitro, including low toxicity, biocompatibility, and long-term retention in the body. N-succinyl-chitosan is useful as a drug carrier because it can be conjugated with various drugs to avoid complications during chemotherapy [29, 32, 33, 51, 65].

C. N-Acetylated Chitosan

N-acetylated chitosan is used as a gene delivery carrier to overcome the gastrointestinal tract's morphological and physiological barriers to target gene expression. Gene delivery to the intestine via N-acetylated chitosan is more efficient than chitosan alone. This finding is especially significant in the duodenum, where the LacZ gene is most effectively expressed using *N*-acetylated chitosan. The IL-10 gene is also successfully transferred to the intestines when mixed with *N*-acetyl chitosan. As a result, plasmid DNA can be delivered to the intestines orally using *N*-acetylated chitosan as a carrier. As a result, we can create a dietary dose system to deliver a DNA vaccine for the treatment of gastrointestinal diseases [28, 29].

D. *N*, *N*-Di carboxymethyl Chitosan

This derivative is produced by the alkylation of chitosan with monochloroacetic acid at 90 °C and pH 8–8.5. This water-soluble derivative is a chelating agent that can be used to treat osteogenesis and other chelating applications [25, 29].

E. Thiolated Chitosan Conjugate

This chitosan derivative was created by covalently attaching isopropyl-S-acetylthioacetimidate to chitosan. This derivative has in situ gelling properties, making it a promising novel tool for a variety of drug delivery systems [15, 29].

F. Polyethylene Glycol-Crosslinked *N*-Methylene Phosphonic Chitosan (NMPC)

By reductive animation, NMPC is modified with polyethylene glycol-aldehyde (PEG-CHO) of varying molecular weight. The cross-linking of NMPC with PEG-COH chains increases hygroscopicity and water swelling. This cross-linked NMPC derivative is suitable as a medical material item due to its film-forming capacity and swelling properties [15, 29].

G. Graft-Copolymerization of Chitosan

Radical polymerization is used to create this derivative with 4-(6-methacryloxyhexyloxy)-4′-nitrobiphenyl. Graft-copolymerization is carried out in a homogeneous environment using Azobisisobutyronitrile (AIBN) as an initiator and 2% acetic acid as a solvent [13, 13].

H. *N*-lauryl-Carboxymethyl-Chitosan

The hydrophobic moieties are provided by chitosan with lauryl groups attached to amino groups, and the hydrophilic moieties are provided by carboxymethyl groups attached to hydroxy groups (N lauryl-carboxymethyl-chitosan (LCC)). Taxol is found to solubilize in LCC by forming micelles with particle sizes less than 100 nm. In vitro hemolysis testing revealed that LCC is less likely to cause membrane damage than polysorbate 80 as an intravenous surfactant [29, 42].

I. Galactosylated, Chitosan-Graft-Poly (Vinyl Pyrrolidone)

GCPVP (galactosylated chitosan-graft-poly (vinyl pyrrolidone)) has been identified as a potential hepatocyte-targeting gene carrier. GCPVP alone and GCPVP/DNA complex have negligible cytotoxicity regardless of GCPVP concentration or charge ratio, but GCPVP/DNA complex has a slightly cytotoxic effect on HepG2 cells when a higher charge ratio and Ca2+ are used. It is established using confocal laser scanning microscopy that the ASGPR of hepatocytes and endocytosis by an interaction between galactose ligands of GCPVP is the major route of transfection of GCPVP/F plasmid complexes [29, 48].

J. Chitosan–Glutathione (GSH) Conjugate

This derivative improves the mucoadhesive and permeation properties of chitosan. The novel thiolated chitosan appears to be a promising multifunctional excipient for various drug delivery systems due to the strong permeation enhancing effect of the chitosan–GSH conjugate/GSH system and the improved cohesive and mucoadhesive properties [29, 44].

4 Methods of Synthesis of Chitosan Nanoparticles

In 1994, Ohya et al., for the first time, described the classification of chitosan NPs for the systemic administration of a chemotherapeutic drug [49]. Following that, the chitosan NPs have been extensively explored by the researchers leading to the development of different methods for their preparation. Various important preparation strategies are mentioned in Fig. 2, and their advantages and drawbacks are depicted in Fig. 3. A brief description of these methods is given below.

4.1 Emulsification and Cross-linking

It is the first method documented in the literature for the preparation of chitosan NPs in which the amino group of chitosan and the aldehyde group of the crosslinking agent has been employed. In this method, an emulsion and an oil phase are prepared. The emulsion consists of an aqueous solution of chitosan while the oil phase comprises span 80 as a stabilizer, toluene, and glutaraldehyde as a crosslinker [27, 49]. On intense mixing of these two phases, the droplets are formed after crosslinking which forms the basis of NPs. The NPs can be separated from the emulsion by centrifugation, multiple washings, and vacuum drying. This approach is, however, no longer employed because of the evident toxicity of glutaraldehyde [75].

4.2 Reversed Micelles Method

It is another method based on covalent crosslinking in which the production of chitosan NPs is assisted by water-in-oil reverse micelle structures (Fig. 4). This method resolved the inconveniences associated with the conventional methods based

Fig. 2 Various methods of preparation of chitosan NPs

on glutaraldehyde as a non-harmful solvent and a crosslinker are employed in this strategy [20, 31]. The aqueous phase consisting of chitosan and glutaraldehyde is mixed with the organic phase which comprises an organic solvent and a lipophilic surfactant. The chitosan NPs are formed via crosslinking within the chitosan-containing core of the micelle.

4.3 *Precipitation-Based Methods*

NPs formed by precipitation methods are generally larger than 600–800 nm. Two precipitation-based methods which have been described in the literature are:

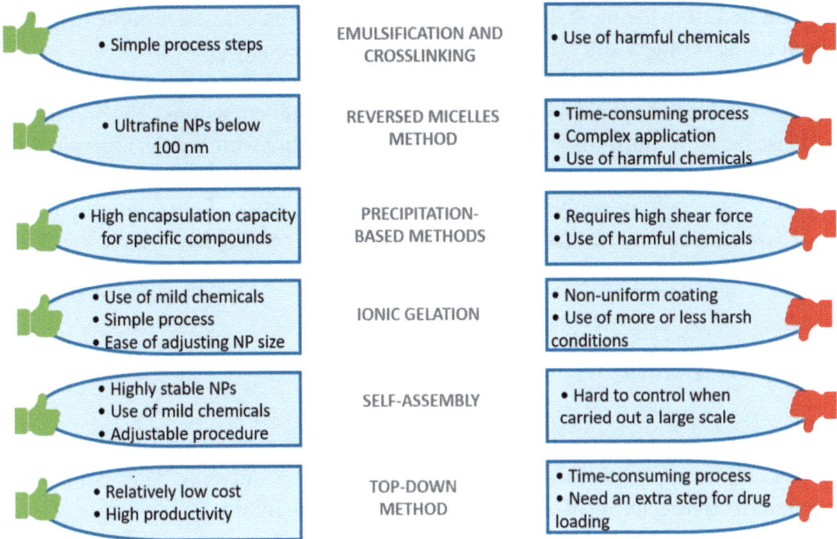

Fig. 3 Advantages and drawbacks of various preparation methods of chitosan NPs

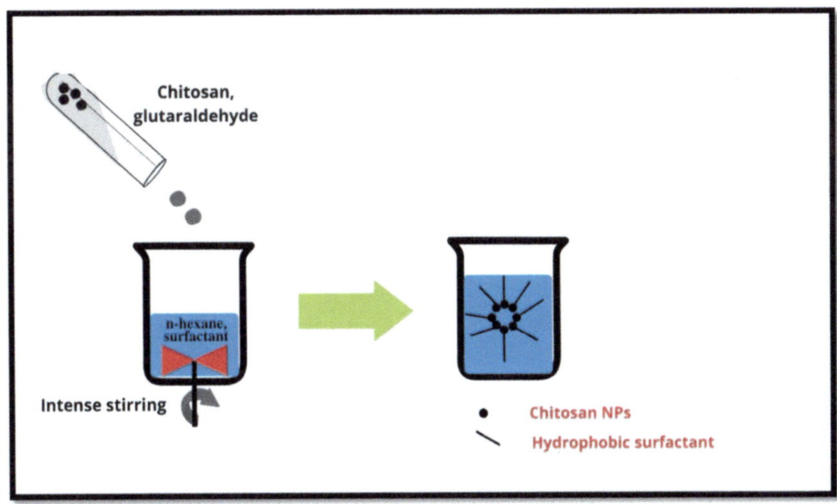

Fig. 4 Schematic illustration of formation of chitosan NPs via reversed micelles method

- The phase inversion precipitation method: This method is based on the combination of emulsification and precipitation. In the presence of a stabilizer, an oil-in-water emulsion is prepared using an organic phase and an aqueous solution of chitosan. After the application of high-pressure homogenization, the emulsion is separated from the methylene chloride by evaporation, causing acetone to diffuse out of the droplets and NPs to precipitate simultaneously [75].

- The emulsion-droplet coalescence method: In this process, the precipitation of NPs is induced by the coalescence of two water-in-oil emulsions. The continuous phase for two emulsions, one with chitosan and another with NaOH, is prepared by mixing liquid paraffin and sorbitan sesquiloleate. The chitosan-containing emulsion is prepared by applying high-speed homogenization. Following the mixing of two emulsions, NaOH diffuses into the ultrafine droplets which lower the solubility of chitosan, inducing NP formation and precipitation [24].

4.4 Ionic Gelation

It is one of the most preferred strategies for the preparation of chitosan NPs primarily because it does not involve the use of toxic crosslinkers or solvents. This method was first described by Calvo et al. in 1997 [8]. This technique follows the principle of ionic crosslinking, which occurs when the oppositely charged groups are present. Firstly, chitosan is added to the aqueous solution of an acid (generally acetic acid) followed by the addition of an aqueous solution of sodium tripolyphosphate (TPP) under intense stirring [75]. The diffusion of anionic molecules and the cationic chitosan molecules takes place leading to crosslinking and ultimately the formation of NPs (Fig. 5). Although various anionic crosslinkers like glutaraldehyde can be used to synthesize chitosan NPs, TPP is preferred because of its biocompatibility and biodegradability [59].

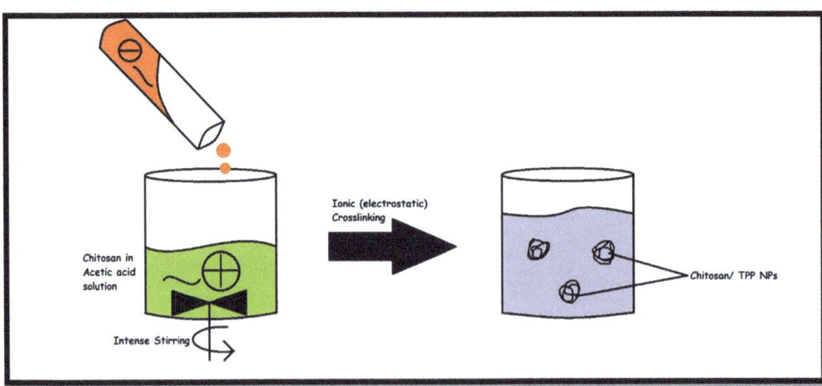

Fig. 5 Schematic illustration of production of chitosan NPs via ionic gelation method

4.5 Self-assembly

This method is extensively employed for the formation of NPs. It is based on numerous interactions which take place at the same time. The nature of interactions can be any of the following [56, 78]:

- Electrostatic
- Hydrophobic
- Hydrogen bonding
- Van der Waals forces

The self-assembly approach may either involve complex formation between chitosan and the natural anionic molecules or the modification of hydrophobicity of chitosan via grafting. NPs produced by this approach is highly suitable for the encapsulation of lipophilic and hydrophilic drugs [56].

4.6 Top-Down Approach

In this approach, two steps are involved in the preparation of nanoparticles:

(i)   Acid hydrolysis: Acid hydrolysis of chitin is carried out in the presence of a strong acid like HCl to form chitin nanocrystals.
(ii) Deacetylation: This step involves treatment with alkali (NaOH) to form chitosan NPs, which are basically the chitin NPs with a deacetylation level of more than 60% [74].

The formation of chitin NPs via the top-down method is illustrated in Fig. 6.

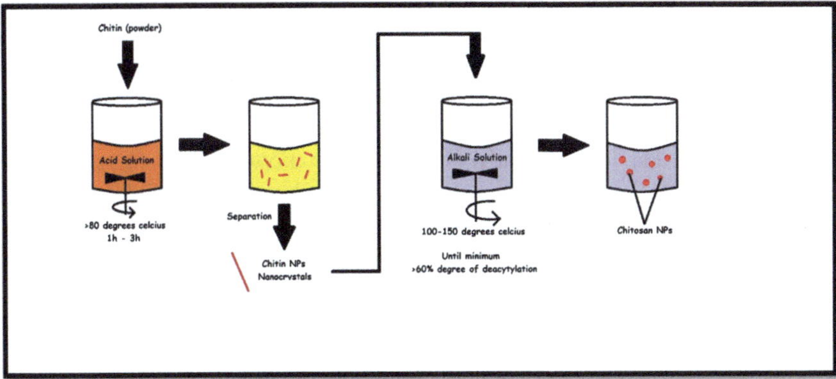

Fig. 6 Schematic illustration of formation of chitosan NPs via top-down method

5   Green Synthesis of Chitosan Nanoparticles

The biological/green methods for the extraction of chitosan are becoming highly prevalent due to the various drawbacks of chemical extraction techniques such as [64]:

- Alteration of the physico-chemical properties of chitin
- High cost of purification processes
- Presence of some chemicals in the wastewater effluents

In biological synthesis, enzymes and micro-organisms are employed for the extraction of chitin and the recovery of chitosan. The various advantages of the biological extraction of chitin are highlighted in Fig. 7. Khanafari and co-workers compared the chemical and biological methods of extraction of chitin from shrimp shells and showed the superiority of the biological method over the chemical one because the structural integrity of chitin was preserved when it was extracted through the biological method [35]. In a study, the extraction of chitin from shrimp shells and fungi was investigated by Teng et al. in which they used a one-pot fermentation process where the proteins were hydrolyzed into amino acids by fungal proteases [68]. Younes et al. extensively reviewed the methods of preparation of chitosan from marine sources and their future outlooks [79].

Spray drying is one of the green preparation routes for the preparation of chitosan NPs. In this method, chitosan is dissolved in the aqueous acetic acid and the resulting solution is passed through a nozzle, keeping the temperature range between 120 and 150 °C, leading to the formation of chitosan NPs. Spray drying can also be used to obtain magnetic chitosan NPs [22].

Supercritical-CO_2-assisted solubilization and atomization (SCASA) is a green method in which only water and CO_2 are used during the preparation process. This technique is devoid of any acid or harmful solvents. After about 48 h of the dissolution of CO_2 in water, the chitosan solution is fed to a fluidized bed, and NPs are formed which are collected by a filter placed on the fluidized bed [75]. The publications associated with the "green synthesis" of chitin/chitosan-based nanomaterials in the literature are very few. As discussed earlier, chitosan nanoparticles are frequently prepared by the ionic gelation method in which the ionic crosslinking takes place between chitosan and TPP. Recently, Gadkari et al. reported the green synthesis of antimicrobial chitosan NPs, which were prepared by chemical crosslinking between chitosan and a cinnamaldehyde, an eco-friendly bactericidal agent [18]. Purified chitin films and nano fibres with high molecular weight were successfully obtained by Qin et al. by employing eco-friendly ionic liquids [55]. Green synthesis of Ag-Ch nanocomposites was reported in which chitosan served as a reducing agent as well as a stabilizing agent, using NaOH as an accelerator. The as-prepared nanocomposite was shown to have antimicrobial activity against *E. coli* and *S. aureus* bacteria [66].

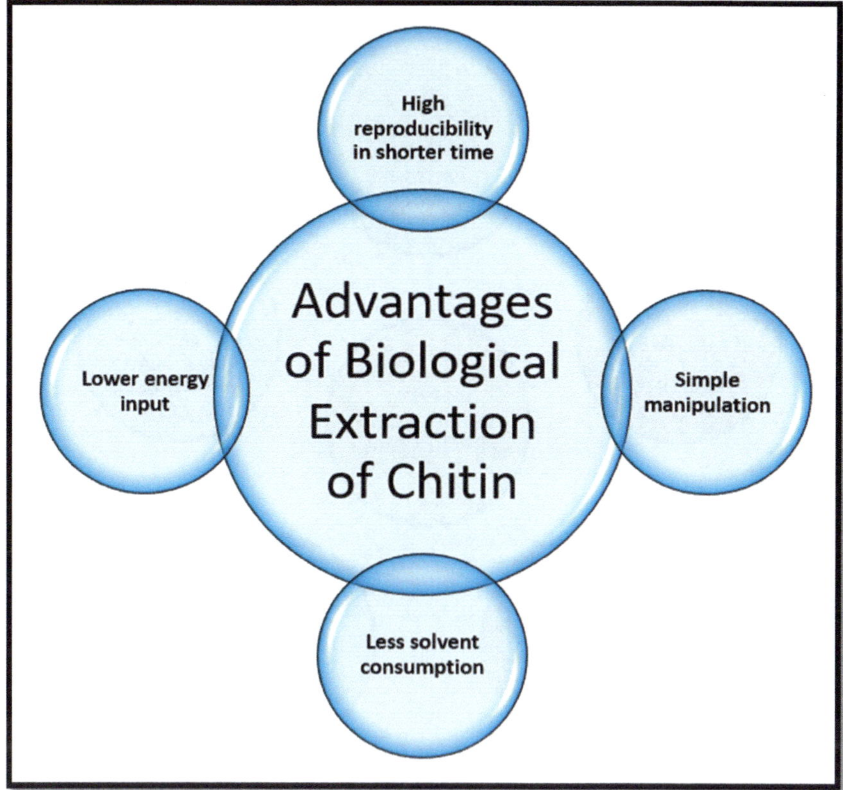

Fig. 7   Advantages of biological extraction of chitin

6   Modification Strategies of Chitosan

There are different scopes of modifications of chitosan that are depicted in Fig. 8.

6.1   Chitosan Modification to Produce Functional Derivatives

There are various types of chitosan derivatives that are produced by modification and are presented in Fig. 9.

6.1.1   Quaternized Chitosan Derivatives

According to numerous publications, it is possible to modify the positive (NH_3^+) charge of chitosan in order to make it soluble over a wide pH range and in a neutral or

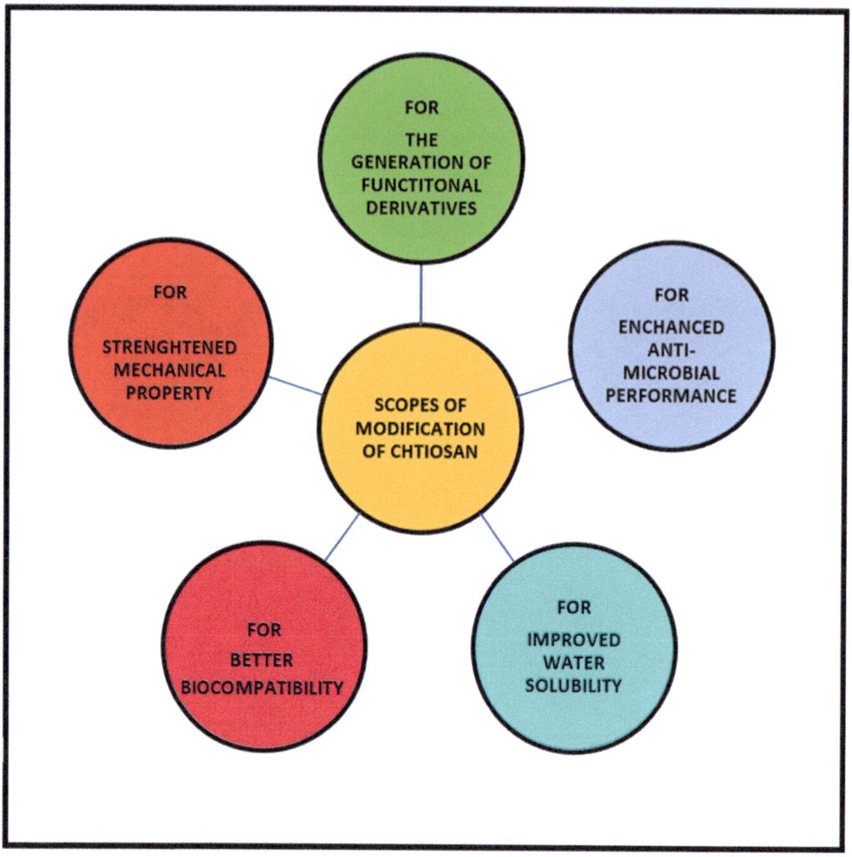

Fig. 8 A schematic diagram depicting a few scopes of modification of chitosan

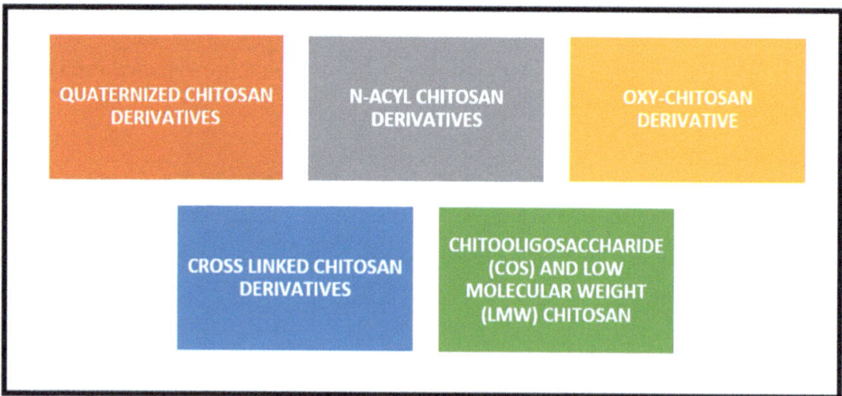

Fig. 9 A schematic representing various types of modified chitosan for the generation of functional derivatives

a little alkaline medium. This is one method for increasing the solubility of chitosan in water. At pH 6.5, chitosan is positively charged, whereas quaternized chitosan remains permanently positively charged at pH > 6.5. Under basic conditions, a quaternization reaction occurs between chitosan and alkyl iodide. The most well-known quaternized chitosan is *N,N,N*-trimethyl chitosan chloride (TMC), which has been described for a variety of applications. TMC is produced through two consecutive reactions. The first involves *N*-methyl-2-pyrrolidinone (NMP), which is used as a solvent in alkaline conditions (NaOH) in the presence of methyl iodide and chitosan, and the second involves the replacement of iodide ions with chloride ions using anionic exchange resin. By varying the carbon length of alkyl halides, different types of quaternized chitosan can be easily obtained [7].

6.1.2 *N*-Acyl Chitosan Derivatives

Chitosan becomes aquaphobic through the *N*-Acylation process, which involves implanting various fatty acids. The reaction is an addition of amide to fatty acid –COOH groups and chitosan $–NH_2$ groups. Acyl halide or acid anhydride is the chemical reagents used in *N*-Acylation. This acylation is typically carried out in pyridine/chloroform, pyridine, and methanol/acetic acid/water. Nonetheless, because the chitosan repeating unit contains two reactive –hydroxyl groups, this reaction can produce O-alkyl chitosan. Many researchers recommend replacement with trityl groups, in place of chitosan's primary –OH groups to avoid O-acylation. Because of the formation of chitosan chloroacyl, this process improves the *N*-Acylation step. *N*-Acyl chitosan has been produced using a variety of acid anhydrides [7].

6.1.3 Oxy-Chitosan Derivatives

Many researchers have looked into the production of chitouronic acid sodium (carboxylated chitin or chitosan) which is soluble in water using TEMPO-an organic catalyst used in the oxidation of –OH functions into –CHO in NaOCl and NaBr conditions. TEMPO is well-known for its use in regioselectively oxidizing the primary hydroxyl groups of various polysaccharides [7]. TEMPO is used to make oxy-chitosan derivatives, specifically 6-oxy-chitosan. Chitouronic sodium salts are typically made from shrimp or fungal cell chitin that has been pre-treated (chemically or enzymatically). A new class of carboxylated chitosan with high biocompatibility on human keratocytes was produced from Trichoderma and Aspergillus fungal biomass. C6-oxy-chitosan, a novel bioactive derivative, has shown to be efficient against activity Leishmania. Recently, an environmentally friendly process has been developed for the C_6 oxidation of chitosan using a TEMPO system to produce chitosan soluble in water has been developed [7].

6.1.4 Chitosan's Cross-linked Derivatives

The cross-linking of chitosan requires the use of very particular chemical agents to get the chains linked together, which ultimately results in a 3-D macromolecular network. Chitosan is cross-linked in the presence of –CHO derivates, such as glyoxal, glutaraldehyde (GTA), or formalin with covalent bonds, to produce chitosan-based hydrogels. The cross-linking reaction with chitosan involves the formation of a Schiff base (imine). The most researched cross-linking agent is GTA [7]. It is cheap, readily available, and synthetic. In the presence of labile hydrogen, the hired response is a condensation response among the aldehyde group and a primary amine group from the chitosan chain. However, GTA is toxic, and natural GTA substitutes are being investigated for use in the production of chitosan hydrogel. Figure 3 depicts the chitosan cross-linking reactions. Citric acid was used as a cross-linking agent in the preparation of chitosan/polyvinyl alcohol (PVA) membranes. This cross-linking strategy was investigated in order to create biomaterials for hemodialysis membranes [7]. Cross-linking citric acid and chitosan was expected to incorporate carboxylate groups (COO) into biomaterials, increasing the bioactive sites on the chitosan membrane for biomolecule transport (urea, creatinine, etc.). PVA was used to improve the crosslinked chitosan membrane's mechanical efficiency and hydrophobicity. Figure 10 depicts the main cross-linking chitosan strategies.

6.1.5 Chitosan with Low Molecular Weight (LMW) and Chito Oligosaccharide (COS)

Because of its excessive viscosity, HMW chitosan is difficult to be applied commercially. Because of the production of COS and the low molecular weight of chitosan, a reduction in its molecular weight is an effective idea for addressing viscosity issues and improving its biological properties [7]. Oligosaccharides are defined as oligomers with a degree of polymerization ranging from 2 to 10, but some higher degrees of polymerization are classified as LMW. COS and LMW chitosan are primarily produced using enzymatic, chemical, and physical methods. The reduction of MW via enzymatic, chemical, or physical processes is linked to improved chitosan water or acetic acid solubility. Depolymerization of chitosan is primarily accomplished through chitosanolysis of acid, which is the most used method for producing COS and LMW chitosan. Using HCl, HNO_2, H_2O_2, and potassium persulfate in chitosanolysis are examples of chemical processes. Sonication, microwave irradiation, gamma irradiation, or a thermal procedure are examples of physical processes. Enzymatic processes employ both specific enzymes and nonspecific enzymes like chitinase, chitosanase and pepsin, cellulase, pronase, lysozyme, papain, hemicellulase, or pectinase, respectively. However, the main issues with enzymatic depolymerization are the costs [7]. High-pressure homogenization (HPH), plasma, and the use of zeolites as adsorbents to purify acid hydrolysis COS and LMW chitosan have all been described as new methods for reducing the molecular mass of chitosan. Furthermore, electrochemical processes for depolymerizing chitosan have been developed.

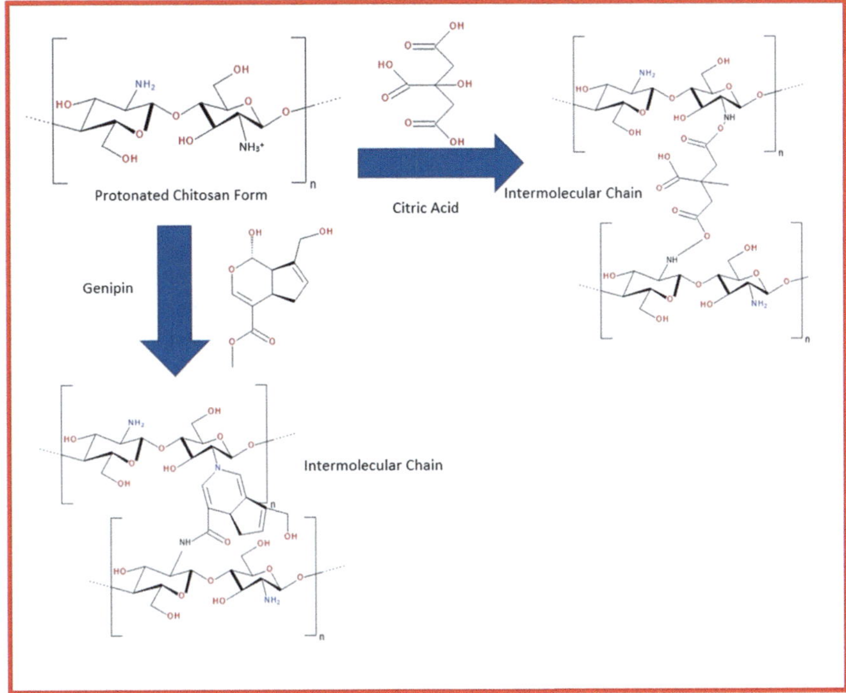

Fig. 10 The cross-linking reactions of chitosan

6.2 *Enhanced Anti-microbial Performance*

6.2.1 Introducing Cationic Compounds

Although the degree of de-acetylation, molecular weight, and content of the NH_2 residual group can influence the anti-microbial activity of chitosan derivatives, the same of chitosan derivatives is primarily dependent on the positively charged NH_2 group, which can interfere with and destroy the negative bacterial membrane, bind bacterial DNA, and block protein synthesis [14]. The higher the positive charge, the greater the microbial activity. Increasing positive electricity can improve the antimicrobial activity of chitosan derivatives:

A. NH_2 Direct Quaternization

Quaternary ammonium salt is the most commonly used anti-microbial group in chitosan derivatives [1]. NH_2's direct quaternization group was carried out primarily to improve the anti-microbial activity of the chitosan derivates, which can be tuned by varying the ratio of the cationic groups present. Following research, it was discovered that the N-trimethyl group was the most important factor in improving the anti-microbial activity of chitosan derivatives [14]. By varying the trimethylation, acetylation, and acylation, the chitosan derivative with the best anti-microbial activity, better than *S. aureus* and *E. coli*, could be obtained.

B. Grafting onto Chitosan

By grafting QAS, and betaine, chitosan derivatives' antimicrobial activity can be improved. Combining acetate chitosan and C_{12}–C_{18} alkyl aminopropyl betaine demonstrated a broader anti-microbial spectrum and superior anti-microbial activity than either compound alone. Despite the fact that glycine betaine contains QAS, its anti-bacterial mechanism differs from traditional QAS due to chelation between the NH_2 group and the trace amount of metal ions (Ca^{2+}, Na^+) that are required for bacteria to maintain their stability and metabolism of the cell, which can destroy the stability and inhibit the growth of cell membrane [16]. As a result, retaining the necessary amounts of NH_2 and OH groups during the chemical modification of chitosan is required [14].

C. Grafting Polymerization from Chitosan

Through grafting copolymerization of chitosan acrylamide and methacrylamido-propyl trimethyl ammonium chloride, PQAS can be incorporated into the chitosan skeleton. The produced chitosan derivatives can destroy the bacteria membrane at just a concentration of 14 mg/L. Similarly, anti-microbial chitosan-based polymerized nanoparticles with positive surface charges were synthesized by grafting *N*-trimethyl aminoethyl methacrylate chloride or methyl methacrylate and *N*-dimethyl aminoethyl methacrylate hydrochloride from chitosan [14].

6.2.2   Neutral Anti-microbial Compound Conjugation

A. Polypeptide

Another way of producing chitosan peptide conjugate (CPC) with improved antimicrobial activity is to integrate it with an antimicrobial peptide (CGGG(KLAKLAK)2) and an enzyme-cleavable peptide (GPLGVRGC). The morphology of CPC can be enhanced due to the peeling off of the protective PEG layer induced by the presence of gelatinase from bacteria, resulting in excellent anti-microbial activity with a minimum inhibitory concentration value of about 7×10^{-6} M based on the disruption of hydrophobic/hydrophilic balance.

B. NO Donor

Chitosan can be used as a loading matrix for NO, an anti-biofilm agent [58] to create a derivative that can slowly release NO by sparging chitosan derivatives with NO gas Planktonic cells treated with NO-loaded chitosan show a decrease in colony-forming units, which could be useful in wound dressing. NO can be loaded on chitosan-poly(amidoamine) via reaction with its secondary NH_2 group, yielding chitosan-poly(amidoamine)/NONOate with a loading amount of 1.7 mol/mg and excellent antimicrobial and wound healing activity.

6.2.3 Metal Coordination

By coordinating chitosan with Ag^+ ions, a construction based on the synergistic impact of chitosan and Ag^+ ions was recently done [9], Moldable hydrogel with an antibacterial action against *S. aureus* and *E. coli*. The Ag^+ ions can be complexed with chitosan and then photo reduced or electro reduced in situ [77]. The anti-microbial effect was greatly enhanced by the synergistic effect of chitosan and metal ions, which stabilized the metal through coordination and by increasing its positive charge, which further promoted its interaction with negatively charged bacteria. Cu(II) complexes formed by chelating O-carboxymethyl chitosan (CMCS) Schiff bases with Cu^{2+} ions have significantly higher antifungal activity against Phytophthora capsici than the original chitosan [54]. The easy release of copper cations from the complex, caused by the benzene ring's space steric hindrance, results in an effective anti-fungal performance.

6.2.4 Physical Integration

The physical blending of chitosan derivatives with antibiotics is another common method to increase the anti-microbial activity of chitosan derivatives. Through the solgel transition process, a polyelectrolyte composite hydrogel based on chitosan/CMchitosan/AgNPs demonstrates long-term antibacterial efficacy against *S. aureus* and *P. aeruginosa* [77]. Even while direct combination with antibiotics can enhance chitosan's anti-microbial performance, slow-release antibiotics may increase the probability of microorganism resistance, which requires careful consideration in future research. Based on the synergistic impact of CNC rod and poly-cationic CS, chitosan can be mixed with cellulose nanocrystals (CNCs) to generate a hydrophobic spray-coating composite with increased antimicrobial action. CNC can cause bacterial membrane rupture, allowing chitosan derivatives with protonated NH_2 groups to enter the cell, resulting in bacterial death [70].

6.3 Improved Water Solubility

For the improvement of solubility in the water of chitosan under a basic and neutral environment, hydrophilic compounds with reactive groups can be conjugated with chitosan through amide and hydroxyl groups [14].

6.3.1 Modification of the Amine Group

Hydrophilic compounds with –COOH, –CHO, –OCN, and epoxy easily bind to the backbone of the chitosan with the aid of coupling reactions with its very reactive amide group. By grafting Brij-S20-succinic anhydride to chitosan, the solubility in the water of chitosan can be enhanced. Brij-S20-succinic anhydride disrupts the inter/intra molecular bonds, lowering crystallinity and increasing solubility. Chitosan complex 1-(4-(2-aminoethyl) phenoxy) zinc(II) phthalocyanine (ZnPcN) [46] on conjugation with acid groups of CMCS, and amide groups of phthalocyanine, improves solubility in the water because the conjugate's absorption intensity ratio is double than that of ZnPcN used alone, and agglomeration in water is reduced. When the amide group of chitosan reacts with the acid group of gluconic acid, the generated sugar-carrying chitosan is soluble under various pH settings due to the hydrophilicity of the gluconyl group and the *N*-acetyl group [14].

6.3.2 Hydroxyl Group Modification

The OH group of chitosan reacts with halo hydrocarbon or sulfonyl chloride to produce CMCS, which has better solubility in the water [40]. The solubility in the water of chitosan can also be improved by the reaction of its hydroxyl group with *N*-acetyl ethylenediamine. It is evident that solubility in water of chitosan can be enhanced by the replacement of the hydrophilic group chemically [14].

6.4 Better Biocompatibility

Biocompatibility is very important, especially in cases where biomaterials have a direct contact with human organs or tissues [14]. It relates to how specific tissues react to different elements.

6.4.1 Organic Modification

A. Introduction of Carboxyl Group

By inserting a carboxyl group into the molecular structure of chitosan derivates, the toxicity of protonated cationic amide or groups in chitosan derivates can be minimized, resulting in CMCS, which is non-toxic, biocompatible, and biodegradable [14].

B. Introduction of Betaine

Zwitterionic betaine can combine huge amounts of water to form a hydration layer through hydrogen bonding and electrostatic contact [21, 34, 80]. The water of hydration between proteins and zwitterionic molecule imposes substantial repulsions against nonspecific protein adsorption because it restricts the movement of water adsorbed at the surface of glycine betaine. With the addition of betaine, the derivative exhibits effective antimicrobial characteristics, anti-nonspecific protein adsorption, and anti-platelet adhesiveness [14].

C. PEGylation

Because of its high hydrophilicity and chain flexibility, PEG has been used to alter chitosan derivatives to minimize toxicity and nonspecific protein adsorption, resulting in a volume exclusion effect [14]. Crosslinking chitosan with aldehyde terminated PEG or grafting activated monomethoxy poly(ethylene glycol) to chlorinated N-phthaloyl chitosan can be used to make the PEG-modified chitosan derivative [43].

D. Introduction of Polycaprolactone

Biocompatible polycaprolactone (PCL) can promote cell proliferation and adhesion [41], as a result, it can be utilized to improve the rate of healing and the quality of neonatal tissue. In comparison to $PHBHH_x$ or chitosan-g-PCL/$PHBHH_x$, the chitosan-g-PCL and poly(3-hydroxybutyrate-co-3-hydroxyhexanoate) ($PHBHH_x$) hybrid have a rougher surface and smaller grain sizes, which can increase cell adhesion and proliferation, resulting in higher initial cell survival.

E. Introduction of Gelatin

Mixed –OH groups modified biocompatible boron nitride nanotubes (BNNTs) with chitosan to form a slow degradative chitosan/BNNT-OH scaffold that showed improved cellular proliferation and adhesion compared to chitosan, based on the increased mechanical performance and pore size of biocompatible boron nitride nanotubes (BNNTs) [6]. The discontinuousness and hydrophilicity of water-swollen gelatin nanospheres are responsible for the deposited derivatives with controlled wettability and surface roughness. The ratio of chitosan and harmless gelatin nanospheres in the derivative can be changed to alter protein adsorption and cell adhesion [14].

6.4.2 Inorganic Modification

A. Carbon/Boron Materials

Some inorganic compounds, such as graphene oxide, can improve chitosan biocompatibility by enhancing its surface shape, roughness, and pole size, which encourages cell attachment and proliferation at the surface of chitosan derivatives [14].

Mixed OH groups modified biocompatible boron nitride nanotubes (BNNTs) with chitosan to form a slow degradative chitosan/BNNT-OH scaffold that showed improved cellular proliferation and adhesion compared to chitosan, based on the increased mechanical performance and pore size of biocompatible boron nitride nanotubes (BNNTs) [14].

B. Metal Ions

By combining biocompatible Zn^{2+} ions with chitosan, the resulting porous CS_2ZnAlg microspheres can further increase [Zn^{2+}] [52]. When compared with the control group, the hemostasis time of the chitosan/Zn^{2+} derivative was drastically reduced (134 5 s) (over 600 s) [14]. Following the termination of bleeding, the chitosan/zinc derivative shapes dark red aggregates of blood cells, platelets, fibrins, and microspheres near the wound site. Chitosan's positive charge can reendow its mucoadhesive activity because it merges with negatively charged red blood cells.

Enhanced Mechanical Property

Chitosan's mechanical performance can be bettered if it forms a great number of networks either by physical or by chemical crosslinking [61] or in combination with metal nanoparticles [39]. The physically crosslinked hydrogel double-network (PCDHN) of chitosan/Na alginate (SA)/calcium ions (Ca^{2+}) is much better than a physical single network of hydrogel cross-linked via electrostatic interactions, and 10 times better than chitosan/SA hydrogel. The creation of a double network through the use of hydrogen bonding can also strengthen chitosan derivatives. Poly(vinyl alcohol) CS [45] (CPH) DN hydrogel has exceptional tensile strength, elongation at break, and compressive strength thanks to physical crosslinking. The adsorption of fracture energy by chitosan's first physical network results in CPH's outstanding excellent mechanical quality, while the second poly vinyl alcohol network maintains the hydrogel's shape. Furthermore, the biocompatible chitosan-based hydrogel with surface mineralization can enhance the differentiation of bone marrow stem cells [14].

7   Conclusions

Chitosan, one of the most available biopolymers, has been extensively studied and used in diverse fields due to its biodegradability, biocompatibility, non-toxicity, antimicrobial, and antifungal properties. However, its broad applications are limited by its poor solubility and insufficient mechanical strength. To this, chemical modification has been demonstrated to be an effective approach. After their first description over two decades ago, Chitosan NPs have acquired a significant position in various industries like pharma and agriculture. Various methods for the modification and preparation of chitosan NPs have been developed. Sometimes, the NP synthesis requires the use of toxic substances which may adversely affect the environment. To overcome this issue, green preparation techniques stand out as an alternative. These methods are eco-friendly as well as cost-effective. Also, these are devoid of toxic chemicals and high energy, pressure, etc., are not required. Moreover, the synthesis of chitosan NPs using biological methods (using microorganisms like bacteria, fungi, and algae) is even more appealing since it is rapid, non-toxic, and environment-friendly. However, there are scarce reports available on the green synthesis of NPs. The development of novel and innovative green methods for the synthesis of chitosan NPs is a domain where more research and developments are expected in the future.

References

1. Abu Elella M (2021) Synthesis and potential applications of modified Xanthan gum. J Chem Eng Res Updates 8:73–97
2. Aranaz I et al (2012) Functional characterization of chitin and chitosan. Curr Chem Biol 3(2):203–230
3. Azmana M et al (2021) A review on chitosan and chitosan-based bionanocomposites: promising material for combatting global issues and its applications. Int J Biol Macromol 185(June):832–848
4. Bagheri-Khoulenjani S, Taghizadeh SM, Mirzadeh H (2009) An investigation on the short-term biodegradability of chitosan with various molecular weights and degrees of deacetylation. Carbohyd Polym 78(4):773–778
5. Bano I et al (2017) Chitosan: a potential biopolymer for wound management. Int J Biol Macromol 102:380–383
6. Boccaccini AR et al (2010) Electrophoretic deposition of biomaterials. J Royal Soc Interface 7(Suppl 5):S581–S613
7. Brasselet C et al (2019) Modification of chitosan for the generation of functional derivatives. Appl Sci (Switzerland) 9(7)
8. Calvo P, Remuñán-López C, Vila-Jato JL, Alonso MJ (1997) Novel hydrophilic chitosan-polyethylene oxide nanoparticles as protein carriers. J Appl Polym Sci 63(1):125–132
9. Cao J et al (2018) Dual physical crosslinking strategy to construct moldable hydrogels with ultrahigh strength and toughness. Adv Func Mater 28:1800739
10. Chang S-H et al (2019) Effect of chitosan molecular weight on anti-inflammatory activity in the RAW 264.7 macrophage model. Int J Biol Macromol 131:167–175
11. Chebotok EN, Yu V, Novikov, Konovalova IN (2006)Depolymerization of chitin and chitosan in the course of base deacetylation.Russ J Appl Chem 79(7):1162–1166
12. Croisier F, Jérôme C (2013) Chitosan-based biomaterials for tissue engineering. Eur Polymer J 49(4):780–792
13. Dal Pozzo A et al (2000)Preparation and characterization of poly(ethyleneglycol)-crosslinked reacetylated chitosans.Carbohyd Polym 42:201–206
14. Ding S, Wang Y, Li J, Chen S (2021) Progress and prospects in chitosan derivatives: modification strategies and medical applications. J Mater Sci Technol 89:209–224
15. Ding W, Lian Q, Samuels RJ, Polk MB (2003)Synthesis and characterization of a novel derivative of chitosan.Polymer 44:547–556
16. Dutta PK, Tripathi S, Mehrotra GK, Dutta J (2009)Perspectives for chitosan based antimicrobial films in food applications.Food Chem 114(4):1173–1182
17. Elsabee MZ, Abdou ES (2013) Chitosan based edible films and coatings: a review. Mater Sci Eng C Mater Biol Appl 33(4):1819–41
18. Gadkari RR et al (2019) Green synthesis of chitosan-cinnamaldehyde cross-linked nanoparticles: characterization and antibacterial activity. Carbohyd Polym 226:115298. https://doi.org/10.1016/j.carbpol.2019.115298
19. Ghosh A, Ali M (2012) Studies on physicochemical characteristics of chitosan derivatives with dicarboxylic acids. J Mater Sci 47:1196–1204
20. Grenha A (2012) Chitosan nanoparticles: a survey of preparation methods. J Drug Target 20(4):291–300
21. Hu Z et al (2018) Investigation of the effects of molecular parameters on the hemostatic properties of chitosan. Molecules (Basel, Switzerland) 23(12)
22. Huang HY, Shieh YT, Shih CM, Twu YK (2010) Magnetic chitosan/iron (II, III) oxide nanoparticles prepared by spray-drying. Carbohyd Polym 81(4):906–910. https://doi.org/10.1016/j.carbpol.2010.04.003
23. Ibañez-Peinado D, Ubeda-Manzanaro M, Martínez A, Rodrigo D (2020) Antimicrobial effect of insect chitosan on Salmonella typhimurium, Escherichia coli O157:H7 and Listeria monocytogenes survival. PLoS ONE 15:1–14, 12 Dec 2020

24. Ichikawa H, Tokumitsu H, Miyamoto M, Fukumori Y (2007) Nanoparticles for neutron capture therapy of cancer. In: 6 Nanotechnologies for the life sciences
25. Inamdar N, Mourya VK, Tiwari A (2010) Carboxymethyl chitosan and its applications. Adv Mater Lett 1:11–33
26. Islam S, Rahman Bhuiyan MA, Islam MN (2017) Chitin and chitosan: structure, properties and applications in biomedical engineering. J Polym Environ 25(3):854–866
27. Jameela SR, Kumary TV, Lal AV, Jayakrishnan A (1998) Progesterone-loaded chitosan microspheres: a long acting biodegradable controlled delivery system. J Control Release 52:17–24. Jameela1998.Pdf
28. Jintapattanakit A et al (2007) Peroral delivery of insulin using chitosan derivatives: a comparative study of polyelectrolyte nanocomplexes and nanoparticles. Int J Pharm 342(1–2):240–249
29. Hemant Yadav KS, Joshi GB, Singh MN, HG Shivakumar (2010) Naturally occurring chitosan and chitosan derivatives: a review.Current Drug Therapy 6(1):2–11
30. Kaczmarek MB et al (2019) Enzymatic modifications of chitin, chitosan, and chitooligosaccharides. Front Bioeng Biotechnol 7(SEP)
31. Kafshgari MH et al (2012) Preparation of alginate and chitosan nanoparticles using a new reverse micellar system. Iran Polym J (English Edition) 21(2):99–107
32. Kato Y, Onishi H, Machida Y (2000) Evaluation of N-succinyl-chitosan as a systemic long-circulating polymer. Biomaterials 21(15):1579–1585
33. Kato Y, Onishi H, Machida Y (2005) Contribution of chitosan and its derivatives to cancer chemotherapy. In Vivo (Athens, Greece) 19(1):301–310
34. Keefe AJ, Jiang S (2011) Poly(zwitterionic)protein conjugates offer increased stability without sacrificing binding affinity or bioactivity. Nat Chem 4(1):59–63
35. Khanafari A, Marandi R, Sanatei S (2008) Recovery of chitin and chitosan from shrimp waste by chemical and microbial methods. Iran J Environ Health Sci Eng 5(1):19–24
36. Kim I-Y et al (2008) Chitosan and its derivatives for tissue engineering applications. Biotechnol Adv 26(1):1–21
37. Kim S (2018) Competitive biological activities of chitosan and its derivatives: antimicrobial, antioxidant, anticancer, and anti-inflammatory activities. Int J Polym Sci 2018:1708172 [Tao Y (ed)]
38. Kong M, Chen XG, Xing K, Park HJ (2010) Antimicrobial properties of chitosan and mode of action: a state of the art review. Int J Food Microbiol 144(1):51–63
39. Krawczak P (2019) Editorial corner—a personal view: polymer composites: evolve towards multifunctionality or perish. Express Polym Lett 13(9):771
40. Kritchenkov AS et al (2020) Efficient reinforcement of chitosan-based coatings for ricotta cheese with non-toxic, active, and smart nanoparticles. Prog Org Coat 145(April):105707
41. Lee K-W et al (2007) Poly(propylene fumarate) bone tissue engineering scaffold fabrication using stereolithography: effects of resin formulations and laser parameters. Biomacromolecules 8(4):1077–1084
42. Leitner VM, Walker GF, Bernkop-Schnürch A (2003) Thiolated polymers: evidence for the formation of disulphide bonds with mucus glycoproteins. Eur J Pharm Biopharm Official Journal of Arbeitsgemeinschaft fur Pharmazeutische Verfahrenstechnik e.V 56(2):207–214
43. Li Y, Wang X, Wei Y, Tao L (2017) Chitosan-based self-healing hydrogel for bioapplications. Chin Chem Lett 28(11):2053–2057
44. Lin Y, Chen Q, Luo H (2007) Preparation and characterization of N-(2-carboxybenzyl)chitosan as a potential PH-sensitive hydrogel for drug delivery. Carbohyd Res 342(1):87–95
45. Lin Z et al (2021) Preparation of chitosan/calcium alginate/bentonite composite hydrogel and its heavy metal ions adsorption properties. Polymers 13(11)
46. Liu S et al (2022) Biocompatible gradient chitosan fibers with controllable swelling and antibacterial properties. Fibers Polym 23(1):1–9
47. Lodhi G et al (2014) Chitooligosaccharide and its derivatives: preparation and biological applications. Biomed Res Int 2014:654913
48. Miwa A et al (1998) Development of novel chitosan derivatives as micellar carriers of taxol. Pharm Res 15(12):1844–1850

49. Ohya Y, Shiratani M, Kobayashi H, Ouchi T (1994) Release behavior of 5-fluorouracil from chitosan-gel nanospheres immobilizing 5-fluorouracil coated with polysaccharides and their cell specific cytotoxicity. J Macromol Sci Part A 31(5):629–642
50. Okamoto Y et al (2002) Analgesic effects of chitin and chitosan. Carbohyd Polym 49(3):249–252
51. Onishi H, Takahashi H, Yoshiyasu M, Machida Y (2001) Preparation and in vitro properties of N-succinylchitosan- or carboxymethylchitin-mitomycin C conjugate microparticles with specified size. Drug Dev Ind Pharm 27(7):659–667
52. Pan M et al (2018) Porous chitosan microspheres containing zinc ion for enhanced thrombosis and hemostasis. Mater Sci Eng C Mater Biol Appl 85:27–36
53. Park BK, Kim M-M (2010) Applications of chitin and its derivatives in biological medicine. Int J Mol Sci 11(12):5152–5164
54. Potara M et al (2011) Synergistic antibacterial activity of chitosan-silver nanocomposites on Staphylococcus aureus. Nanotechnology 22(13):135101
55. Qin Y, Xingmei L, Sun N, Rogers RD (2010)Dissolution or extraction of Crustacean shells using ionic liquids to obtain high molecular weight purified chitin and direct production of chitin films and fibers.Green Chem 12(6):968–997
56. Quiñones JP, Peniche H, Peniche C (2018) Chitosan based self-assembled nanoparticles in drug delivery. Polymers 10(3):1–32
57. Rahman Bhuiyan MA et al (2017) Chitosan coated cotton fiber: physical and antimicrobial properties for apparel use. J Polym Environ 25(2):334–342
58. Reighard KP et al (2017) Role of nitric oxide-releasing chitosan oligosaccharides on mucus viscoelasticity. ACS Biomater Sci Eng 3(6):1017–1026
59. Riegger BR et al (2018) Chitosan nanoparticles via high-pressure homogenization-assisted miniemulsion crosslinking for mixed-matrix membrane adsorbers. Carbohyd Polym 201:172–181. https://doi.org/10.1016/j.carbpol.2018.07.059
60. Sannan T, Kurita K, Iwakura Y (2003) Studies on chitin, 2. Effect of deacetylation on solubility. Die Makromolekulare Chemie 177:3589–3600
61. Serhan M et al (2019) Total iron measurement in human serum with a smartphone. In: AIChE annual meeting, conference proceedings, Nov 2019
62. Shariatinia Z (2019) Pharmaceutical applications of chitosan. Adv Coll Interface Sci 263:131–194
63. Singh R, Shitiz K, Singh A (2017) Chitin and chitosan: biopolymers for wound management. Int Wound J 14(6):1276–1289
64. Sivanesan I et al (2021) Nanoforms/nanocomposites for drug delivery applications, 1–21
65. Song Y, Onishi H, Nagai T (1993) Pharmacokinetic characteristics and antitumor activity of the N-succinyl-chitosan-mitomycin C conjugate and the carboxymethyl-chitin-mitomycin C conjugate. Biol Pharm Bull 16(1):48–54
66. Susilowati E, Maryani, Ashadi (2019) Green synthesis of silver-chitosan nanocomposite and their application as antibacterial material. J Phys Conf Ser 1153(1)
67. Tan H, Chu C, Payne K, Marra K (2009) Injectable in situ forming biodegradable chitosan-hyaluronic acid based hydrogels for cartilage tissue engineering. Biomaterials 30:2499–2506
68. Teng WL et al (2001) Concurrent production of chitin from shrimp shells and fungi. Carbohyd Res 332(3):305–316
69. Thanou M, Verhoef JC, Junginger HE (2001) Oral drug absorption enhancement by chitosan and its derivatives. Adv Drug Deliv Rev 52(2):117–126
70. Tyagi P et al (2019) High-strength antibacterial chitosan-cellulose nanocrystal composite tissue paper. Langmuir ACS J Surf Colloids 35(1):104–112
71. Verlee A, Mincke S, Stevens CV (2017) Recent developments in antibacterial and antifungal chitosan and its derivatives. Carbohyd Polym 164:268–283
72. Wang N et al (2020) Antibacterial effect of chitosan and its derivative on Enterococcus faecalis associated with endodontic infection. Exp Ther Med 19(6):3805–3813
73. Wang YW et al (2019) Biological effects of chitosan-based dressing on hemostasis mechanism. Polymers 11(11)

74. Wijesena RN et al (2015) A method for top down preparation of chitosan nanoparticles and nanofibers. Carbohyd Polym 117:731–738. https://doi.org/10.1016/j.carbpol.2014.10.055
75. Yanat M, Schroën K (2021) Preparation methods and applications of chitosan nanoparticles; with an outlook toward reinforcement of biodegradable packaging. React Funct Polym 161(Feb)
76. Yang J et al (2008) Effect of chitosan molecular weight and deacetylation degree on hemostasis. J Biomed Mater Res Part B Appl Biomater 84(1):131–137
77. Yang J et al (2020) Preparation of a chitosan/carboxymethyl chitosan/AgNPs polyelectrolyte composite physical hydrogel with self-healing ability, antibacterial properties, and good biosafety simultaneously, and its application as a wound dressing. Compos B Eng 197:108139
78. Yang Y et al (2014) Advances in self-assembled chitosan nanomaterials for drug delivery. Biotechnol Adv 32(7):1301–1316.https://doi.org/10.1016/j.biotechadv.2014.07.007
79. Younes I et al (2012) Chitin and chitosan preparation from shrimp shells using optimized enzymatic deproteinization. Process Biochem 47(12):2032–2039. https://doi.org/10.1016/j.procbio.2012.07.017
80. Zhang L et al (2013) Zwitterionic hydrogels implanted in mice resist the foreign-body reaction. Nat Biotechnol 31(6):553–556

Chapter 3
Characterization Techniques for Chitosan and Its Based Nanocomposites

Gunjan Purohit and Diwan S. Rawat

Abstract Chitosan nanocomposites/nanoparticles (NPs) are biobased polymeric materials that have gained booming intent due to their versatile physicochemical characteristics and properties. The processed chitosan, i.e., chitosan nanoparticles in comparison to bulk counterparts possesses versatile biological/biodegradable applications because of their smaller sizes, higher surface area, and cationic nature. Various morphologies of chitosan are reported in the literature, viz., nano vehicles, nanoparticles, nanocomposites, fibers, meshes, nanocapsules, and so on for a variety of applications. The underlying chapter critically reviews the series of characterization techniques that determine chitosan nanocomposites' morphological and physicochemical characteristics such as surface properties, charges, particle size, particle appearance, elemental composition, and surface interactions. The current chapter is aimed to discuss the popular techniques, namely, transmission electron microscopy (TEM), scanning electron microscopy (SEM), energy-dispersive X-ray spectroscopy (EDX), X-ray photoelectron spectroscopy (XPS), X-ray diffraction (XRD), inductively coupled plasma mass spectroscopy (ICP-MS), fourier-transform infrared spectroscopy (FT-IR), atomic emission spectroscopy (AES), dynamic light scattering (DLS), atomic force microscopy (AFM), nuclear magnetic resonance (NMR), etc., which helps in determining and developing a consistent, precise, and reliable characterization of chitosan nanocomposites.

Keywords Chitosan nanoparticles · Morphological · Properties · Solid-state properties · Interaction analysis · Surface charge

Abbreviations

AAS	Atomic absorption spectroscopy
AES	Atomic emission spectroscopy
AFM	Atomic force microscopy

G. Purohit · D. S. Rawat (✉)
Department of Chemistry, University of Delhi, Delhi 110007, India
e-mail: dsrawat@chemistry.du.ac.in

© The Author(s), under exclusive license to Springer Nature Singapore Pte Ltd. 2022
S. Gulati (ed.), *Chitosan-Based Nanocomposite Materials*,
https://doi.org/10.1007/978-981-19-5338-5_3

CHI   Chitosan (CHI)
DLS   Dynamic light scattering
DMC/TPP   *N,N*-Dimethyl chitosan/tripolyphosphate
EDX   Energy-dispersive X-ray spectroscopy
ESCA   Electron spectroscopy for chemical analysis
ETAs   Electrothermal atomizers
FESEM   Field emitter scanning electron microscopy
FT-IR   Fourier-transform infrared spectroscopy
HCL   Hollow cathode lamp
ICDD   International Centre for Diffraction Data
ICP-MS   Inductively coupled plasma mass spectroscopy
JCPDS   Joint Committee on Powder Diffraction Standards
MW   Molecular weight
NMR   Nuclear magnetic resonance
NPs   Nanoparticles
PCS   Photon correlation spectroscopy
PLA/CS   Poly(lactic acid)/chitosan
QELS   Quasi-elastic light scattering
SEM   Scanning electron microscopy
TEM   Transmission electron microscopy
TMC/TPP   *N,N,N*-Trimethyl chitosan/tripolyphosphate
XPS   X-ray photoelectron spectroscopy
XRD   X-ray diffraction

1 Introduction

Chitosan (CHI), i.e., "cationic (1-4)-2-amino-2-deoxy-β-d-glucan" is a linear semicrystalline naturally occurring biopolymer/polysaccharide that is produced by deacetylation of homopolysaccharide "chitin" [1]. The primary source of chitosan is marine organisms, fungi-cell walls, and insects-exoskeletons which have major components containing glucosamine and N-acetyl glucosamine associated via β-(1-4) linkages. During the chitin-deacetylation process, a segment of the N-acetyl groups is lost in an alkaline medium, and this polysaccharide is named chitosan (CHI) possessing D-glucosamine-derived chitin units. The process of chitin-deacetylation can be performed either by using enzymatic hydrolysis (chitin deacetylase) or under alkaline conditions using a blend of anhydrous hydrazine-hydrazine sulfate or potassium/sodium hydroxide solutions. Only in an acidic medium, this CHI is soluble and positively charged. When the pH of CHI crosses its pKa value (≈6.5), the polymeric chitosan dissipates its positive charge and precipitates, making it a pH-responsive material [2]. The presence of abundant amino and highly reactive hydroxyl functionalization on the chitosan backbone enhances its chelating ability and polyelectrolytic nature. Such elusive properties of chitosan make it soluble only in a dilute acidic solution and insoluble in water or any other organic solvents. Chitosan also exhibits antifungal, antimicrobial, analgesic, hemostatic, and mucoadhesive properties. The origin of the chitin source and deacetylation method alters the structural properties of chitosan, viz., molecular weight (MW), and the degree of deacetylation. The low toxicity, biocompatibility, and biodegradability of chitosan and its derived nanoparticles have attracted the attention of academicians/researchers due to its potential applicability in the renowned areas, viz., agriculture, engineering, biotechnology, pharmaceuticals, etc. Chitosan nanomaterials in comparison to bulk counterparts exhibit an array of applications due to their high surface-to-volume ratio, smaller size, cationic nature, etc. The processed metal encapsulated, or naked chitosan nanomaterials could be of various shapes and sizes such as nano vehicles, nanoparticles (NPs), nanofibers/meshes, 3D scaffolds, nanocapsules, etc. [3]. These altered chitosan-based nanoparticles (NPs) have been studied for disease control and growth-promoting agents in numerous plant species, thereby displaying superior biological activity compared to their bulk/micron-sized chitosan counterparts/substrates. The multi functionalities of chitosan biopolymers make them special as for other biopolymers such vast applicability has not yet been reported in the literature. There are several synthetic strategies are reported in the literature using which chitosan nanoparticles or nanocomposites can be achieved such as solvent casting, ionic gelation method (cross-linking reaction), electrospinning, layer-by-layer (LbL) depositions, freeze-drying, etc. Before making use of synthesized chitosan NPs for various applications it is required to characterize the material fully. The topological properties, surface/particle appearance, charges, particle size, elemental composition, interactions among each other, etc., can be determined by making use of physicochemical and morphological characterization techniques [1–3]. This chapter broadly and exclusively discusses a series of techniques, viz., transmission electron microscopy (TEM), scanning electron microscopy (SEM), energy-dispersive X-ray spectroscopy (EDX), dynamic light scattering (DLS), X-ray diffraction (XRD), X-ray photo-electron spectroscopy (XPS), atomic absorption spectroscopy (AAS), inductively coupled plasma mass spectroscopy (ICP), fourier-transform infrared spectroscopy (FTIR), atomic force microscopy (AFM), nuclear magnetic resonance (NMR), etc., used for precise characterization of chitosan-based nanomaterials/nanoparticles.

2 Nanomaterial Properties

2.1 *Morphological and Topological Properties of Chitosan Nanomaterials*

The morphological, internal arrangement/architecture, surface topography, degree of aggregation, chemical identification, and particle size of chitosan NPs can be evaluated by making use of transmission electron microscopy (TEM), or cryo-TEM or scanning electron microscopy (SEM).

2.1.1 Scanning Electron Microscopy (SEM)

Scanning electron microscopy (SEM) is a valuable tool that determines the surface visualizations, in-depth study of functionalized/agglomerated chitosan (CHI) nanoparticles (NPs), and estimates the sample composition via energy-dispersive X-ray spectroscopy EDX. SEM uses an energized electron beam (typically 1–30 eV) which scans the surface of the sample in a raster pattern wherein the secondary emitted electrons or backscattered electrons are detected, thereby achieving resolutions in the nanometre range.

Depending on the nature of the sample (chitosan nanomaterials), the electronic interaction with the sample varies, resulting in various types of emitted electrons at or on the sample surface. The detected electrons of different energies are processed and displayed as a pixel in the monitor, thereby visualizing 3D images or composition of nanomaterials (refer to Fig. 1). When these electrons are beamed under a high electric field, it is then popularly known as field emitter scanning electron microscopy (FE-SEM) [4–6]. Conventionally, to improve the electrical conductivity and contrast of nanomaterials, ultrathin coating with noble metals, viz, gold (Au), silver (Ag), and platinum (Pt) are used under high vacuum surroundings [7].

The application of SEM analysis for having a pictorial visualization is extensively used in literature for CHI-based nanomaterials/NPs [7]. In 2010, Dev et al. reported the synthesis of poly (lactic acid)/chitosan "PLA/CS" NPs using emulsion and solvent evaporation techniques [8]. Figure 2a shows the SEM images of PLA/CS NPs at different magnifications, wherein these particles' surface morphology is spherical [8]. Hosseini et. al 2013 prepared the essential oil encapsulated chitosan NPs using a two-step process of oil-in-water emulsion preceded by ionic gelation [9]. The morphology of prepared chitosan NPs found to be spherical in nature which is nicely intact and had a clear distribution among themselves as evident from Fig. 2b [9].

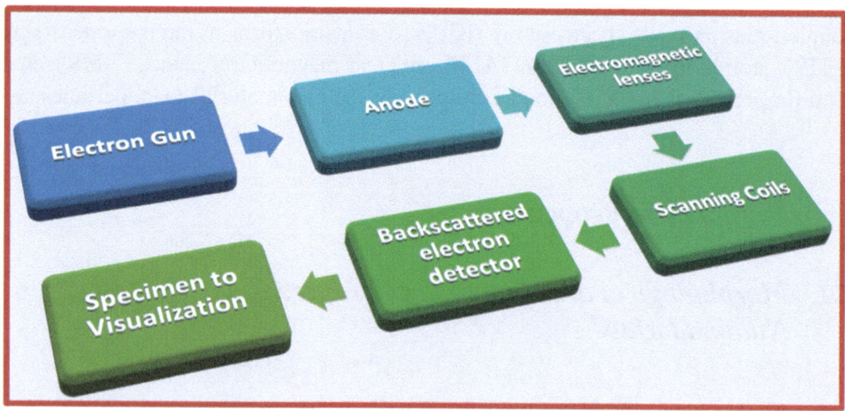

Fig. 1 Flowchart representing an overview of the SEM analysis process [4–6]

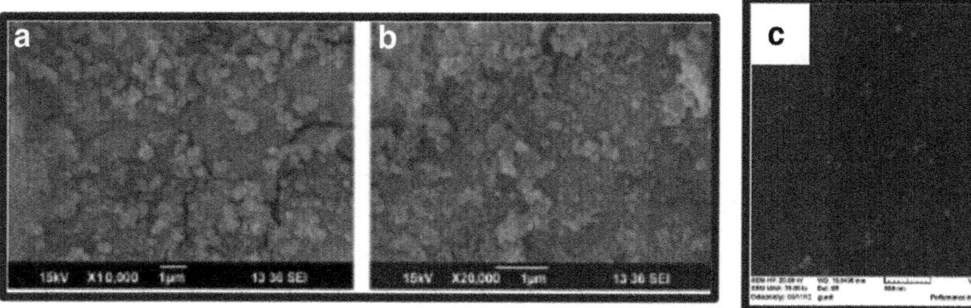

Fig. 2 SEM images of **a** poly-lactic acid/chitosan "PLA/CS NPs" [8] Reproduced from Dev et al. with permission from Elsevier **b** chitosan NPs [9] Reproduced from Hosseini et al. with permission from Elsevier

2.1.2 Transmission Electron Microscopy (TEM)

The higher spatial resolution, i.e., equivalence with the atomic dimensions makes transmission electron microscopy "TEM" quite a popular technique to characterize chitosan-based nanomaterials, which provides chemical information and internal structural arrangements in an image form. In TEM, a highly energized beam of electrons passes via a thin foil wherein the electrons are transformed into elastic/inelastic electrons which then interact with the specimen containing a sample of chitosan nanomaterial. The interactions of those inelastic/elastic electrons with the specimen gave rise to the emission of reflected/transmitted particles. The difference in energies is then detected and used to generate the higher resolution magnified images captured by the camera followed by visualization in a monitor. It is noteworthy to mention that ratio between the specimen, image plane, and the objective lens is taken into account which has to be magnified by the lens (Fig. 3a) [10, 11].

Although both SEM and TEM depicts the visualization of nanomaterial, i.e., how it looks, how the atoms are arranged, the extent of aggregation, and information on particle size/dimension but when compared to each other TEM is advantageous in proving higher resolution (as low as 0.2 nm) with good quality analytical measurements. TEM provides detailed information as it utilizes the energetic electrons which can be used to infer the composition, morphological, imaging, and crystallographic analysis. Therefore, TEM used three main techniques, i.e., electron microscopy, imaging, and diffraction pattern (Fig. 3b). Sample preparation in the case of TEM aialysis is quite simple, wherein before analysis a drop or two of suspension of chitosan-based nanomaterials (after sonication) is placed into a carbon-coated copper grid followed by drying or sometimes IR irradiation. TEM also distinguishes between the monocrystalline, polycrystalline, and amorphous chitosan-based nanomaterials/NPs [10]. It is noteworthy to mention that the size distribution or size estimation using TEM of nanomaterials is precise, not accurate as the chances of aggregation can occur during the evaporation of the sample preparation process. Cryo-TEM is superior to usual TEM analysis as in the former case, by using cryogenic temperature

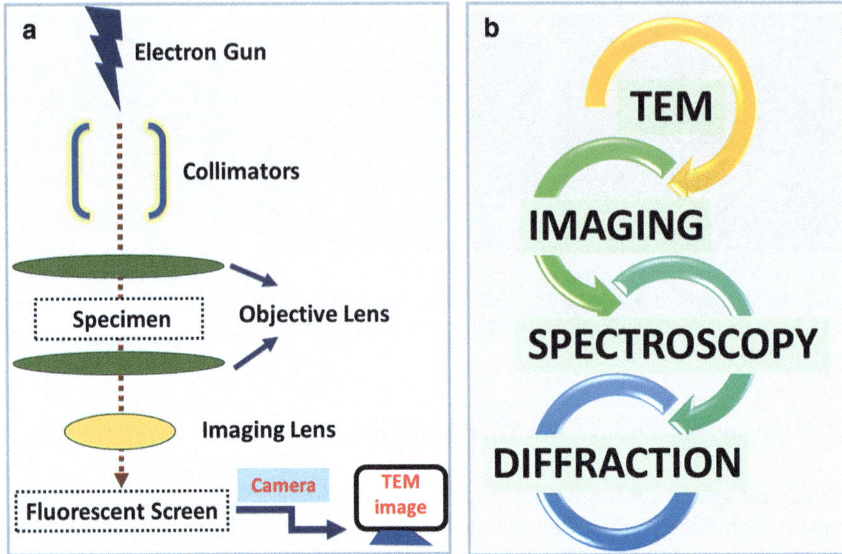

Fig. 3 **a** Flow chart representing steps involved in working principle of TEM analysis [10, 11]; **b** Three main techniques of TEM [11]

(-100 to -175 °C) a suspension of nanomaterial is solidified, and the visualization is done when the specimen is in the frozen state. The aforementioned method bypassed the problem of nanoparticle aggregation, usage of heavy-metal contrasting agents, and solvation thus giving not only precise but also predicting an accurate size estimation [11]. Cryo-TEM is particularly useful when one must distinguish whether the chitosan-based nanomaterials have the natural tendency of self-assembly or if it only occurs during sample preparation wherein solvation/evaporation procedures came into the picture [11–13]. In the literature, numerous reports have mentioned the use of TEM analysis for chitosan-based nanomaterials/NPs. Facchi et al., (2016) successfully reported the synthesis of N-modified chitosan NPs for curcumin delivery using a benzyl alcohol/water emulsion system (Fig. 4) [14]. Figure 4a–d shows the TEM analysis of N,N,N-Trimethyl chitosan/tripolyphosphate (TMC/TPP) and N,N-dimethyl chitosan/tripolyphosphate (DMC/TPP) NPs which have irregular spherical geometries attributed to vigorous stirring while using the benzyl alcohol/water emulsion system [14]. TEM analysis also tells that the average particle size for DMC/TPP NPs is up to 317 nm, while for TMC/TPP NPs it is ~99 nm [14].

In the year 2002, Banerjee et al. reported the synthesis of ultrafine chitosan nanoparticles wherein, the amine groups are cross-linked (10% or 100%) using a reverse micellar system containing surfactant, i.e., sodium bis(ethylhexyl) sulfosuccinate (AOT) and n-hexane [15]. The TEM analysis depicts the particles are shaped spherically and are aggregated with the dimensions of 30 nm when 10% amine functionalization (for chitosan NPs) is cross-linked whereas with 100% cross-linking the particle size shoots up to 110 nm (Fig. 4e, f [15]. Numerous other reports too

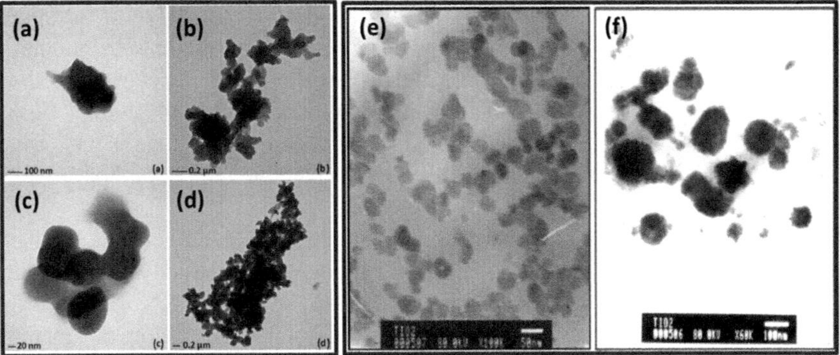

Fig. 4 TEM images of **a, b** DMC/TPP [14] Reproduced from Facchi et al. with permission from Elsevier; **c, d** TMC/TPP chitosan nanoparticles [14] Reproduced from Facchi et al. with permission from Elsevier; **e** 10% [15] and **f** 100% cross-linked chitosan nanoparticles [15] Reproduced from Banerjee et al. with permission from Elsevier

showed spherical kind morphologies of chitosan-based NPs [16–18] and also with metal encapsulation such as Cu with chitosan [19, 20].

2.2 Solid-State Properties of Chitosan-Based Nanomaterials/NPs

Like every other nanoparticle system, chitosan-based nanomaterials/NPs are not elementary molecules rather they are configured as three main layers. The first and foremost is the surface layer wherein a suitable functionalization can be done, encapsulating metal ions within, coating with surfactants, etc. The second layer, i.e., the middle layer/shell layer is chemically distinct followed by the third layer which is essentially the core and is the central/innermost segment of chitosan-based NPs. To identify the crystallinity, composition, defect structure, grain size, etc., all are solid-state properties that can be effectively useful to elucidate the molecular dynamics of chitosan-based nanomaterials/NPs. X-ray diffraction (XRD) is a powerful tool that can be used to study the solid-state properties of chitosan-based nanomaterials/NPs [1, 21, 22].

2.2.1 X-ray Diffraction (XRD)

In material science, it is an extremely important aspect to determine the crystallographic structure of nanomaterials. In the year of 1912, Max Von Laue made an important discovery that the two-dimensional diffraction gratings of a substance (nanomaterial/NPs) in presence of X-ray wavelength act in a similar fashion to that of plane spacing in a crystal lattice. XRD is a non-destructive analytical technique that solves the purpose of surface/phase identification of chitosan-based nanomaterials, thereby giving information on cell dimensions and atomic spacings as well. In XRD (Please refer to Fig. 5), the incident X-rays (generated by cathode ray "Cu X-ray tube" with 1.5418 Å) are filtered to produce monochromatic radiation which was then collimated to concentrate and are used to irradiate the sample (chitosan-based nanomaterial) [23–25].

Upon satisfying Bragg's law ($n\lambda = 2d \sin\theta$), the interaction between the incident rays and chitosan-based nanomaterial produces a constructive interference followed by a diffracted ray. Bragg's law correlates the wavelength of an incident ray "λ" proportional to that of diffraction angle "θ" and lattice spacing "d", thereby measuring the respective intensities and scattering angle that is diffracted from the sample [23, 24]. The sample material which is chitosan-based nanomaterial must be homogenized be it in a film form or finely grounded powdered form. By changing the 2θ (2 theta) angles to all plausible ranges, a finite diffraction direction of the lattice can be attained. Each material has a specific unique set of d-spacings whose corresponding positions/intensities are available from International Centre for Diffraction Data (ICDD) or in earlier times known as Joint Committee on Powder Diffraction Standards JCPDS data as a reference pattern/database [23]. Thus, by converting diffraction peaks to d-spacings one can easily identify the grain size, composition, shape of a unit cell, crystalline nature, etc., of chitosan-based nanomaterials/NPs [24–27]. It is noteworthy to mention that the diffraction angle direction depends on the nature, shape, and size of a unit cell of chitosan-based nanomaterials/NPs, whereas the intensities of diffraction patterns depend on the structural internal arrangements of the atoms.

Rhim et al. (2006) successfully reported the XRD spectra of chitosan powder wherein, a characteristics peaks corresponding to 2θ values of 10.9° and 19.8° which corresponds to the amorphous structure of chitosan (Fig. 6d, black graph) [28].

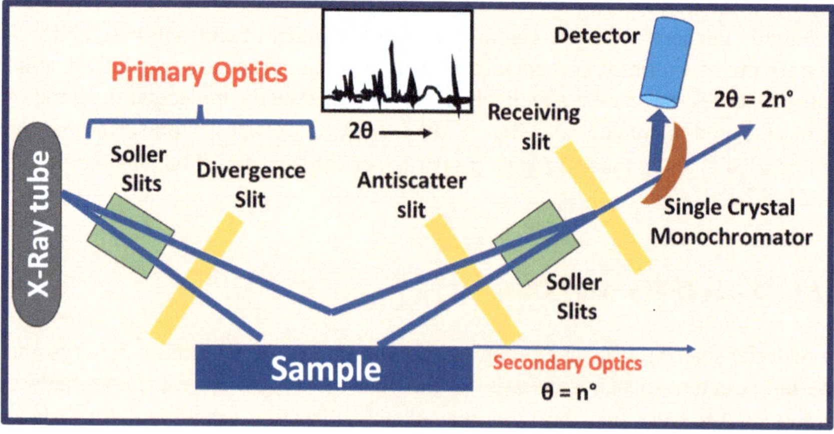

Fig. 5 Working principle of X-ray spectroscopy [23–25]

Whereas, after chemical processes, the as-synthesized chitosan nanofilm showed a characteristic peak of 2θ values of 8°, 11° corresponds to the hydrated crystalline structure of chitosan nanofilm and 18° to amorphous characteristics (Fig. 6d, blue graph) [28] It is well known in the literature that the structural modifications of chitosan alter its internal lattice structure arrangements. Factors such as the dissolution process, drying, precipitation, processing chemical treatment, molecular weight, degree of deacetylation, etc., cause the changes in chitosan structural arrangements [28]. Furthermore, Hosseini in 2013, recorded the XRD patterns of OEO-loaded chitosan nanoparticles over a 2θ range of 5–50°, and it showed the characteristic peak corresponding to 2θ of 25° indicating a high degree of crystallinity (Fig. 6a–c) [29]. As shown in Fig. 6a–c, the broad peak in the XRD spectrum indicated the cross-linking reactions between chitosan and TPP. This proves that the chitosan NPs are having a densely packed network structure corresponding to the interpenetrating polymeric chain of chitosan with TPP counterparts [29]. XRD diffractograms of chitosan-based nanomaterials are reported in literature many times [30–33] For example, metal encapsulated on chitosan, i.e., Cu-chitosan NPs shows a characteristic 2θ range between 19.5 and 21.0 which corresponds to the crystalline nature of Cu-chitosan NPs system [34].

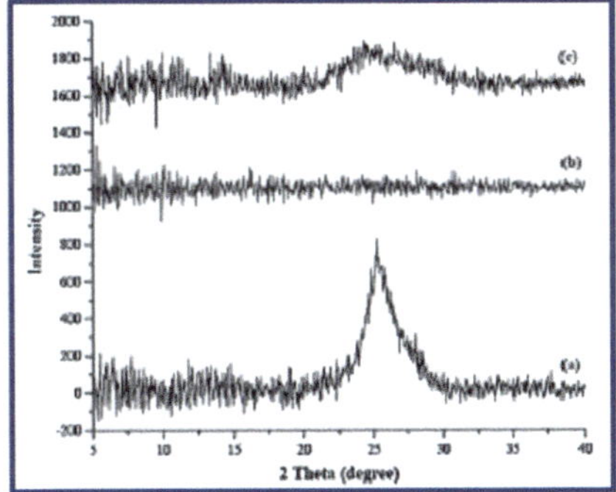
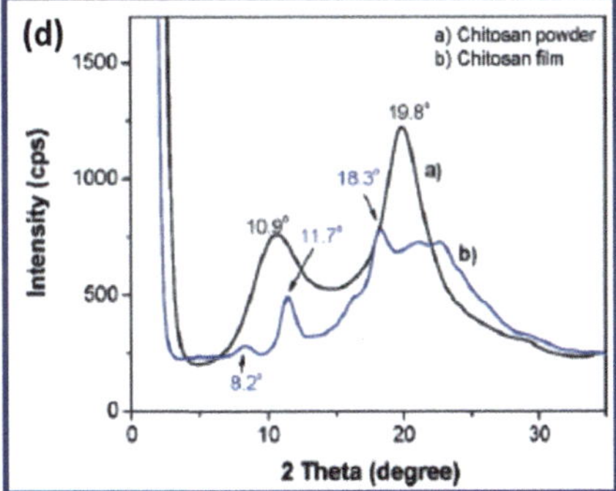

Fig. 6 XRD spectra of **a** chitosan powder [29] Reproduced from Hosseini et al. with permission from Elsevier **b** chitosan nanoparticles [29] Reproduced from Hosseini et al. with permission from Elsevier **c** OEO-loaded chitosan NPs [29] Reproduced from Hosseini et al. with permission from Elsevier **d** comparison between chitosan powder (black graph) and chitosan film (blue graph) [28] Reproduced from Rhim et al. with permission from Elsevier

2.3 Elemental Analysis/Properties of Chitosan-Based Nanomaterials/NPs

2.3.1 X-ray Photoelectron Spectroscopy (XPS)

X-ray photoelectron spectroscopy "XPS" is one of the prominent surface science techniques with which one can infer/analyze the chemical state, electronic states, and elemental composition/empirical formula of nanomaterials. In the year 1887 Hz, discovered the photoelectric effect which is the basis of the XPS working principle. Later in the 1960s, Siegbahn and his research group at Uppsala University, Sweden extended this work to surface analysis and coined the term XPS or ESCA (electron spectroscopy for chemical analysis) for which Siegbahn won a noble prize in physics in 1981 [35, 39, 43].

As shown in Fig. 7a, XPS, a qualitative analysis involves the bombardment of the sample with single energy or monochromatic or monoenergetic X-ray photons (Mg Kα 1253.6 eV or Al Kα 1486.6 eV, line width $\approx$0.7–0.85 eV), followed by computing the kinetic energy of emitted electrons from the topmost layer of the sample (1–10 nm) [35]. The electrons which are emitted from outermost/surface atoms have characteristic peaks in the XPS spectrum, thereby enabling one to identify and quantify the surface elements present (except hydrogen and helium). The chemical states of elements present within can be identified and quantified simply by measuring/investigating the minute variations occurring in binding energies of emitted photoelectrons, auger electrons, multiple splitting, satellite peaks, etc. (Please refer to Fig. 7b). It is important to mention that an XPS spectrum is plotted as the relative number of electrons against their respective binding energies (eV). XPS is an ultra-high vacuum and sensitive technique that changes in the electronic configuration of the atoms or chemical bonds can be easily detected thereby the existence of certain elements or species can be identified. In the case of chitosan-based nanomaterials, the determination of elements such as C, N, O and P can be easily identified as these elements constitute the majority and have characteristic peaks corresponding to their respective binding energies [36]. Boufi et al. in 2013 reported the XPS for the mild-wet synthesized gold/silver NPs in an aqueous chitosan solution [35]. Figure 8a–b shows the survey XPS spectra and N 1a XPS region of chitosan nanoparticles (blue spectrum) which clearly indicates the presence of N 1s whose peak corresponds to the binding energy of 399.4 $\pm$ 0.2 eV. The binding energy value of 399.4 is characteristic of nitrogen in NH_2 groups or NH-CO groups indicating a mixed chitosan-chitin structure [35].

Trapani et al. (2011) demonstrated the XPS spectra of chitosan nanoparticles [36]. Figure 9a–b shows the high-resolution C1s XPS spectra wherein the typically observed ratio of 3.8 of binding energies values of 286.5 and 288.0 eV corresponds to CS (chitosan bulk, Fig. 9a) and CSNPs (chitosan nanoparticles, Fig. 9b) [36]. Moreover, Fig. 9f–g shows the curve fitted N 1s high-resolution regions of XPS spectra for chitosan and chitosan NPs wherein the lower binding energy value of 399.6 eV, 401.1 and 402.0 eV corresponds to the aminic group, amide group, and

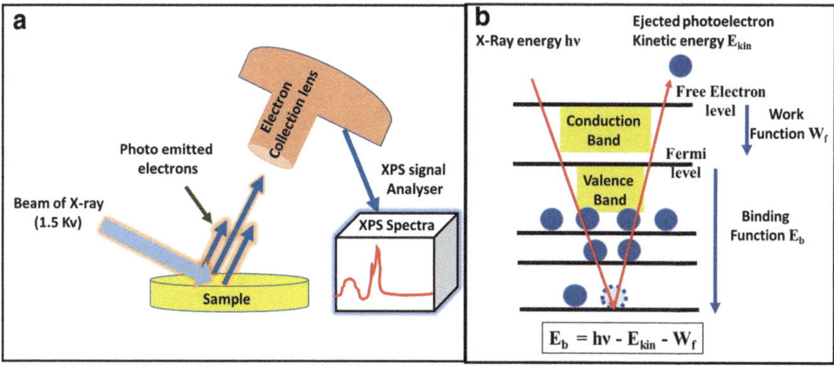

Fig. 7 Schematic representation of **a** XPS process and **b** basic working principle of XPS involving auger process and photoemission [35, 36]

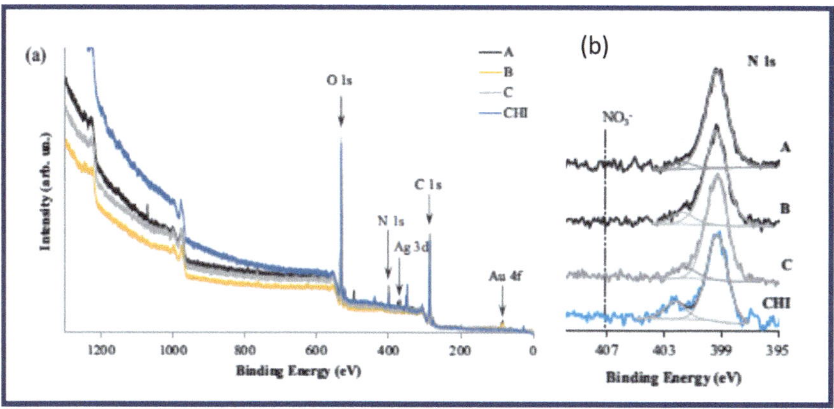

Fig. 8 a XPS survey spectra of chitosan NPs (blue graph, –CHI); **b** XPS N1s regions of CHI NPs (blue graph) [35] Reproduced from Boufi et al. with permission from Elsevier

protonated quaternary nitrogen, respectively [36]. Many other prominent research groups reported the XPS data for chitosan and chitosan-based nanoparticles which are modified by encapsulating such as copper or chitosan aerogels [37–43].

2.3.2 Atomic Absorption Spectroscopy (AAS)

To determine the metal concentration at the pictogram level in a variety of samples, atomic absorption spectroscopy "AAS" is used which is an extremely sensitive technique for elemental analysis [44]. In AAS, the reduction in the intensity of optical radiation of cell containing gaseous atoms of samples are measured. Typically, in AAS, an analyte absorbs specific wavelengths emitted by a hollow cathode lamp

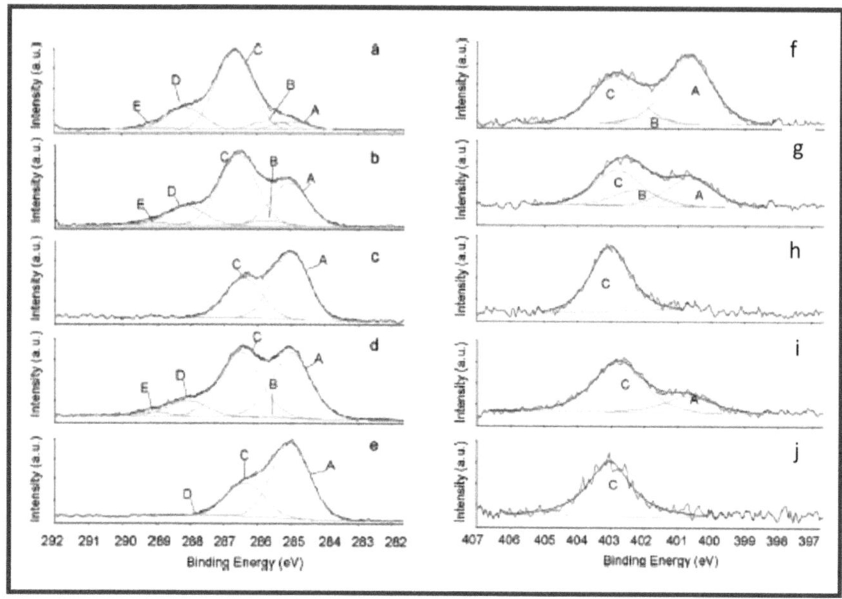

Fig. 9 XPS of **a** C1s regions of pure CS; **b** C1s regions of CSNPs; **f** N1s regions of pure CS; **g** N1s regions of CSNPs [36] Reproduced from Trapani et al. with permission from Elsevier

"HCL" which is used as a light source as shown in Fig. 10 [45, 46]. In AAS, either the flames or graphitic-furnaces/electrothermal atomizers "ETAs" are commonly used atom-cells. The sensitivity of flames is relatively less than that of ETAs as in the case of ETAs the temperature can be supervised by a power supply, whereas flames consist of a meticulously controlled combustion environment. To prevent the process of combustion at elevated temperatures the use of inert gas such as argon is employed. On absorbing the particle wavelength, the sample/analyte in an "atom-cell" turns into a gaseous state which then travels to a detector whose job is to quantify and isolate the wavelengths of interest followed by processing the data in a computer/control instrumentation operation. To provide optimal accuracy, precision, and minimal interferences, the analyte or samples are made by digestion procedures and are usually of specific concentration in an aqueous phase [46].

It is important to mention that every single element absorbs wavelengths of electromagnetic radiation for HCL source differently. In other words, the absorbance of elements is extremely specific and particular for the absorbing wavelength of interest and is measured against the standards. When the standards are measured, it implies that the instrument is calibrated for particular elements of interest, and hence the unknown sample can be processed, and the concentration can be obtained from the digital output display unit. The basic phenomenon involved in AAS is the concept of ionization energy which is basically the energy required to excite an electron and is particular for each element. In the case of chitosan-based nanomaterials/NPs,

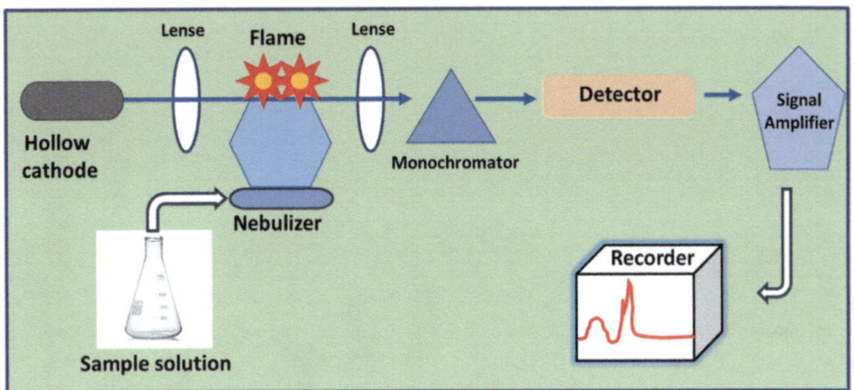

Fig. 10 Schematic representation of atomic absorption spectroscopy (AAS) [45, 46]

every atom has its own distinctive fingerprint regions, enabling the qualitative AAS analysis of the material [47–49].

Chitosan-based nanomaterials have been widely used to encapsulate the metal ions and hence the AAS analysis can be done against different reaction parameters too such as agitation time, rate of reaction, pH, etc. [19, 50, 51]. In general, the Concentration of encapsulated metals (%) can be calculated by taking the ratio of the amount of released metal ions to the total amount of metals present in a nanomaterial followed by multiplying it by 100. Liu et al. in 2002, demonstrated the use of chitosan-based nanomaterials for the metal encapsulation, and the amount of metal ions was then evaluated by making use of AAS technique [52]. Figure 11 shows the relative percentage of absorbance of metal ions such as Ca^{2+}, Zn^{2+}, Ag^+, Cr^{3+}, and Mg^{2+}, on the chitosan-modified glass beads using the AAS spectroscopy [54].

2.3.3 Inductively Coupled Plasma Mass Spectroscopy (ICP)

In inductively coupled plasma mass spectroscopy "ICP-MS" an inductively coupled plasma is used to atomize the sample. Upon atomization, it creates atomic and small-polyatomic ions which were then detected. Figure 12 shows a schematic representation of a single quadrupole of IC-MS which has six major slots/compartments [53]. It starts with a sample solution in an introduction system which is then processed into a nebulizer followed by an inductively coupled plasma in the presence of inert gas for say Argon. The highly ionizable plasma then atomizes the sample thereby generating the polyatomic ions (fine aerosol form) which then using ion optics were extracted to an interface region. The electrostatic lens of ion optic focuses the polyatomic ions into a quadrupole mass analyzer.

In accordance with the mass by charge ratio "m/z ratio" the detection and separation of ions take place. For instance in ICP-MS, the samples which are to be analyzed are digested in a similar fashion in an aqueous phase as it is done in the case of AAS

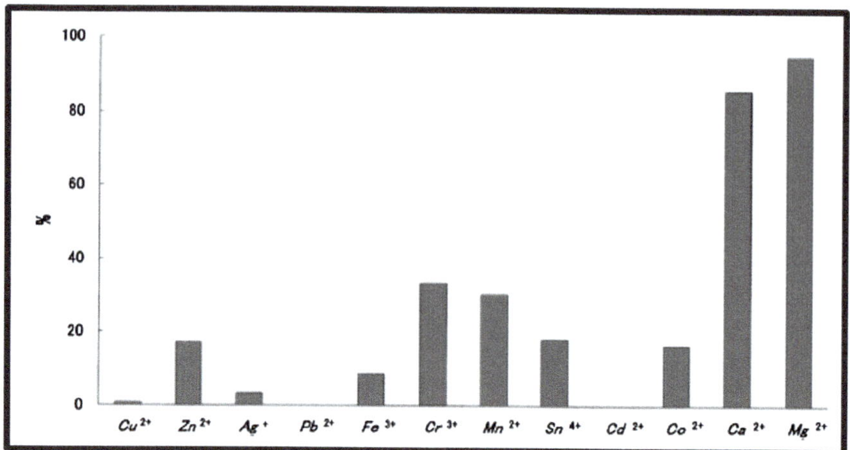

Fig. 11 The metal ion adsorption rate of the chitosan-modified glass beads using AAS [54] Reproduced from Liu et al. with permission from Elsevier

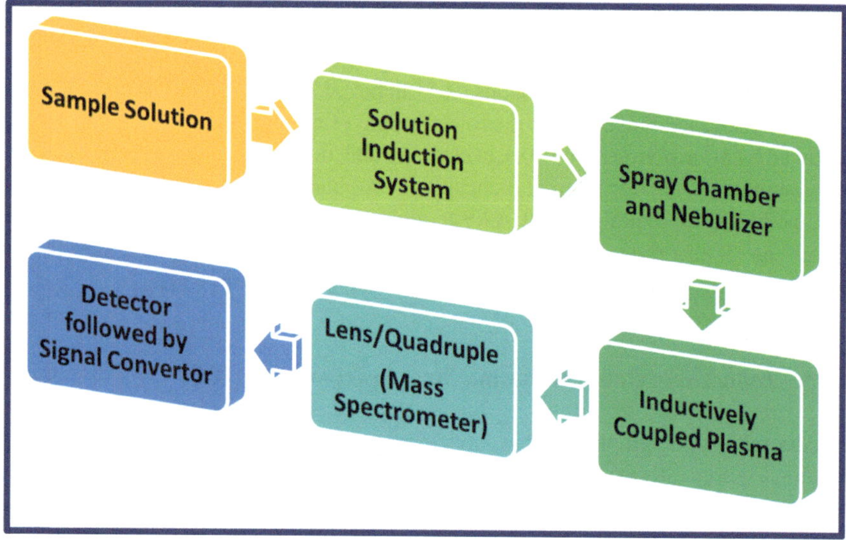

Fig. 12 Schematic representation demonstrating steps involved in ICP analysis [53]

sample preparation case too. The sample preparation includes the use of hydrochloric or nitric acid or in some cases alkaline solutions too and diluent is deionized water [53]

There are not many reports in the literature that explain the use of ICP-MS on chitosan-based nanoparticles. Although, ICP-MS proves to be an effective and supportive technique for the estimation of metal ions bound or formed using

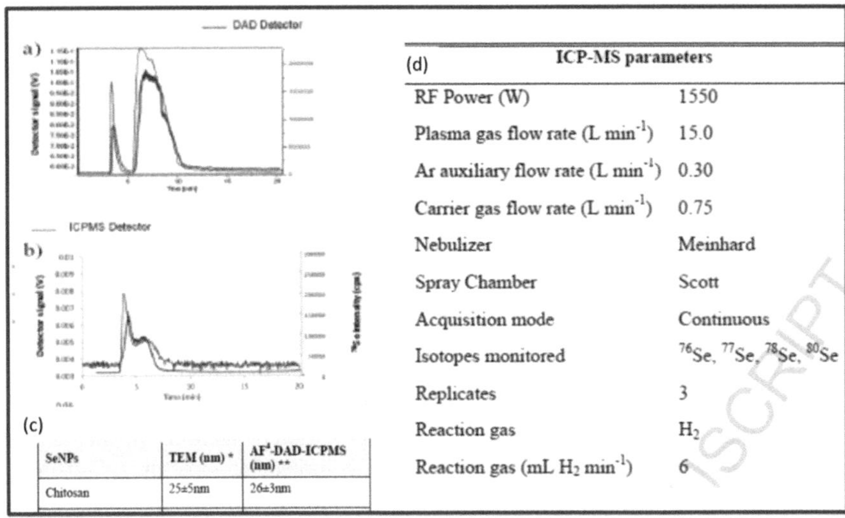

Fig. 13 **a, b** The fractograms showing the fractionation of Se NPs prepared in chitosan and Triton X-100 **c** TEM size estimation of SE NPs and AF^4-DAD-ICP-MS **d** Operating conditions for AF^4-DAD-ICP-MS [58] Reproduced from Palomo-Siguero et al. with permission from Elsevier

chitosan-based nanomaterials [54–57]. Palomo-Siguero et al. in 2017 synthesized the Selenium NPs using a solution-phase strategy in presence of stabilizers such as chitosan(polysaccharides) or non-ionic surfactant (Triton X-100) [58]. Figure 13a–d shows the ICP-MS fractograms of fractionization of Se NPs synthesized by using chitosan and Triton X-100 stabilizers [58]. The selenium peaks were identified by using ICP-MS whose operating conditions for running ICP-MS are shown in Fig. 13d [58].

2.4   *Interaction Analysis/Properties of Chitosan-Based Nanomaterials/NPs*

The type of bonds present within the chitosan-based nanoparticles and their interaction/functional group analysis is an important aspect that tells the kinds of specific bonds present within the system. To measure such properties that include the guided alterations and modifications, fourier-transform infrared spectroscopy is an indispensable tool to study the same.

2.4.1 Fourier-Transform Infrared Spectroscopy (FT-IR)

Fourier-transform infrared spectroscopy "FTIR" is a chemical identification technique that gives insightful information on the interaction of bonds with their corresponding functionalities.

The underlying principle of the FTIR technique is based on the absorption of electromagnetic radiation of the infrared region (4000–400 cm^{-1}) by a molecule (inorganic/organic). For a molecule to be IR active, the dipole moment should change on absorbing IR radiation, and it becomes IR active. Figure 14 shows the schematic representation demonstrating steps involved in FT-IR analysis wherein the frequency is measured in terms of wavenumbers, and the data is recorded in the form of an interference pattern which is then converted into a spectrum (transmittance/absorbance form) [59]. The spectrum is usually in the form of distinct lines that could be narrow or broad and corresponds to a specific frequency thereby helping in identifying the nature of bonds, functionalities pertaining to molecular structures, and interactions [59]. The samples could be of any type (liquid, solid, or gaseous), but in general solid or liquid samples are prepared using KBr (100: 1), and the pellet is made by using a hydraulic press machine. Initially, at room temperature, the instrument is calibrated by recording a blank KBr background followed by a KBr pellet containing the samples of our interest let's say chitosan nanomaterial/NPs in an absorbance/transmittance mode. It is worth mentioning that the FTIR technique is quite an impressive analytical tool that helped in understanding the functionalities of chitosan-based nanomaterials/NPs [8, 60–65].

Banerjee et al. in 2002 characterized the chitosan NPs using FTIR analysis. The chitosan NPs were synthesized by cross-linking reaction between the chitosan polymer and glutaraldehyde using the reverse micellar technique [66]. Figure 15a–b shows the FTIR spectra for chitosan and chitosan NPs. It may be noted that in Fig. 5b,

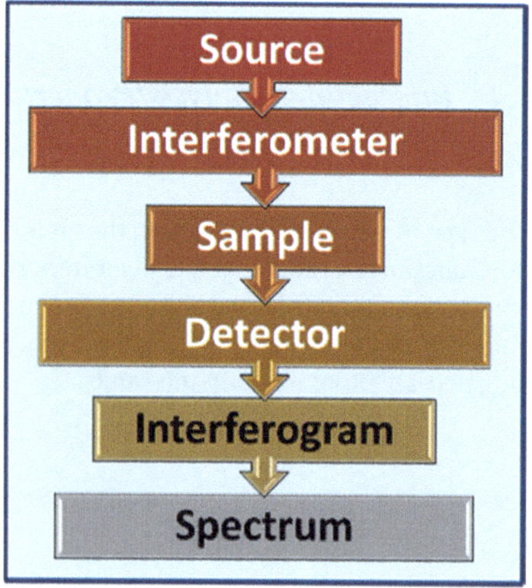

Fig. 14 Schematic representation demonstrating steps involved in FT-IR analysis [59]

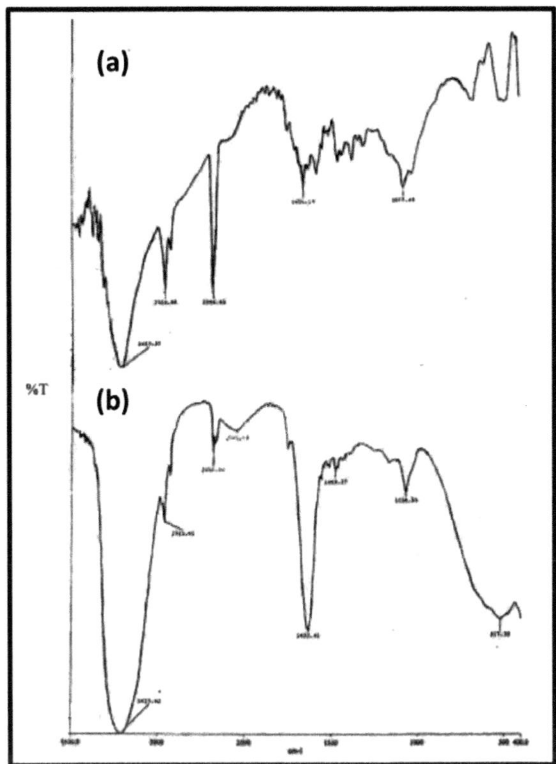

Fig. 15 FTIR spectra of **a** chitosan polymer and **b** cross-linked chitosan nanoparticles [66] Reproduced from Banerjee et al. with permission from Elsevier

an additional peak at 1634 cm^{-1} is in agreement with the stretching vibrations of $C = N$, elucidating cross-linked chitosan NPs [66]. Moreover, a peak at 1650 cm^{-1} corresponds to the scissoring vibrations of NH_2 of primary amines in chitosan chains. The peaks around 1020–1075 cm^{-1} imply the symmetric stretch of C–O–C bonds and 2926 cm^{-1} manifested the string polymeric backbone C–H vibrations [66].

2.5 Determination of Chitosan-Based Nanomaterials/NPs

For chitosan nanoparticles determining topological properties such as surface charge, particle size distribution, as well as particle size, can be evaluated by the dynamic light scattering "DLS" technique.

2.5.1 Dynamic Light Scattering (DLS)

Dynamic light scattering (DLS) or photon correlation spectroscopy (PCS) or quasi-elastic light scattering (QELS) is a versatile and powerful tool to examine the

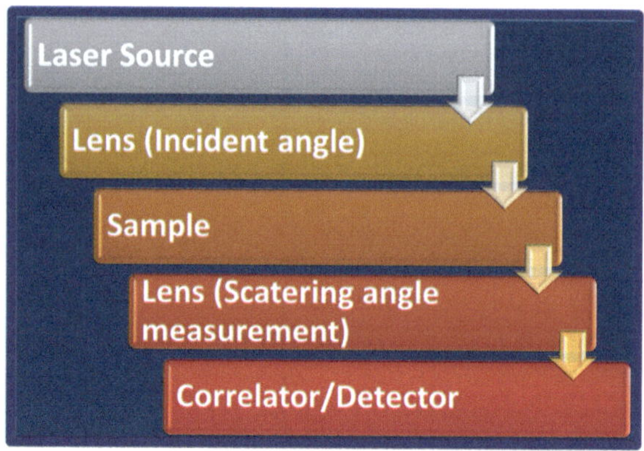

Fig. 16 Schematic representation of dynamic light scattering [67]

diffusion behavior of molecules in the solution phase. The hydrodynamic radii or diffusion coefficient is calculated in the DLS or PCS technique, which defers the size distribution, particle size, surface charge, and shapes of particles (nanoparticles/macromolecules) [67]. Typically, in DLS, the sample to be analyzed has to be in a solution phase, and hence the size is estimated by studying the Brownian motion of suspended particles having different scattered angles 'θ' (Refer Fig. 16) [67]. The random motion of suspended particles/molecules is because of the continuous bombardment by the solvent molecules around them. Using the DLS technique, one can measure the particles in the submicron region and of the dimension of <1 nm. In DLS, the size of the NPs was calculated using Stoke-Einstein equation. The size of chitosan NPs/nanomaterials which is hydrodynamic radii "d_h" can be determined from the diffusion of particles "D", i.e., $d_h = kT/3\pi\eta D$ wherein k is Boltzmann constant, η is the viscosity of the medium, and T is the absolute temperature [67].

Much scientific work is already reported in the literature where the DLS technique proves to be an important aspect to determine the size distribution as well as surface charge properties [14, 68, 69]. To get reliable data, it is important to combine the more sophisticated techniques such as microscopy with the DLS technique. For fully dispersed chitosan nanoparticles/nanomaterials reliable as well as reproducible data can be obtained wherein the concept of aggregation can be eliminated. Banerjee et al. in 2002 reported the DLS data of chitosan nanoparticles as shown in Fig. 17a [66]. Since the underlying principle of the DLS technique is based on the Brownian motion of particles and chitosan is polymeric in nature hence the actual size is always lesser than the observed average particle size. Li et al. in 2008 estimated the average particle size distribution of ferrite-coated chitosan NPs using the DLS technique wherein the size varies from 10 to 100 nm as shown in Fig. 17b [70].

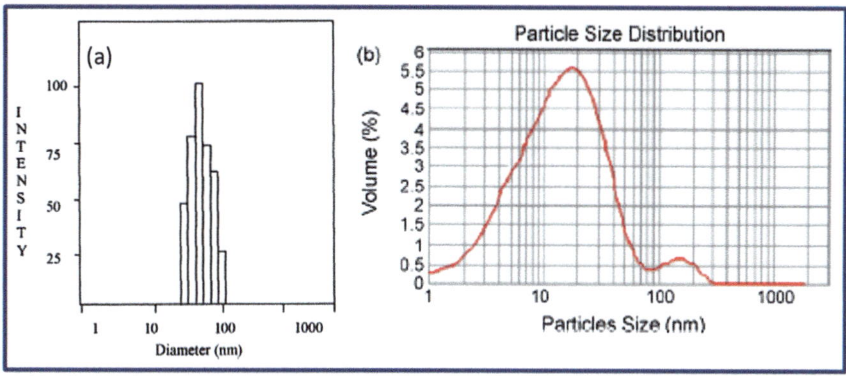

Fig. 17 **a** Size distribution of chitosan nanoparticles by QELS [66] Reproduced from Banerjee et al. with permission from Elsevier **b** Particle size distribution of the Fe_3O_4–chitosan nanoparticles [70] Reproduced from Li et al. with permission from Elsevier

3 Conclusion

Chitosan-based nanomaterials/NPs have gained attention in recent times and have shown exponential growth specifically in the area of medical sciences and agriculture. Characterizing the chitosan-based nanomaterials/NPs provides reliability, accuracy, and consistent results especially to understand the nanomaterials and their sustainability. The important characterizing tools such as SEM, TEM, Cryo-TEM, XRD, XPS, AAS, ICP, FTIR, and DLS help in understanding the chemical and physical properties of chitosan-based nanomaterials/NPs. Therefore, representing the role of each technique in a conclusive manner will help the scientific community to understand the properties of chitosan-based nanomaterials/NPs. In this way, it will be beneficial for researchers, academicians as well as students to select and understand the most suitable technique of their choice to assess their uses in a precise manner.

References

1. Mourdikoudis S, Pallares RM, Thanh NTK (2018) Characterization techniques for nanoparticles: comparison and complementarity upon studying nanoparticle properties. Nanoscale 10:12871
2. Qu B, Luo Y (2021) A review on the preparation and characterization of chitosan-clay nanocomposite films and coatings for food packaging applications. Carbohydr Polym Technol Appl 2:10010
3. Choudhary RC, Kumari S, Kumaraswamy RV, Pal A, Raliya R, Biswas P, Saharan V (2019) Characterization methods for chitosan-based nanomaterials. In: Prasad R (eds) Plant nanobionics. Nanotechnology in the life sciences. Springer, Cham. https://doi.org/10.1007/978-3-030-12496-0_5
4. McMullan D (1995) Scanning electron microscopy 1928–1965, Scanning 17, pp175–185. https://doi.org/10.1002/sca.4950170309

5. Ponomarev A, Yudovich M, Nikitin V, Nikitin D, Barchenko V, Sobol' V (2000) Some features of analysis of solutions of fullerenes C60 and C70 by their absorption spectra. Optics Spectrosc 88:195–196

6. Saharan V, Pal A (2016) Chitosan based nanomaterials in plant growth and protection. Plant Sci. https://doi.org/10.1007/978-81-322-3601-6_3

7. Choudhary RC, Kumaraswamy RV, Kumari S, Sharma SS, Pal A, Raliya R, Biswas P, Saharan V (2017a) Synthesis, characterization, and application of chitosan nanomaterials loaded with zinc and copper for plant growth and protection. In: Prasad et al (eds) Nanotechnology. Springer Nature Singapore Pte Ltd, Singapore

8. Dev A, Binulal NS, Anitha A, Nair SV, Furuike T, Tamura H, Jayakumar R (2010) Preparation of poly(lactic acid)/chitosan nanoparticles for anti-HIV drug delivery applications. Carbohydr Polym 80:833–838

9. Hosseinia SF, Zandib M, Rezaeia M, Farahmandghavi F (2013) Two-step method for encapsulation of oregano essential oil in chitosan nanoparticles: preparation, characterization and in vitro release study. Carbohydr Polym 95:50–56

10. Mourdikoudis RM, Pallares NTK, Thanh (2018) Characterization techniques for nanoparticles: comparison and complementarity upon studying nanoparticle properties. Nanoscale 10:12871–12934

11. Parupudi A, Mulagapati SHR, Subramony JA (2022) Chapter 1—Nanoparticle technologies: recent state of the art and emerging opportunities. In: Kesharwani P, Singh KK (eds) Nanoparticle therapeutics. Academic Press, pp 3–46. ISBN 9780128207574. https://doi.org/10.1016/B978-0-12-820757-4.00009-0

12. Kumar PS, Pavithra KG, Naushad M (2019) Chapter 4—Characterization techniques for nanomaterials. In: Thomas S, Sakho EHM, Kalarikkal N, Oluwafemi SO, Wu J (eds) Nanomaterials for solar cell applications. Elsevier, pp 97–124. ISBN 9780128133378. https://doi.org/10.1016/B978-0-12-813337-8.00004-7

13. Pennycook SJ (2005) Transmission electron microscopy. In: Bassani F, Liedl GL, Wyder P (eds) Encyclopedia of condensed matter physics. Elsevier, pp 240–247. ISBN 9780123694010. https://doi.org/10.1016/B0-12-369401-9/00582-9

14. Facchi SP, Scariot DB, Bueno PVA, Souza PR, Figueiredo LC, Follmann HDM, Nunes CS, Monteiro JP, Bonafé EG, Nakamura CV, Muniz EC, Martins AF (2016) Preparation and cytotoxicity of N-modified chitosan nanoparticles applied in curcumin delivery. Int J Biol Macromol 87:237–245

15. Banerjee T, Mitra S, Singh AK, Sharma RK, Maitra A (2002) Preparation, characterization and biodistribution of ultrafine chitosan nanoparticles. Int J Pharm 243:93–105

16. Liu H, Gao C (2009) Preparation and properties of ionically cross-linked chitosan nanoparticles. Polym Adv Technol 20:613–619

17. Saharan V, Mehrotra A, Khatik R, Rawal P, Sharma SS, Pal A (2013) Synthesis of chitosan-based nanoparticles and their in vitro evaluation against phytopathogenic fungi. Int J Biol Macromol 62:677–683

18. Yuwei C, Jianlong W (2011) Preparation and characterization of magnetic chitosan nanoparticles and its application for Cu(II) removal. Chem Eng Sci 168:286–292

19. Choudhary RC, Kumaraswamy RV, Kumari S, Sharma SS, Pal A, Raliya R, Biswas P, Saharan V (2017b) Cu-chitosan nanoparticle boost defense responses and plant growth in maize (Zeamays L.). Sci Report 7:9754

20. Brunel F, Gueddari NEE, Moerschbacher BM (2013) Complexation of copper (II) with chitosan nanogels: toward control of microbial growth. Carbohydr Polym 92:1348–1356

21. Khan I, Saeed K, Khan I (2019) Nanoparticles: properties, applications and toxicities. Arab J Chem 12:908–931

22. Boateng E, Chen A (2020) Recent advances in nanomaterial-based solid-state hydrogen storage. Mater Today Adv 6:100022

23. Wong-Ng W, McMurdie HF, Hubbard CR, Mighell AD (2001) JCPDS-ICDD research associateship (cooperative program with NBS/NIST). J Res Natl Inst Stand Technol 106:1013–1028

24. Gonon M (2021) Case studies in the X-ray diffraction of ceramics. In: Pomeroy M (ed) Encyclopedia of materials: technical ceramics and glasses. Elsevier, pp 560–577
25. Vishwakarma V, Uthaman S (2020) 9-Environmental impact of sustainable green concrete, In: Liew MS, Nguyen-Tri P, Nguyen TA, Kakooei S (eds) Micro and nano technologies, smart nanoconcretes and cement-based materials. Elsevier, pp 241–255
26. Sima F, Ristoscu C, Duta L, Gallet O, Anselme K, Mihailescu IN (2016) 3-Laser thin films deposition and characterization for biomedical applications. In: Vilar R (ed) Laser surface modification of biomaterials. Woodhead Publishing, pp 77–125
27. Bunaciu AA, Udriştioiu EG, Aboul-Enein HY (2015) X-ray diffraction: instrumentation and applications. Anal Chem 45:289–299
28. Rhim JW, Hong SI, Park HM, Ng PKW (2006) Preparation and characterization of chitosan-based nanocomposite films with antimicrobial activity. J Agric Food Chem 54:5814–5822
29. Hosseini SF, Zandi M, Rezaei M, Farahmandghavi F (2013) Two-step method for encapsulation of oregano essential oil in chitosan nanoparticles: preparation, characterization and in vitro release study. Carbohydr Polym 95:50–56
30. Ieva E, Trapani A, Cioffi N, Ditaranto N, Monopoli A, Sabbatini L (2009) Analytical characterization of chitosan nanoparticles for peptide drug delivery applications. Anal Bioanal Chem 393:207–215
31. Kalaivani R, Maruthupandy M, Muneeswaran T, Hameedha Beevi A, Anand M, Ramakritinan CM, Kumaraguru AK (2018) ynthesis of chitosan mediated silver nanoparticles (Ag NPs) for potential antimicrobial applications. Front Lab Med 2:30–35
32. Manikandan A, Sathiyabama M (2015) Green synthesis of copper-chitosan nanoparticles and study of its antibacterial activity. J Nanomed Nanotechnol 6:1000251
33. Chae KS, Shin CS, Shin WS (2018) Characteristics of cricket (Gryllus bimaculatus) chitosan and chitosan-based nanoparticles. Food Sci Biotechnol 27:631–639
34. Li LH, Deng JC, Deng HR, Liu ZL, Xin L (2010) Synthesis and characterization of chitosan/ZnO nanoparticle composite membranes. Carbohydr Res 345:994–998
35. Boufi S, Vilar MR, Ferraria AM, Botelho-do-Rego AM (2013) In situ photochemical generation of silver and gold nanoparticles on chitosan. Colloids Surf A: Physicochem Eng Asp 439:151–158
36. Trapani A, Giglio ED, Cafagna D, Denora N, Agrimi G, Cassano T, Gaetani S, Cuomo V, Trapani G (2011) Characterization and evaluation of chitosan nanoparticles for dopamine brain delivery. Int J Pharm 419:296–307
37. Matienzo LJ, Winnacker SK (2002) Dry processes for surface modification of a biopolymer: chitosan. Macromol Mater Eng 287:871–880
38. Zhang S, Xiao Q, Xiao Y, Li Z, Xiong S, Ding F, He J (2022) Chitosan based aerogels with low shrinkage by chemical cross-linking and supramolecular interaction. Gels (Basel, Switzerland) 8:131
39. Huang Z, Li Z, Zheng L, Zhou L, Chai Z, Wang X, Shi W (2017) Interaction mechanism of uranium(VI) with three-dimensional graphene oxide-chitosan composite: insights from batch experiments, IR, XPS, and EXAFS spectroscopy. Chem Eng Sci 328:1066–1074
40. Taketa TB, dos Santos DM, Fiamingo A, Vaz JM, Beppu MM, Campana-Filho SP, Cohen RE, Rubner MF (2018) Investigation of the internal chemical composition of chitosan based LbL films by depth-profiling X-ray Photoelectron Spectroscopy (XPS) analysis. Langmuir 34:1429–1440
41. Rashid S, Shen C, Chen X, Li S, Chen Y, Wen Y, Liu J (2015) Enhanced catalytic ability of chitosan-Cu-Fe bimetal complex for the removal of dyes in aqueous solution. RSC Adv 5:90731–90741
42. Zheng Z, Wei Y, Wang G, Wang A, Gong Y, Zhang X (2009) Surface properties of chitosan films modified with polycations and their effects on the behavior of PC12 cells. J Bioact Compat Polym 24:63–82
43. Daniyal W, Fen YW, Saleviter S, Chanlek N, Nakajima H, Abdullah J, Yusof NA (2021) X-ray photoelectron spectroscopy analysis of chitosan-graphene oxide-based composite thin films for potential optical sensing applications. Polymers 13:478

44. Vale MGR, Oleszczuk N (2006) Current status of direct solid sampling for electrothermal atomic absorption spectrometry—a critical review of the development between 1995 and 2005. Appl Spectrosc Rev 41:377–400
45. Ozbeka N, Baysal A (2017) Determination of sulfur by high-resolution continuum source atomic absorption spectrometry: review of studies over the last 10 years. TrAC—Trends Anal Chem 88:62–76
46. Cal-Prieto MJ, Felipe-Sotelo M, Carlosena A, Andrade JM, Lo´pez-Mahı´a P, Muniategui S, Prada D (2002) Slurry sampling for direct analysis of solid materials by electrothermal atomic absorption spectrometry (ETAAS). A literature review from 1990 to 2000 Talanta 56:1–51
47. Rubio R, Sahuquillo A, Rauret G, Quevauviller P (1992) Determination of chromium in environmental and biological samples by atomic absorption spectroscopy: a review. Int J Environ Anal Chem 47:99–128
48. Lagalante AF (1999) Atomic absorption spectroscopy: a tutorial review. Appl Spectrosc Rev 34:173–189
49. Butcher DJ (2005) Atomic absorption spectrometry; interferences and background correction. In: Worsfold P, Townshend A, Poole C (eds) Encyclopedia of analytical science, 2nd edn. Elsevier, pp 157–163
50. Maruca R, Suder BJ, Wightman JP (1982) Interaction of heavy metals with chitin and chitosan. III. Chromium. J Appl Polym Sci 27:4827–4837
51. Azarova YA, Pestov AV, Ustinov AY, Bratskaya SY (2015) Application of chitosan and its N-heterocyclic derivatives for preconcentration of noble metal ions and their determination using atomic absorption spectrometry. Carbohydr Polym 134:680–686
52. Liu XD, Tokura S, Haruki M, Nishi N, Sakairi N (2002) Surface modification of nonporous glass beads with chitosan and their adsorption property for transition metal ions. Carbohydr Polym 49:103–108
53. Wilschefski SC, Baxter MR (2019) Inductively coupled plasma mass spectrometry: introduction to analytical aspects. Clin Biochem Rev 40:115–133
54. Galazzi1 RM, Chacón-Madrid K, Freitas DC, da Costa1 LF, Arruda MAZ (2020) Inductively coupled plasma mass spectrometry based platforms for studies involving nanoparticle effects in biological samples. Rapid Commun Mass Spectrom 34:e8726
55. Geertsen V, Barruet E, Gobeaux F, Lacour J, Taché O (2018) Contribution to accurate spherical gold nanoparticle size determination by single-particle inductively coupled mass spectrometry: a comparison with small-angle X-ray scattering. Anal Chem 90:9742–9750
56. Palomo-Siguero M, López-Heras MI, Cámaraa C, Madrid Y (2015) Accumulation and biotransformation of chitosan-modified selenium nanoparticles in exposed radish (*Raphanus sativus*). J Anal At Spectrom 30:1237–1244
57. Cui C, He M, Chen B, Hu B (2014) Chitosan modified magnetic nanoparticles based solid phase extraction combined with ICP-OES for the speciation of Cr(III) and Cr(VI). Anal Methods 6:8577–8583
58. Palomo-Siguero M, Vera P, Echegoyen Y, Nerin C, Cámara C, Madrid Y (2017) Asymmetrical flow field-flow fractionation coupled to inductively coupled plasma mass spectrometry for sizing SeNPs for packaging applications. Spectrochim Acta B At Spectrosc 132:19–25
59. Blanco-Andujar C (2014) Ph.D. Thesis, Sodium carbonate mediated synthesis of iron oxide NPs to improve magnetic hyperthermia efficiency and induce apoptosis. University College London. Coates J (2006) Interpretation of infrared spectra, a practical approach. In: Meyers RA (ed) Encyclopedia of analytical chemistry. Wiley
60. Qi L, Xu Z, Jiang X, Hu C, Zou X (2004) Preparation and antibacterial activity of chitosan nanoparticles. Carbohydr Res 339:2693–2700
61. Xu Y, Du Y (2003) Effect of molecular structure of chitosan on protein delivery properties of chitosan nanoparticles. Int J Pharm 250:215–226
62. Tang ESK, Huang M, Lim LY (2003) Ultrasonication of chitosan and chitosan nanoparticles. Int J Pharm 265:103–114
63. Varma R, Vasudevan S (2020) Extraction, characterization, and antimicrobial activity of chitosan from horse mussel modiolus modiolus. ACS Omega 5:20224–20230

64. El-Ahmady El-Naggar N, Saber WIA, Zwei AM, Bashir SI ()2022 An innovative green synthesis approach of chitosan nanoparticles and their inhibitory activity against phytopathogenic Botrytis cinerea on strawberry leaves. Sci Rep 12:3515
65. Mondéjar-López M, Rubio-Moraga A, López-Jimenez AJ, Martínez JCG, Ahrazem O, Gómez-Gómez L, Niza E (2022) Chitosan nanoparticles loaded with garlic essential oil: a new alternative to tebuconazole as seed dressing agent. Carbohydr Polym 277:118815
66. Banerjee T, Mitra S, Singh AK, Sharma RK, Maitra A (2002) Preparation, characterization, and biodistribution of ultrafine chitosan nanoparticles. Int J Pharm 243:93–105
67. Harding SE (1994) Determination of diffusion coefficients of biological macromolecules by dynamic light scattering. In: Jones C, Mulloy B, Thomas AH (eds) Microscopy, optical spectroscopy, and macroscopic techniques. Methods in molecular biology, vol 22. Humana Press. https://doi.org/10.1385/0-89603-232-9:97
68. Stetefeld J, McKenna SA, Patel TR (2016) Dynamic light scattering: a practical guide and applications in biomedical sciences. Biophys Rev 8:409–427
69. Qi L, Xu Z, Jiang X, Hu C, Zou X (2004) Preparation and antibacterial activity of chitosan nanoparticles. Carbohydr Res 16:2693–2700
70. Li G, Jiang Y, Huang K, Ding P, Chen J (2008) Preparation and properties of magnetic Fe_3O_4–chitosan nanoparticles. J Alloys Compd 466:451–456

Chapter 4
Implementation of Chitosan-Based Nanocomposites for Drug Delivery System

Gyanendra Kumar, Mohd Ehtesham, and Dhanraj T. Masram

Abstract Chitosan (CS) is a biopolymer with many unique characteristic's properties including physicochemical, biological/pharmacological, biodegradability, nontoxicity, and biocompatibility. Among its many biomedical applications, chitosan has led to the inhibition of several lethal diseases. However, chitosan is produced from chitin via hydrolytic deacetylation. After the addition of carbon-based nanocomposite, the physical and mechanical properties of chitosan have been greatly enhanced. Composites of chitosan with additional noble metals, carbon-based nanocomposites comparable to carbon nanotubes (CNTs), graphene or graphene oxide, and many other moieties were found to have higher the activity or drug encapsulation ability of chitosan. In recent years, numerous carbon-based nanocomposite materials, also including multi-walled carbon nanotubes (MWCNT) as well as graphene oxide (GO), were employed to enhance the drug carrier capabilities of chitosan. Graphene oxide is combined by chemical interactions with the biocompatible polymer chitosan to form a carbon-based nanocomposite of biopolymer, which are improved the drug loading capacity. This chapter examines the applications of chitosan as well as chitosan-based bio nanocomposites for the delivery of drugs. The continuous and prolonged pharmacological administration facilitated by chitosan-based nanocomposites enables the successful treatment of infections and illnesses.

Keywords Nanobiocomposites · Chitosan polymer · Nanotechnology · Drug delivery

G. Kumar
Swami Shraddhanand College, University of Delhi, Alipur, Delhi 110036, India

M. Ehtesham
Department of Chemistry, Jamia Millia Islami University, Delhi 110025, India

G. Kumar · D. T. Masram (✉)
Department of Chemistry, University of Delhi, Delhi 110007, India
e-mail: dhanraj_masram27@rediffmail.com

© The Author(s), under exclusive license to Springer Nature Singapore Pte Ltd. 2022
S. Gulati (ed.), *Chitosan-Based Nanocomposite Materials*,
https://doi.org/10.1007/978-981-19-5338-5_4

Abbreviations

5-FU	5-Fluorouracil
CNTs	Carbon nanotubes
CS	Chitosan
CTMS	Chloro trimethyl silane
CU	Curcumin
DD	Drug delivery
GO	Graphene Oxide
MOF-5	Metal organic framework
MWCNTs	Multiwalled Carbon nanotube
PVA	Poly vinyl alcohol
rGO	Reduced graphene oxide

1 Introduction

Chitosan (CS) is getting a lot of attention because it has numerous good abilities, such as biodegradability, biocompatibility, nontoxicity, growth regulation, antimicrobial activity (Fig. 1), and stress-inhibiting activity in plants [1]. It is the second most prevalent natural polymer in the world [2]. Especially in comparison to chitin, chitosan has an outstanding capacity to form complexes, which can be attributed primarily to the presence of unbound $-NH_2$ units distributed over the chitosan chain [3]. Chitin and chitosan seem to be economically valuable polymers and they can be extracted from marine waste such as crustaceans, crabs, shrimp, arthropods, and microorganisms [4]. In plants, chitin is a polymer made up of a disaccharide of two-acetamido-two-deoxy-β-D-glucose that is allied together through a β (1 $\rightarrow$ 4) bond. Furthermore, chitosan was found in the 1900s after the deacetylation of chitin, which had a lot of acetyl glucosamine units [5]. Chitin was first extracted from mushrooms in 1811 by a French professor, Henni Braconnot [6]. Chitosan is a cationic polymer by nature, making it unstable in environments with different pH and ionic strength. For example, the chitosan polymer, unlike chitin, is dissolvable in weak acids similar to acetic or uric acid [7]. Biopolymers such as chitosan are commonly used in biomedical products because of their many beneficial properties, antibacterial actions, hemostasis, pro- and anti-inflammatory cytokine stimulation, lipids, numerous advanced factors, and chemokines polyelectrolyte compound development, increased mucoadhesive strong point, as well as increased tablet crumbliness [8]. Consequently, chitosan has made its way into various therapeutic applications, such as tissue regeneration, medicine transfer, biosensors, and wound dressing.

Chitosan-based nanomaterials have been used in a variety of drug delivery processes including nanoparticles, microbeads, and microfibres [7]. Recently, nanoparticles have become a popular choice for nanocomposites, specifically for therapeutic applications like battered delivery of medicines and genes, bioimaging,

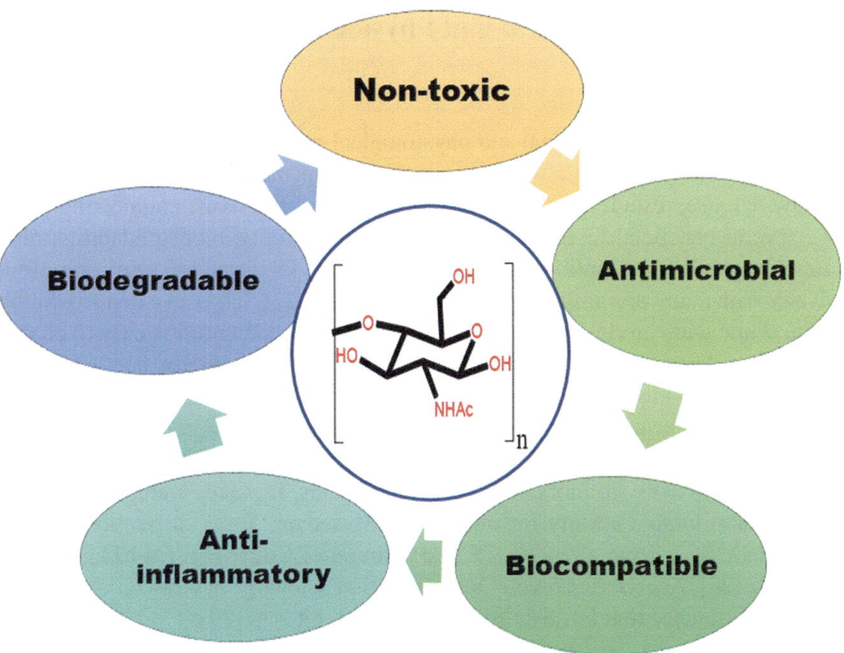

Fig. 1 Numerous biomedical uses are possible due to chitosan's specific biological characteristics

and tissue regeneration, like skin and bone regeneration (Fig. 1). Nanoparticles have a lot of exceptional properties that make them a good choice for making nanocomposites. Further, cells were grown on chitosan-based bio composites increase, and these substrates encourage cell differentiation without the use of advanced factors, suitable for tissue renewal [9, 10]. Multiple characteristics, including cytotoxicity, motorized properties, and healing efficacy, must be careful when developing a framework or foundation for tissue rebirth. To make chitosan-based nanocomposites into films, a number of current procedures have been implemented [11], fiber-meshes [12], and hydrogels [13, 14]. Moreover, chitosan nanocomposite films were used in a variety of therapeutic implementations, including drug delivery [15, 16]. There are different supporting materials have been used such as graphene, [17] nano clay, [18], silver, [19], and titania [20] to prepare chitosan nanocomposites films/materials for such delivery of drug applications. Throughout this chapter, we shall examine chitosan-based nanocomposites for drug delivery.

2 Important Biochemical and Physiological Properties of Chitosan

CS has exceptional biochemical and physiological properties, finding it suited for a variety of uses such as food, cosmetics, water handling, membranes, environmentally friendly defense, resource development, biomedicine, and tissue engineering. They have several beneficial features including biocompatibility, biodegradability, film-forming capability, and antibacterial activity [21, 22]. Chitosan is a semi-crystalline polymer with many dissimilar shapes when it is solid [23]. It has an extended double helical shape with varying packing densities and water content. It is easy to change the hydrated "tendon" CS to the anhydrous crystalline procedure by heating it up or by changing it to monocarboxylic salt [24, 25]. Aside from amino groups, chitosan has 2-hydroxyl clusters for necessary chemical variations. CS can be subjected to a variety of processes including etherification, esterification, and crosslinking [26]. When CS is mixed with inorganic and organic acids, it makes water-soluble salts [27]. It has a durable affinity for metal ions. Moreover, X-ray diffraction patterns can be made by putting a tendon CS in a solution of Cd21, Hg21, Pb21, Zn21, or Cu21 ions [28]. Acetylation level, ionic strength, and charge neutralization of -NH_2 groups all influence that intrinsic pKa number for chitosan [29].

CS as well as its derivatives of the compound through enhanced antibacterial activity have also formed [30]. The antibacterial capabilities of chitosan mixed with nylon-six chelated with silver ions were demonstrated compared to gram-positive and gram-negative microorganisms [31]. Devlieghere et al. looked studied the antibacterial activity of profitable chitosan with a high degree of deacetylation (94%) and a low molar mass against psychrotrophic spoiling microorganisms including food pathogens [32]. They discovered as gram-negative microorganisms were much more sensitive to chitosan than gram-positive bacteria, while susceptibilities within gram-positive bacteria were highly variable. At the same time, CS-zinc compounds demonstrated a broad range of efficient antimicrobial properties, with antibacterial treatment and antifungal action [33]. Silva et al. used buriti oil in chitosan films to create a 100% microbial barrier [34]. The antibacterial movement of chitosan films and coatings was also examined. It was discovered that quaternized chitosan, carboxymethyl chitosan, and quaternized carboxymethyl chitosan presented significant antimicrobial activity compared to Escherichia coli and Staphylococcus aureus [35]. Seyfarth et al. investigated CS for antifungal activity against Candida albicans, Candida krusei, and Candida glabrata [36]. They concluded that when molar mass declined, so did the antifungal activity. Another remarkable attribute of CS is its ability to heal wounds. Ueno et al. [37] explored the habit of chitosan as a wound healer in dogs. They thought that Chitosan could speed up the infusion of polymorphonuclear cells (PMN) and help fibroblasts make more collagen in the initial stages of wound healing. Four drug delivery applications of Nanocomposite Resources Howling et al. revealed that CS stimulates fibroblast proliferation in vitro, but only when it is significantly deacetylated [38]. Also, they found that extremely deacetylated CSs were more energetic than chitin and other less deacetylated chitosan in the body. Lim et al. also found that

CS was cytocompatibility and could help cells grow in the lab [39]. Okamoto et al. [40] studied dogs to see if open wound healing speeds up the healing process. They initiate that the size of the inflammatory cells in the controller group rose significantly more than in the chitin and CS groups. On the extra hand, the chitin and CS groups had a higher rate of re-epithelialization than the control group. Mi et al. investigated an asymmetric chitosan membrane in vivo in rats [41]. They concluded that the membrane-covered wound was hemostatic and healed rapidly. In addition, histological examinations revealed that in wounds coated with the membrane, epithelialization has been accelerated as collagen accumulation within the dermis became highly organized. The physicochemical qualities of chitosan are used in drug delivery, and efforts/research are ongoing to enhance the physicochemical characteristics of chitosan to maximize its applicability for much more efficient drug administration. Further research into chitosan's physical and chemical modifications to facilitate macromolecule transport to the drug's target is predicted. Both toxicities, as well as immunology using chitosan-based target network, must be thoroughly investigated after drug administration to assess the long-term ramifications of this delivery technology, particularly when employed systemically.

Recently, De Souza et al. have also recommended using chitosan in the production of nanocomposite vaccines to treat arbovirus infection [42]. Additionally, such observations have led to the exploration of chitosan as a possible therapy for SARS-CoV-2 (the aetiological agent of COVID-19) [43, 44]. Our long-standing concern about chitosan as a pharmaceutical agent, together with recent findings relating to chitosan and SARS-CoV-19 has prompted us to reconsider chitosan's antiviral characteristics in this chapter. Figure 2 summarizes the possible function of chitosan in the treatment of viral infections.

3 Carbon-Based Bionanocomposites

Two important and widely used procedures for fabricating CS-based membranes as adsorbents are solution casting and solvent disappearance as well as phase downturn (followed by chemical stimulation of the resulting membrane). It has also been shown that making CS composites on your own can be a good and safe idea. Hydrogels made with graphene oxide (GO) and carbon (CS) macromolecules were made by putting them together on their own [46]. Additionally, numerous training has established that self-assembly is a simple process for generating novel complex resources within height adsorption volumes [47, 48].

4 Carbon-Based Nanocomposites

Carbon can be referred to as a unique and hidden element in the periodic table due to its chemical abilities. Carbon compounds are made up of many different things,

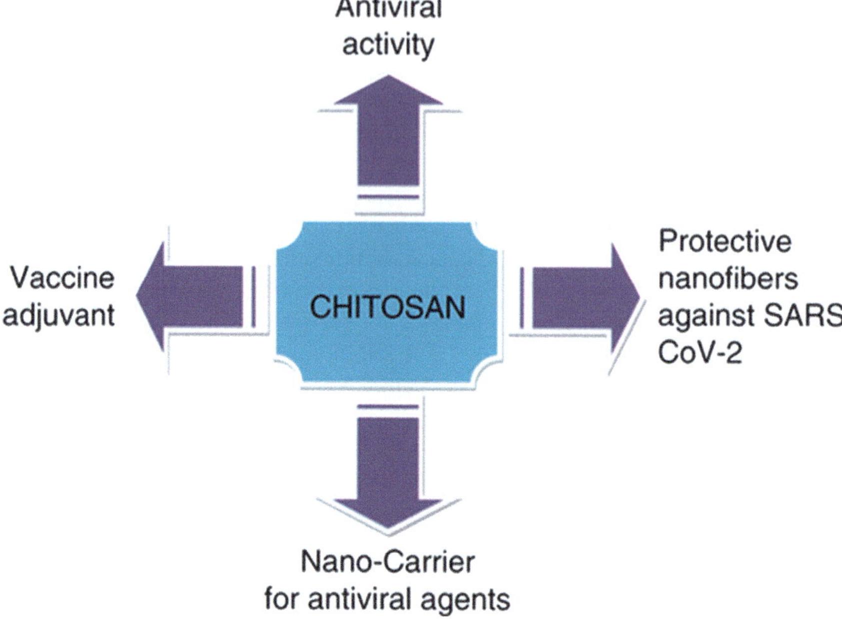

Fig. 2 Schematic illustration of chitosan's possible involvement in viral infection prevention [45]

from petroleum complexes to medications and polymers, but they all reduce into this category. Further, carbon nanostructures are a very important and very specific type of nanotechnology. Moreover, carbon nanostructures are made of carbon compounds. It has exceptional physical and chemical attributes, and they play a big role in the development of innovative and better types of machinery. Carbon comes in many dissimilar forms, such as graphite, diamond, fullerene, amorphous graphene, carbon nanotubes, and carbon nanofibers. In nanochemistry, these forms have lately been employed in the production of nanocomposite materials.

5   Graphene/Graphene Oxide

Graphene is a highly desirable material. Polymers with graphene are said to be very durable and have good thermal conductivity, and electrical conductivity [49]. On the other hand, graphene has a lot of surface area, making it a good adsorbent for metal ions [50]. Furthermore, graphene with oxygen-containing functional groups is polar and has a good adsorptive affinity for metal ions [51]. They can make CS membranes more smooth and dense by adding graphene nanosheets. In this case, when GOs are added to the polymer matrix, wrinkles and microchannels are formed, giving it the appearance of a graphitized and semi-layered structure.

6 Carbon Nanotubes (CNTs)

CNTs were discovered for the first time in 1991 and have since concerned an excessive deal of devotion due to their exceptional and one-of-a-kind properties [51]. CNTs are hollow cylindrical buildings that resemble tubular graphene plates. They are available in two configurations: single-walled and multi-walled. Because it has a lot of clear surface area, the right structure for tubes, and electrical and semiconductor conductivity, it can be functionalized by dissimilar types of surface functional groups [52].

CS is one of the most mutual polymers used to adapt CNTs. Multiwalled Carbon nanotubes are good additions to polymer membranes because they can make them more durable and absorb more. The surface functional groups of new adsorbents show a big part in how the adsorbents work. This is because many novel adsorbents use surface interactions to remove chemicals. According to Salehi et al. could change the features of the CS/PVA membranes with the help of MWCNTs that were amine-functionalized [53]. The main problem with CNTs is that they do not spread evenly in the matrix of polymers. As a result of interactions with Van der Waal, carbon nanotubes tend to form groups and are bundled together. Different procedures and treatments have been used [54]. The most frequently used process is a chemical variation of the surface via acid behavior. When acid-functionalized CNTs are added to the Chitosan/polyvinyl alcohol membrane, they can change the sorption volume and antifouling properties [55].

7 Applications of Chitosan-Based Nanocomposites in Drug Delivery

Table 1 shows and sums up a list of drug delivery applications that use chitosan nanocomposites. Chitosan nanocomposites are often used to deliver drugs to treat diseases like osteoarthritis and cancer [56]. Drug-embedded nanocomposites have a lot of good qualities, like good pharmacokinetics and the ability to deliver drugs to specific target sites. Stimuli-responsiveness aids in the fine-tuning of medicine release rates and the avoidance of burst medicine releases. Medicines can be effectively embedded in chitosan-based resources [57]. Organic chitosan variants were one possibility for modifying that amorphous form of a resource, which impacts drug loading proficiency and release features [58, 59]. According to the literature, many nanoparticles, such as compact GO, [60] clay minerals, [61] gold nanomaterials, [62] highly porous zeolites, [63] layer dual hydroxide, [64] and iron nanoparticles, [65], as well as SiO_2 nanostructured [66], are employing to make numerous categories of chitosan-based drug delivery carriers. Because of the generation of tortuous pathways for clay, 2D layered nanosheets such as graphene or carbon nanotubes (CNT) exhibit better issue features [67]. In addition, some researchers have tried to add mesoporous silica and extra nanoparticles, like hydroxyapatite, to chitosan matrices to improve drug loading and release rates [68–71]. Their findings suggest that the nanoparticles' pore size aided in encapsulating larger amounts of drugs (up to 90%) and in regulating medicine release at the chosen site and time. Additionally, numerous nanoparticles like carbon dots, CNTs, and GO were also used as photothermal agents in NIR-I and NIR-II windows to treat a variety of tumours [72, 73]. They demonstrated high photothermal renovation effectiveness and rapid tumour ablation. Chitosan nanocomposite substrates' photothermal effectiveness facilitated the start of the medicine release (percent) at the target tissue because they were so good at it.

For another thing, CTMS have primarily mixed with nanohydroxyapatite before being mixed with chitosan polyacrylamide to make the nanohydroxyapatite compound (Cs/PAAm/nHA). The sum of CTMS added to the Cs/PAAm/nHA composite can be changed to achieve the desired amount of deprivation of the compound and the amount of medicine released [83]. A reduction process was used to create chitosan/rGO nanomaterial, which has then employed in a transdermal carrier of drug scheme. Because of the reduced graphene, chitosan is more electrically conductive, making the nanocomposites suitable for drug administration by electroporation or iontophoresis [84]. The hybrid $Au\text{-}Fe_3O_4$ nanoparticles were useful in the formation of continuous delivery of drug arrangements based on chitosan that has been grafted with glycolic acid to make it more effective. Despite the fact that the nano scaffold was designed to be stable, cell growth, adhesion, and movement were all affected by the pH of the medium. This is even though the nano scaffold was found to be stable. The nanoparticles that have been made control how much cyclophosphamide comes out of a phosphate buffer (pH 7.0) solution that has the drug in it [85]. These nanoparticles had a wide range of properties in the water, including biocompatibility, antibacterial movement, magnetic properties, decent dispersibility, and a suitable hydrodynamic size. This study looked at how well FA-PEG-chitosan-coated Fe_3O_4 nanoparticles could be targeted. We compared them to chitosan-coated Fe_3O_4 nanoparticles and the FA-PEG-chitosan-coated Fe_3O_4 nanoparticles. Nanoparticles made with exact targeting properties, super magnetism, and extensive blood circulation holes were found to be very good for drug delivery and hyperthermia treatment [86]. Various methods have been used to make nanocomposites that don't separate the solvents from the nanocomposites, such as casting and evaporating the solvent to make chitosan organic rectorite nanocomposite films. It was discovered that dissimilar weight ratios of chitosan to living rectorite were effective in drug application when loaded with medicines simultaneously [87]. Crosslinking agents like dicarboxylic acid were used to make hydrophilic nanoparticles based on chitosan. They were used to make the nanoparticles stick together. The material that was made is biodegradable, can be mixed with water, and has been used as an injectable drug [88]. CS nanocarriers that had 5-FU inside of them were made because the nanocarriers could be used for general drug delivery because the nanomaterial is pH sensitive [89]. The improved chitosan-g-glycolic acid and $P\text{-}Fe_3O_4$ nanoparticles had many holes, which caused phase separation. As time goes on, the porous structure releases less and less drug, but it still releases a lot at first. The Fe_3O_4-gold-chitosan core–shell nanostructure was created with chitosan and the crosslinking agent formaldehyde

Table 1 Nanocomposite substrates are made of chitosan for drug delivery

Additives for nanocomposites	Chitosan details and fabrication technique (if available)	Characteristics of pure chitosan materials	References
The MTD medication was encapsulated in Zn-based (MOF-5)	Using the Stirring Method	At pH 2, the MTD drug has been estimated to be 539 mg/g	Kumar et al. [74]
Co-encapsulated (5-FU) and (CU) in chitosan/rGO composite material	A solution casting/low-molecular mass of ~48,000–180,000 g/mol is having 85% DDA	There was no burst release, and medication loading productivity was 93% for 5-FU and 95% for CU. The dual drug-packed rGO nanocomposites increased anticancer movement, killing 80% of cancer cells while being compatible with fibroblasts	Dhanavel et al. [75]
Chitosan, silver nanoparticles, and MWCNT are encapsulated in 5-fluorouracil (5-FU)	Blending a low-molecular-mass solution having 85% DDA	Overall drug encapsulation productivity of nanocomposites films was 96%, with sustained release lasting up to 72 h	Nivethaa et al. [76]
An alginate-chitosan/montmorillonite type clay called alginate-chitosan/montmorillonite clay is used to treat cancer	85% DDA in a solution/medium-average molecular mass blend	5-FU-encapsulated nanocomposite films containing 30% MMT demonstrated good loading and release efficiency- The nanocomposite films had a substantially greater burst release of medication	Azhar and Olad [77]

(continued)

Table 1 (continued)

Additives for nanocomposites	Chitosan details and fabrication technique (if available)	Characteristics of pure chitosan materials	References
5-Fluorouracil (5-FU) is inside nanocomposites made of chitosan and gold nanoparticles that respond to electric fields	The average MW of solution casting is 270,000 g/mol, with an 85% DDA content	The drug release features of nanocomposites were altered using an external electric field (DC) within the electrolyte solution. When the pH was 5.3 and an external electric field was introduced, the nanocomposite films released more drugs (63%)	Chandran and Sandhyarani [78]
Incorporation of silica nanoparticles in thermosensitive hydrogels	With 75–85% DDA, the average MW of solution approach/viscosity is 50,000–190,000 g/mol	The addition of silica nanoparticles to chitosan hydrogels increased humoral immunity and dramatically increased the growth of CD4+ T cells	Gordon et al. [79]
Iron nanoparticle-based magnetic hydrogels	in situ hybridization approach that is simple to use	Magnetic hydrogels have been developed. Under a low-frequency irregular magnetic field, drug release was changed from passive to pulsatile, and the addition of iron nanoparticles improved biocompatibility and automatic stability, notably elastic modulus, when compared to pristine chitosan hydrogels	Li et al. [80]

(continued)

Table 1 (continued)

Additives for nanocomposites	Chitosan details and fabrication technique (if available)	Characteristics of pure chitosan materials	References
	Solution casting	A nanocomposite film with a pH of 5.5 in the tumor tissue had a loading of drug productivity of 95% and a cumulative release rate of 88% when the pH was that high At pH 7.4, drug release was just 49%. The drug-encapsulated nanocomposites exhibited a high degree of cell compatibility	Yang et al. [81]
Bionanocomposite beads of chitosan and GO-MTD	Gelation Method	In this case, this is the first time that Chi/GO beads have been used to deliver metronidazole (MTD) orally for 84 h	Kumar et al. [82]

via glucose reduction of Au(III). The resulting Fe_3O_4-gold-chitosan nanostructures are effective for medication delivery [90, 91].

The nanocomposites through bone-bioactivity also drug-eluting capability are being evaluated as suitable coating resources for metallic bone implantations. Patel et al. demonstrated the transport of ampicillin using chitosan–bioactive glass as nanoparticles. For 10–11 weeks, the therapeutic medication that was well-integrated during the coating process was shown to last. An antibacterial trial against Streptococcus mutants also demonstrated the effect of medication release. As a result, it has been found that CS–BGn could be used to coat metallic inserts and scaffolds for bone restoration and renewal [92]. In the same way, Jayalekshmi et al. [93], Demonstrated the habit of a golden nanoparticle integrated polymer/bioactive glass complex for regulated drug delivery. At the end of this study, the system was used to deliver doxorubicin. Further, at pH 7.4, 18% mostly was apparently released from the scheme on the eighth day. This means that it was found that the nanoscale of the composite was a factor in the system's high drug delivery efficiency and that the system's nontoxicity, biocompatibility, and controlled release of doxorubicin were proof that the system could be used for chemotherapy.

8 Conclusion and Future Prospects

It is essential to discuss their flaws to improve material-based research as well as innovative implementations. Regarding chitosan-based nanocomposites, a rising variety of studies indicate that CS has significant biomedical applications for potential in drug delivery. The studies in this chapter show that tissues react to chitosan nanocomposites differently from how they respond to each component. Nanoparticle surface modification and chitosan modifications have a big impact on cell activation, adhesion, growth, and differentiation. Moieties like small biomolecules, peptides, proteins, and polymers are being used to change surface characteristics toward improved biological effects, permitting improved cell-substrate interaction and enhancing substrate mechanical properties. Chitosan nanocomposites with customized features produce positive results, such as faster bone regeneration, focused drug/cell delivery, improved wound healing, angiogenesis, and osteogenesis. Furthermore, the molecular mass and drug delivery (DD) have a significant effect on the usage of chitosan. Just, a study looked at the biological properties of a library of chitosan with different percentages of DD, acetylation patterns, and molecular mass. It was shown that chitosan at concentrations of more than 30,000 g/mol inhibits glucosamine-induced macrophage cytokines via lysosomal rupturing. Tiny particles, such as carbon nanotubes, silver sulfadiazine, and bioactive glass, have been shown to be harmful. Despite their shortcomings, chitosan nanocomposite substrates have a high potential for biological implementations. The current chapter outlines the drug's activity and applications in the systemic delivery of the drug. Following this is a quick summary of biodegradable polymers for drug delivery, with an emphasis on chitosan as a prospective delivery drug carrier. As a result, the goal of the chapter is to emphasize the difficulties in developing chitosan-based drug delivery for local diseases and provides an overview of chitosan and its derivatives.

References

1. Mujtaba M, Morsi RE, Kerch G, Elsabee MZ, Kaya M, Labidi J, Khawar KM (2019) Current advancements in chitosan-based film production for food technology; a review. Int J Biol Macromol 121:889–904
2. Kaya M, Baran T, Erdoğan S, Menteş A, Özüsağlam MA, Çakmak YS (2014) Physicochemical comparison of chitin and chitosan obtained from larvae and adult Colorado potato beetle (Leptinotarsa decemlineata). Mater Sci Eng: C 45:72–81
3. Shukla SK, Mishra AK, Arotiba OA, Mamba BB (2013) Chitosan-based nanomaterials: a state-of-the-art review. Int J of Biol Macromol 59:46–58

4. Abdou ES, Nagy KS, Elsabee MZ (2008) Extraction and characterization of chitin and chitosan from local sources. Biores Technol 99(5):1359–1367
5. Tokuyasu K, Ono H, Hayashi K, Mori Y (1999) Reverse hydrolysis reaction of chitin deacetylase and enzymatic synthesis of β-d-GlcNAc-(1→4)-GlcN from chitobiose. Carbohydr Res 322:26–31
6. Muzzarelli RA, Boudrant J, Meyer D, Manno N, DeMarchis M, Paoletti MG (2012) Current views on fungal chitin/chitosan, human chitinases, food preservation, glucans, pectins and inulin: a tribute to Henri Braconnot, precursor of the carbohydrate polymers science, on the chitin bicentennial. Carbohydr Polym 87:995–1012
7. Juntong Y, Depeng W, Nagaraja G, Mahmood KK, Sudisha J, Mujtaba M (2021) Current trends and challenges in the synthesis and applications of chitosan-based nanocomposites for plants: a review. Carbohydr Polym 261:117904
8. Fong D, Ariganello MB, Girard-Lauzière J, Hoemann CD (2015) Biodegradable chitosan microparticles induce delayed STAT-1 activation and lead to distinct cytokine responses in differentially polarized human macrophages in vitro. Acta Biomater 12:183–194
9. Wang Y, Gu H (2015) Core–shell-type magnetic mesoporous silica nanocomposites for bioimaging and therapeutic agent delivery. Adv Mater 27:576–585
10. Polini A, Pisignano D, Parodi M, Quarto R, Scaglione S (2011) Osteoinduction of human mesenchymal stem cells by bioactive composite scaffolds without supplemental osteogenic growth factors. PLoS ONE 6:e26211
11. Kumar S, Krishnakumar B, Sobral A, Koh J (2019) Bio-based (chitosan/PVA/ZnO) nanocomposites film: thermally stable and photoluminescence material for removal of organic dye. Carbohydr Polym 205:559
12. Naseri N, Algan C, Jacobs V, John M, Oksman K, Mathew AP (2014) Electrospun chitosan-based nanocomposite mats reinforced with chitin nanocrystals for wound dressing. Carbohydr Polym 109:7
13. Yin M, Wan S, Ren X, Chu CC (2021) Development of inherently antibacterial, biodegradable, and biologically active chitosan/pseudo-protein hybrid hydrogels as biofunctional wound dressings. ACS Appl Mater Interfaces 13:14688
14. Qianqian O, Songzhi K, Yongmei H, Xianghong J, Sidong L, Puwang L, Hui L (2021) Preparation of nano-hydroxyapatite/chitosan/tilapia skin peptides hydrogels and its burn wound treatment. Int J Biol Macromol 181:369
15. Soe MT, Pongjanyakul T, Limpongsa E, Jaipakdee N (2020) Modified glutinous rice starch-chitosan composite films for buccal delivery of hydrophilic drug. Carbohydr Polym 245:116556
16. Abilova GK, Kaldybekov DB, Irmukhametova GS, Kazybayeva DS, Iskakbayeva ZA, Kudaibergenov SE, Khutoryanskiy VV (2020) Chitosan/poly(2-ethyl-2-oxazoline) films with ciprofloxacin for application in vaginal drug delivery. Materials 13:1709
17. He L, Wang H, Xia G, Sun J, Song R (2014) Chitosan/graphene oxide nanocomposite films with enhanced interfacial interaction and their electrochemical applications. Appl Surf Sci 314:510
18. Han YS, Lee SH, Choi KH, Park I (2010) Preparation and characterization of chitosan–clay nanocomposites with antimicrobial activity. J Phys Chem Solids 71:464
19. Shah A, Hussain I, Murtaza G (2018) Chemical synthesis and characterization of chitosan/silver nanocomposites films and their potential antibacterial activity. Int J Biol Macromol 116:520–529
20. Kaewklin P, Siripatrawan U, Suwanagul A, Lee YS (2018) Active packaging from chitosan-titanium dioxide nanocomposite film for prolonging storage life of tomato fruit. Int J Biol Macromol 112:523–529
21. Honarkar H, Barikani M (2009) Applications of biopolymers I: chitosan. Monatshefte fu¨r Chemie-Chemical Monthly 140(12):1403
22. Rinaudo M (2006) Chitin and chitosan: properties and applications. Prog Polym Sci 31(7):603632
23. Okuyama K, Noguchi K, Miyazawa T, Yui T, Ogawa K (1997) Molecular and crystal structure of hydrated chitosan. Macromolecules 30(19):58495855

24. Ogawa K (2014) Effect of heating an aqueous suspension of chitosan on the crystallinity and polymorphs. Agric Biol Chem 55(9):2375–2379
25. Okuyama K, Noguchi K, Hanafusa Y, Osawa K, Ogawa K (1999) Structural study of anhydrous tendon chitosan obtained via chitosan/acetic acid complex. Int J Biol Macromol 26(4):285293
26. Hon D (1996) Chitin and chitosan: medical applications. Marcel Dekker Inc., New York, p 631649
27. Mishra RK, Tiwari SK, Mohapatra S, Thomas S (2019) Chapter 1—Efficient Nanocarriers for drug-delivery systems: types and fabrication. In: Nanocarriers for drug delivery, pp 1–41
28. Maniukiewicz W (2010) 8 X-ray diffraction studies of chitin, chitosan, and their derivatives, chitin, chitosan, oligosaccharides and their derivatives: biological activities and applications. p 83
29. Guibal E (2005) Heterogeneous catalysis on chitosan-based materials: a review. Prog Polym Sci 30(1):71109
30. Zhong Z, Xing R, Liu S, Wang L, Cai S, Li P (2008) Synthesis of acyl thiourea derivatives of chitosan and their antimicrobial activities in vitro. Carbohydr Res 343(3):566570
31. Ma Y, Zhou T, Zhao C (2008) Preparation of chitosannylon-6 blended membranes containing silver ions as antibacterial materials. Carbohydr Res 343(2):230237
32. Devlieghere F, Vermeulen A, Debevere J (2004) Chitosan: antimicrobial activity, interactions with food components and applicability as a coating on fruit and vegetables. Food Microbiol 21(6):703714
33. Wang X, Du Y, Liu H (2004) Preparation, characterization and antimicrobial activity of chitosan–Zn complex. Carbohydr Polym 56(1):2126
34. Silva MF, Lopes PS, da Silva CF, Yoshida CMP (2016) Active packaging material based on buriti oil–Mauritia flexuosa Lf (Arecaceae) incorporated into chitosan films. J Appl Polym Sci 133(12)
35. Dutta J, Dutta PK (2010) 15 Antimicrobial activity of chitin, chitosan, and their oligosaccharides, chitin, chitosan, oligosaccharides and their derivatives: biological activities and applications. pp 195
36. Seyfarth F, Schliemann S, Elsner P, Hipler U-C (2008) Antifungal effect of high-and lowmolecular-weight chitosan hydrochloride, carboxymethyl chitosan, chitosan oligosaccharide and N-acetyl-D-glucosamine against Candida albicans, Candida krusei and Candida glabrata. Int J Pharm 353(1):139148
37. Ueno H, Yamada H, Tanaka I, Kaba N, Matsuura M, Okumura M et al (1999) Accelerating effects of chitosan for healing at early phase of experimental open wound in dogs. Biomaterials 20(15):14071414
38. Howling GI, Dettmar PW, Goddard PA, Hampson FC, Dornish M, Wood EJ (2001) The effect of chitin and chitosan on the proliferation of human skin fibroblasts and keratinocytes in vitro. Biomaterials 22(22):29592966
39. Lim C, Yaacob N, Ismail Z, Halim A (2010) In vitro biocompatibility of chitosan porous skin regenerating templates (PSRTs) using primary human skin keratinocytes. Toxicol In Vitro 24(3):721727
40. Okamoto Y, Shibazaki K, Minami S, Matsuhashi A, Tanioka SI, Shigemasa Y (1995) Evaluation of chitin and chitosan on open wound healing in dogs. J Vet Med Sci 57(5):851854
41. Mi FL, Shyu SS, Wu YB, Lee ST, Shyong JY, Huang RN (2001) Fabrication and characterization of a sponge-like asymmetric chitosan membrane as a wound dressing. Biomaterials 22(2):165173
42. De Souza GAP, Rocha RP, Gonçalves RL, Ferreira CS, de Mello Silva B, de Castro, RFG et al (2021) Nanoparticles as vaccines to prevent arbovirus infection: a long road ahead. Pathogens 10:36
43. Gurunathan S, Qasim M, Choi Y, Do JT, Park C, Hong K et al (2020) Antiviral potential of nanoparticles—can nanoparticles fight against coronaviruses? Nanomaterials 10(9):1645
44. Pyrć K, Milewska A, Duran EB, Botwina P, Lopes R, Arenas-Pinto A et al (2021) (2020) SARS-CoV-2 inhibition in human airway epithelial cells using a mucoadhesive, amphiphilic chitosan that may serve as an anti-viral nasal spray. Sci Rep 11:20012

45. Jaber N, Al-Remawi M, Al-Akayleh F, Al-Muhtaseb N, Al-Adham ISI, Collier PJ (2022) A review of the antiviral activity of chitosan, including patented applications and its potential use against COVID-19. J Appl Microbiol 132:41–58

46. Zhao H, Jiao T, Zhang L, Zhou J, Zhang Q, Peng Q, Yan X (2015) Preparation and adsorption capacity evaluation of graphene oxide-chitosan composite hydrogels. Sci China Mater 58:811–818

47. Xing R, Wang W, Jiao T, Ma K, Zhang Q, Hong W, Qiu H, Zhou J, Zhang L, Peng Q (2017) Bioinspired polydopamine sheathed nanofibers containing carboxylate graphene oxide nanosheet for high-efficient dyes scavenger ACS Sustain Chem 5:4948–4956

48. Jiao T, Zhao H, Zhou J, Zhang Q, Luo X, Hu J, Peng Q, Yan X (2015) Self-assembly reduced graphene oxide nanosheet hydrogel fabrication by anchorage of chitosan/silver and its potential efficient application toward dye degradation for wastewater treatments. ACS Sustain Chem 3:3130–3139

49. Fan H, Wang L, Zhao K, Li N, Shi Z, Ge Z, Jin ZJB (2010) Fabrication, mechanical properties, and biocompatibility of graphene-reinforced chitosan composites. Biomacromolecules 11:2345–2351

50. Yu G, Lu Y, Guo J, Patel M, Bafana A, Wang X, Qiu B, Jeffryes C, Wei S, Guo Z (2018) Carbon nanotubes, graphene, and their derivatives for heavy metal removal. Adv Compos Hybrid Mater 1:56–78

51. Thostenson T, Ren Z, Chou TW (2001) Multiscale modeling of compressive behavior of carbon nanotube/polymer composites. Compos Sci Technol 61:1899

52. Stobinski L, Lesiak B, Kövér L, Tóth J, Biniak S, Trykowski G, Judek J (2010) Multiwall carbon nanotubes purification and oxidation by nitric acid studied by the FTIR and electron spectroscopy methods. J Alloy Compd 501:77–84

53. Salehi E, Madaeni SS, Rajabi L, Vatanpour V, Derakhshan AA, Zinadini S, Ghorabi S, Monfared HA (2012) Novel chitosan/poly (vinyl) alcohol thin adsorptive membranes modified with amino functionalized multi-walled carbon nanotubes for Cu (II) removal from water: preparation, characterization, adsorption kinetics and thermodynamics. Sep Purif Technol 89:309–319

54. Ma PC, Siddiqui NA, Marom G, Kim JK (2010) Manufacturing, dispersion and functionalization of carbon nanotubes for polymer-based nanocomposites: a review. Compos Part A Appl Sci Manuf 41:1345–1367

55. Zarghami S, Tofighy MA, Mohammadi T (2015) Adsorption of zinc and lead ions from aqueous solutions using chitosan/polyvinyl alcohol membrane incorporated via acid-functionalized carbon nanotubes. J Dispers Sci Technol 36:1793–1798

56. Sharma B, Shekhar S, Jain P, Sharma R, Chauhan K (2021) Graphene based biopolymer nanocomposites. Springer, Singapore, p 135

57. Pradhan S, Brooks AK, Yadavalli VK (2020) Nature-derived materials for the fabrication of functional biodevices. Mater Today Bio 7:100065

58. Kumar MR, Muzzarelli RA, Muzzarelli C, Sashiwa H, Domb A (2004) Chitosan chemistry and pharmaceutical perspectives. Chem Rev 104:6017

59. Ahmed F, Soliman FM, Adly MA, Soliman HAM, El-Matbouli M, Saleh M (2019) Recent progress in biomedical applications of chitosan and its nanocomposites in aquaculture: a review. Res Vet Sci 126:68–82

60. Baktash MS, Zarrabi A, Avazverdi E, Reis NM (2021) Development and optimization of a new hybrid chitosan-grafted graphene oxide/magnetic nanoparticle system for theranostic applications. J Mol Liq 322:114515

61. Jauković V, Krajišnik D, Daković A, Damjanović A, Krstić J, Stojanović J, Čalija B (2021) Influence of selective acid-etching on functionality of halloysite-chitosan nanocontainers for sustained drug release. Mater Sci Eng: C 123:112029

62. Horo H, Bhattacharyya S, Mandal B, Kundu LM (2021) Synthesis of functionalized silk-coated chitosan-gold nanoparticles and microparticles for target-directed delivery of antitumor agents. Carbohydr Polym 258:117659

63. Abdelhamid HN, Dowaidar M, Langel Ü (2020) Carbonized chitosan encapsulated hierarchical porous zeolitic imidazolate frameworks nanoparticles for gene delivery. Microporous Mesoporous Mater 302:110200

64. Chen YX, Zhu R, Xu ZL, Ke QF, Zhang CQ, Guo Y-P (2017) Self-assembly of pifithrin-α-loaded layered double hydroxide/chitosan nanohybrid composites as a drug delivery system for bone repair materials. J Mater Chem B 5:2245

65. Zhang YG, Zhu YJ, Chen F, Sun TW (2017) A novel composite scaffold comprising ultralong hydroxyapatite microtubes and chitosan: preparation and application in drug delivery. J Mater Chem B 5:3898

66. Liu W, Wang F, Zhu Y, Li X, Liu X, Pang J, Pan W (2018) Galactosylated chitosan-functionalized mesoporous silica nanoparticle loading by calcium leucovorin for colon cancer cell-targeted drug delivery. Molecules 23:3082

67. Murugesan S, Scheibel T (2020) Copolymer/clay nanocomposites for biomedical applications. Adv Funct Mater 30:1908101

68. Kozakevych RB, Bolbukh YM, Tertykh VA (2013) Controlled release of diclofenac sodium from silica-chitosan composites. World J Nano Sci Eng 3:69

69. Ahmed A, Hearn J, Abdelmagid W, Zhang H (2012) Dual-tuned drug release by nanofibrous scaffolds of chitosan and mesoporous silica microspheres. J Mater Chem 22:25027

70. Zhang J, Liu G, Wu Q, Zuo J, Qin Y, Wang J (2012) Novel mesoporous hydroxyapatite/chitosan composite for bone repair. J Bionic Eng 9:243

71. Pourshahrestani S, Zeimaran E, Kadri NA, Gargiulo N, Jindal HM, Naveen SV, Sekaran SD, Kamarul T, Towler MR (2017) Potency and cytotoxicity of a novel gallium-containing mesoporous bioactive glass/chitosan composite scaffold as hemostatic agents. ACS Appl Mater Interfaces 9:31381

72. Su Z, Sun D, Zhang L, He M, Jiang Y, Millar B, Douglas P, Mariotti D, Maguire P, Sun D (2021) Chitosan/silver nanoparticle/graphene oxide nanocomposites with multi-drug release antimicrobial, and photothermal conversion functions. Materials 14:2351

73. Wang H, Mukherjee S, Yi J, Banerjee P, Chen Q, Zhou S (2017) Biocompatible chitosan-carbon dot hybrid nanogels for nir-imaging-guided synergistic photothermal-chemo therapy. ACS Appl Mater Interfaces 9:18639

74. Kumar G, Kant A, Kumar M, Masram DT (2019) Synthesis, characterizations and kinetic study of metal organic framework nanocomposite excipient used as extended-release delivery vehicle for an antibiotic drug. Inorg Chim Acta 496:119036

75. Dhanavel S, Revathy T, Sivaranjani T, Sivakumar K, Palani P, Narayanan V, Stephen A (2020) 5-Fluorouracil and curcumin co-encapsulated chitosan/reduced graphene oxide nanocomposites against human colon cancer cell lines. Polym Bull 77:213

76. Nivethaa EA, Dhanavel S, Rebekah A, Narayanan V, Stephen A (2016) A comparative study of 5-Fluorouracil release from chitosan/silver and chitosan/silver/MWCNT nanocomposites and their cytotoxicity towards MCF-7. Mater Sci Eng C 66:244–250

77. Azhar FF, Olad A (2014) A study on sustained release formulations for oral delivery of 5-fluorouracil based on alginate–chitosan/montmorillonite nanocomposite systems. Appl Clay Sci 101:288–296

78. Chandran PR, Sandhyarani N (2014) An electric field responsive drug delivery system based on chitosan–gold nanocomposites for site specific and controlled delivery of 5-fluorouracil. RSC Adv 4:44922–44929

79. Gordon S, Teichmann E, Young K, Finnie K, Rades T, Hook S (2010) In vitro and in vivo investigation of thermosensitive chitosan hydrogels containing silica nanoparticles for vaccine delivery. Eur J Pharm Sci 41:360

80. Li B, Jia D, Zhou Y, Hu Q, Cai W (2006) In situ hybridization to chitosan/magnetite nanocomposite induced by the magnetic field. J Magn Magn Mater 306:223

81. Yang F, Wen X, Ke QF, Xie XT, Guo YP (2018) pH-responsive mesoporous ZSM-5 zeolites/chitosan core-shell nanodisks loaded with doxorubicin against osteosarcoma. Mater Sci Eng C 85:142–153

82. Kumar G, Chaudhary K, Mogha NK, Kant A, Masram DT (2021) Extended release of metronidazole drug using chitosan/graphene oxide bionanocomposite beads as the drug carrier. ACS Omega 6(31):20433–20444

83. Huanbutta K, Sriamornsak P, Luangtana-Anan M, Limmatvapirat S, Puttipipatkhachorn S, Lim L-Y, Terada K, Nunthanid J (2013) Application of multiple stepwise spinning diskprocessing for the synthesis of poly(methyl acrylates) coated chitosan-diclofenacsodium nanoparticles for colonic drug delivery. Eur J Pharm Sci 50(3–4):303–311

84. Justin R, Chen B (2014) Strong and conductive chitosan-reduced graphene oxide nanocomposites for transdermal drug delivery. J Mater Chem B 2:3759–3770

85. Depan D, Kumar AP, Singh RP (2009) Structureprocessproperty relationship of the polar graphene oxide-mediated cellular response and stimulated growth of osteoblasts on hybrid chitosan network structure nanocomposite scaffolds. Acta Biomater 5:93–100

86. Zhou S, Li Y, Cui F, Jia M, Yang X, Wang Y et al (2013) Development of multifunctional folate-poly(ethylene glycol)-chitosan-coated Fe_3O_4 nanoparticles for biomedical applications. Macromol Res 22:58–66

87. Wang X, Du Y, Luo J, Lin B, Kennedy JF (2007) Chitosan/organic rectorite nanocomposite films: structure, characteristic and drug delivery behavior. Carbohydr Polym 69:41–49

88. Bodnar M, Hartmann JF, Borbely J (2005) Nanoparticles from chitosan. Macromol Symp 321–326

89. Aydin RST, Pulat M (2012) 5-Fluorouracil encapsulated chitosan nanoparticles for pHstimulated drug delivery: evaluation of controlled release kinetics. Nanomater 42:1–10

90. Kumari S, Singh RP (2012) Glycolic acid-g-chitosan-gold nanoflower nanocomposite scaffolds for drug delivery and tissue engineering. Int J Biol Macromol 50(3):878–883

91. Salehizadeh H, Hekmatian E, Sadeghi M, Kennedy K (2012) Synthesis and characterization of core-shell Fe_3O_4-gold-chitosan nanostructure. J Nanobiotechnol 10(1):3

92. Patel KD, El-Fiqi A, Lee H-Y, Singh RK, Kim D-A, Lee H-H, Kim H-W (2012) Chitosan-nanobioactive glass electrophoretic coatings with bone regenerative and drug delivering potential. J Mater Chem 22(47):24945–24956

93. Jayalekshmi AC (2015) Sharma CP Gold nanoparticle incorporated polymer/bioactive glass composite for controlled drug delivery application. Colloids Surf B 126:280–287

Chapter 5
Potential of Chitosan-Based Nanocomposites for Biomedical Application in Gene Therapy

Manoj Trivedi and Sanjay Kumar

Abstract The application of gene therapy in the field of molecular medicine is an extremely promising approach to curing distinct varieties of illnesses and disorders of the human race. Currently, challenges of the gene therapy are to find secure and effective vectors which might be capable of delivering genes to the specific cells and getting them to express inside the cells. Because of safety concerns, artificial delivery systems are desired in comparison to viral vectors for gene delivery so numerous attention has been centered on the development of the effective vectors. However, Researchers are confronted with numerous problems consisting of low gene transfer efficiency, cytotoxicity, and lack of cell-targeting capability for the usage of these synthetic vectors. Chitosan, which is the biodegradable and non-toxic cationic polysaccharide, is generally preferred to the other cationic polymers as a non-viral vector mainly due to its properties of chemical versatility, excellence in transcellular transport, effectiveness as a DNA-condensing agent, and efficient and permanent transfection. The objective of this chapter is to indicate the importance and give an overview of the applications of chitosan and its derivatives as novel non-viral vectors for gene delivery.

Keywords Chitosan · Polymer · Carrier · Cationic · Gene · Transfection · Ligand

Abbreviations

DNA	Deoxyribonucleic acid
mPEG	Methoxy Polyethylene glycol
mV	Millivolts
PDMAEA	Poly[2-(dimethylamino) ethyl acrylate]
PDMAEMA	Poly[2-(dimethylamino) ethyl methacrylate]
PEC	Polyelectrolyte complex

M. Trivedi (✉) · S. Kumar
Department of Chemistry, Sri Venkateswara College, University of Delhi, New Delhi 110021, India
e-mail: mtrivedi@svc.ac.in

© The Author(s), under exclusive license to Springer Nature Singapore Pte Ltd. 2022   121
S. Gulati (ed.), *Chitosan-Based Nanocomposite Materials*,
https://doi.org/10.1007/978-981-19-5338-5_5

PEG	Polyethylene glycol
PEI	Polyethylenimine
PEI	Poly(L-lysine)
pHPMA	Poly[N-(2-hydroxypropyl methacrylamide]
p-NIPAM	Poly(N-isopropylacrylamide)
RNA	Ribonucleic acid

1 Introduction

Since the discovery of DNA's structure, its functions, and gene transfer applications have gotten a lot of interest, from cell transfection to the manufacture of transgenic animals by getting transgenic embryos to gene therapy. Several studies were conducted, as follows:

- Changing the gene code and looking into the roles of genes.
- Transfer of DNA to the organism or its cells (transgenesis).
- Gene therapy is used to treat gene mutations or deficiencies.
- Therapeutics are produced using transgenic prokaryotes and eukaryotes, particularly pharmaceutical animals such as goats and cattle (bio-pharming of therapeutics).
- Model laboratory animals are created to study genetic illnesses caused by genetic damage or mutations.
- Farm animals, particularly pigs, are being studied for use as a tissue bank for human transplants (xenotransplantation).

In these investigations, many successful outcomes have been obtained, as well as innovative methodologies. Apart from in vitro experiments, in vivo applications have also been carried out, but due to some unsolved problems and restrictions or difficulties, such as targeting of gene carrier particles, undesirable acute or late side effects of genes, and their carrier systems, obstacles relating to human applications have yet to be overcome. Transfection (gene transfer) is a process in which a gene is transferred to the nucleus of another cell and implanted in its DNA. Many approaches and protocols for transgenesis applications have been created by researchers. The gene is transferred to the tissue/cells using a variety of ways.

- Electroporation.
- Direct injection of genes into the nucleus or pronucleus.
- Using viral vectors to transfer genes.
- Non-viral vectors are used to deliver genes.

Because the electroporation technique is only utilized in cell suspension, it has limited use. Furthermore, microinjection of the gene directly into the nucleus/pronucleus necessitates specialized and costly equipment, and this technique can be utilized in cell culture systems and after egg fertilization. The reliability of viral vectors in vivo experiments is still an open question, and the results of these studies are generally poor. As a result, researchers created a novel targeted gene carrier system, and non-viral gene carrier systems have become more commonly used in transgenesis and gene therapy applications. However, these methods have issues such as difficulty in targeting the gene carrier system, early cytoplasmic enzymatic activity degradation of the particles and the gene in the cells, and non-viral agent cytotoxicity. Currently, research is focused on resolving these issues. By delivering genetic material to the patient, gene therapy has been utilized to prevent genetic abnormalities. With this technique, the patient's therapy, which is the regeneration of damaged biological activities or the restoration of homeostasis, is carried out at the molecular level. The basic goal is to overcome biological barriers that prevent therapeutic genes from reaching the desired location. Preclinical and clinical gene therapy research has advanced considerably in the last 15 years. Gene therapy has recently gained popularity as a promising treatment option for genetic illnesses, cancer, cardiovascular disease, and viral infection. Gene therapy not only tries to treat diseases but also to transfer recombinant genetic material to the nucleus, where gene expression, which activates or deactivates protein synthesis, occurs. It is clear that well-targeted, non-toxic gene carrier mechanisms are required to transport the gene to the nucleus. Chitosan is a non-viral gene carrier that is commonly used in gene therapy.

2 Chitosan as a Gene Carrier

The carrier system that delivers a gene for gene expression is called a vector. Vectors are mainly divided into two groups: non-viral vectors and viral vectors. Effective transfection and gene expression of viral vector therapy genes are of great clinical importance for gene therapy. Due to its structure, the virus performs gene transfection very effectively. Due to this property of viral vectors, the required carrier system is preserved and improved. Recently, viral vectors with different genomic characteristics such as retrovirus, adenovirus, adeno-associated virus, herpesvirus, and poxvirus have been commonly used for the efficient provision of gene transport capacity and gene expression. Retroviruses are the most popular of these viral vectors because of their highest gene transfection efficiency and highest expression of therapeutic genes. While these functions emphasize the importance of safe RNA and DNA viral carrier systems, the same report reveals the difficulty of using clinical viral vectors [1, 2]. For example, adenovirus provides the highest gene expression and can infect dividing and non-dividing cells, but it elicits an immune response through viral protein and transient gene expression [3–5]. Similarly, retroviruses facilitate the manipulation of the viral genome. Although easy to combine with DNA, their advantages are difficult to target, complex combinations within the genome, and instability [5]. The first clinical study deals with viral vector compliance and reliability. There are numerous viral vector systems that have been tested in ex vivo and in vivo studies. In recent years, studies have focused on virus targeting, cell type expression, and time of expression to enhance existing effects [5]. Researchers prefer viral vectors for their effectiveness in transfection studies, but the use of these vectors has characteristics such as cytotoxicity, immunosensitivity to viral antigens, and possible viral combinations. Is limited because it is low. Because of these negative features, researchers have been drawn to non-viral vectors. Today, gene therapy is clinically effective in treating many illnesses. Since 1989, when gene therapy was first introduced, this approach has been used in more than 3000 patients with approximately 600 interventions, but no gene therapy product raises toxicity concerns. In addition, there are significant usage differences between viral and non-viral vectors. Viral vectors are used in about 75% of clinical treatment cases, while non-viral vectors are used in less than 25% [6, 7]. Some highlights of the gene transfer application are shown in chronological order in Fig. 5.1 [8–15]. Due to these instabilities, researchers not only compared viral vectors with non-viral vectors but also sought to understand the potential and limitations of non-viral systems. In many of these studies, the researchers of gene therapy and pharmaceutic technologies recognize that viral vectors are merged with the genome and the cell is mutated, so cancer occurs in vivo. Besides that, the immune response is activated, and that's why the treatment becomes more difficult [7, 16]. Especially, although the numerous non-viral vectors are synthesized and their features designed, these systems are not effective enough for gene transfer, so there are not any commercial products. The non-viral vectors are divided into two groups lipophilic vectors and polymeric vectors.

Non-viral vectors containing cationic liposomes are commonly used before and during clinical trials, but an important part of non-viral lipophilic vectors has the same toxicity as viral vectors. And although there are clinical problems, the main advantage of these polymers is that they can create various modified cations in the structure of the polymer to enhance their potency. As a result, plasmid DNA is released in a controlled manner, increasing stability to enzymes in the blood such as nucleases, reducing non-specific uptake, regulating interactions with cells and plasma molecules, and simultaneously eliminating the immunizing of the system. These modifications are applied to improve the release properties of the polymer, allowing the genetic material to be released in a controlled manner at the target site [17]. On the other hand, non-viral gene transfer systems with different polymer structures are

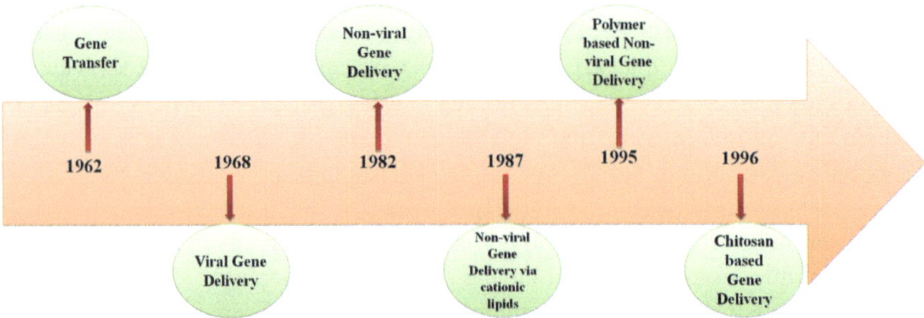

Fig. 5.1 The Continuous growth of gene transfer applications

safe and economical gene transfer systems with improved various synthetic vectors, some of which are commercially available. In addition, non-viral vectors are not as efficient as viral vectors and generally solve as some of the cationic carrier systems used with anionic genes to generate ion complexes capable of exhibiting cytotoxic effects [18, 19]. The common goal of these studies is the synthesis of polymer carrier systems that are as effective and less toxic as viral vectors.

3   Application of Polymers in Gene Delivery

3.1   Artificial Polymers

There are several non-viral gene delivery systems for transferring genetic material to the cell nucleus. The non-viral gene delivery system consists of polymers and lipids or liposomes. Cationic polymers have many advantages for use in gene delivery such as low toxicity and immune response, being easy to handle, and being stable [20, 21]. However, some issues need to be resolved, including toxicity, reduced transfection efficiency, and lack of biodegradability. These properties of the polymer need to be modified in different ways. These biodegradable non-viral polymer gene delivery systems are called transgenic polymers. Transgenic polymers can be divided into two groups: natural transgenic polymers and synthetic transgenic polymers. Synthetic transgenic polymers are generally preferred for gene delivery because of their ease of modification. According to Amiji [6], synthetic transgenic polymers include non-biodegradable transgenic synthetic polymers [polyethylenimine (PEI), polyethylene glycol (PEG) conjugates, etc.], biodegradable transgenic synthetic polymers (poly-β-aminoesters, polyamido amines, poly-imidazoles, etc.), polyethylene oxide/polypropylene oxide copolymers and polymeric polyethylene oxide, polyalkylcyanoacrylate nanomicrospheres. However, the main drawbacks of these polymers are their high toxicity. The main reasons for toxicity are the polymer skeleton and the density and distribution of positive charges along with the molecular weight. PEI's with a molecular weight of 22–25 kDa are used for gene therapy because of their reduced cytotoxicity and high transfection efficiency [22, 23]. Other non-biodegradable transgenic polymers are used to deliver genetic material such as DNA, oligonucleotides, and small interfering RNAs. Some of these polymers are poly[2-(dimethylamino) ethyl methacrylate] (PDMAEMA), poly[2-(dimethylamino) ethyl acrylate] (PDMAEA), and N-vinyl pyrrolidone. They show high transfection efficiency like PEI, but due to their negative properties that reduce stability in blood, high interaction with serum components, etc., these polymers are PEG and poly[N-(2-hydroxypropyl methacrylamide)] (pHPMA) is bound [24]. PEG and pHPMA materials mask the instability of nanoparticles in serum due to the presence of nanoparticles in the hydrophilic layer on the polyplex. However, the PEG and pHPMA groups interfere with complex: DNA complex formation, thereby reducing the efficiency of polyplex transfection. Researchers are working on improving various ways to solve these problems [24–26].

3.2 Natural Polymers

Cationic polymers (polycations) are one of the most commonly used carrier systems in molecular gene transfer systems. The polyelectrolyte complex (PEC) obtained by the interaction of DNA and polycations protects DNA from enzymes such as DNAse. In addition, transgenic polycations interact with serum DNA due to their cationic properties at physiological pH.PEC systems are easier to manufacture and have a lower immune response than viral vectors, but researchers are pursuing further research due to negative properties such as biodegradability problems and reduced transfection efficiency. The main transgenic polycations are PEI, poly(L-lysine), dendrimers, gelatin, and chitosan, a natural cationic polysaccharide. As mentioned above, one of these polycations, chitosan, is natural, biodegradable, biocompatible, non-toxic, and does not contain negative charges in the PEC system [6], so it is used in transfection studies. It is attracting the attention of researchers. Natural polymers commonly used in gene delivery are poly (amino acids) such as poly-L-lysine (PLL), polyornithine, polyarginine, chitosan, dextran, collagen, gelatin, and their modified derivatives. PLLs and other polys (amino acids) are important polymers for use in gene delivery systems due to their biodegradability, but these polymers are highly toxic [22]. Other polycations, namely, dextran [27], collagen [28], gelatin [29, 30], and their modified derivatives are used in gene delivery systems, but researchers have observed sufficient transfection efficiency so did not do it. Therefore, chitosan has been widely used in many studies due to its characteristic properties.

4 General Characteristics of Chitosan

Chitosan is a linear polysaccharide composed of glucosamine and N-acetyl-glucosamine units bound by β (1–4) glycosidic bonds and a partially deacetylated product of the natural polysaccharide chitin. Chitin, a biopolymer, is the most abundant organic compound in nature and is an important component of the exoskeleton of animals, mainly found in the shells of crustaceans such as crabs, shrimp, and krill. Since chitosan is an N-deacetylated derivative of chitin, the degree of acetylation determines whether the biopolymer is chitin or chitosan. When the degree of deacetylation of chitin, which is the content of glucosamine, exceeds about 50%, it dissolves in acidic aqueous solutions such as acetic acid, lactic acid, hydrochloric acid, and aspartic acid, and is called chitosan [31, 32]. Chitosan does not dissolve at basic pH values. Chitosan exhibits varying degrees of solubility in dilute aqueous media, depending on the free amine content of the chain. The molar ratio of acetylated amine groups to deacetylated amine groups in chitosan also determines the sensitivity or biodegradability of the enzyme. Chitosan is an inexpensive, biocompatible, animal/human biodegradable, non-toxic cationic polymer. In addition to these properties, it exhibits other excellent biological properties such as immunological, antibacterial, and wound healing activity. Chitosan degradation products are also non-toxic, non-immunogenic, and non-carcinogenic. The chemical modification of chitosan that results in a variety of derivatives is easy to apply. A variety of possible modification reactions can be applied, including nitration, phosphorylation, sulfation, thiolation, acylation, hydroxyalkylation, graft polymerization, amination, and combinations of chitosan derivatives with cyclodextrin. In particular, the physical, mechanical, chemical, bioactive properties, and commercial availability of chitosan make chitosan a very attractive biomaterial in biomedicine. Therefore, since the nineteenth century [33–35], chitosan derivatives have found widespread use in many areas, including biotechnology (especially biomedicine) and environmental applications. For example, a future and important use of chitosan in biomedicine for gene delivery. Chitosan has a high positive charge density due to the D-glucosamine unit in the structure. It exhibits polycationic properties at acidic and neutral pH. The amine group of chitosan is protonated and chitosan forms a PEC with negatively charged DNA [6, 36, 37]. Currently, positively charged groups (amino groups) and negatively charged ones, nucleic acids that form stable complexes or biological membranes and in vivo targets (e.g., PEI or polyamide amine dendrimers) due to their low toxicity and immunogenicity [33, 34]. Therefore, chitosan and its derivatives have recently been recognized as safe and efficient cationic carriers for gene delivery [6, 33, 34, 38–41].

5 Factors Affecting Gene Transfer in Chitosan Delivery System

Many delivery systems have been developed to maximize transfection efficiency and minimize side effects. For this purpose, Kasyua and Karudahave been used in vivo, versatile payload acceptability, low toxicity or non-toxicity, low immunogenicity, stealth, active targeting, proper size, and proper surface charge. It provided information on the optimized properties of these carrier systems for efficient cellular penetration: activity, intracellular targeting mechanism and high productivity [42]. Bhavsar and Amiji pointed out other properties like reactivity, biocompatibility, non-heat resistance, impurities, availability in medicinal grade, load capacity, permeability, swelling, viscoelasticity, and local environment sensitivity [43]. As defined above, we have looked at these two classifications for rectification of the non-viral vectors compared to viral vectors. Size and charge density are very playing a crucial role in vitro and in vivo overall performance of polymeric gene delivery systems. The polymeric gene delivery systems have a positive charge which is complex with the DNA having a negative charge. These cationic densities boom the encapsulation performance and enhance the uptake of DNA into the cells through the interplay of the negatively charged cell membrane [18, 44–46].

5.1 *Surface Charge and Zeta Potential*

The polymeric gene delivery systems that are complexed with DNA are called polyplex. The positive charge density of the polymer increases the DNA encapsulation efficiency and stability as well as immunity. Therefore, in various studies, PEG or the same other molecule forms a complex with the polymer carrier to provide charge balance, and these modifications make these carriers safe [47, 48]. Charges are also important for cell penetration. Cationic nanoparticles enter cells via endocytosis. Positively charged particles react with the negatively charged sugar coating on the outside of the cell membrane (on the extracellular polymer material on the surface of the cell membrane) and are taken up by cells by various intracellular mechanisms. When particles (which did not invade cells by passive and active transport) are taken up by endocytosis, multiple acidic groups cause endosome destabilization at low pKa values, resulting in a proton sponge effect. The polymer is then protected by moving protons to the endosome and increasing the ion charge density until the endosome becomes unstable [10]. From the formation of polymer support-DNA complexes to cell invasion and gene transfer to the nucleus, the required surface charge is called the zeta potential [49, 50]. Expressed as colloidal stability of particles distributed in a liquid (usually water) and when an electric field is applied to the liquid, it moves to the negative or positive electrode according to the surface charge ratio. As is well-known, in the case of colloidal systems, when a net surface charge is formed, a reverse charge begins to be generated in the outer layer, thereby forming an electric double layer. The innermost layer that surrounds the opposite layer for each surface charge of a particle is called the star layer. Each particle acts as a single entity consisting of this bilayer whose positive charge is equal to its negative charge. The potential difference between this field and the surface charge of the particle is called the zeta potential (ζ). The unit is millivolts (mV) and is measured with a zeta meter [51]. Particles are stable below -30 mV and above $+30$ mV. For stable nanoparticles used in gene delivery applications, the average zeta potential is up to $+30$ mV. The ionization of the terminal groups of the non-viral polymer carrier depends on the degree of surface ionization proportional to the pH of the dispersion. For zwitterionic particles, the surface charge is positive at low pH and the surface charge is negative at high pH. These charges equilibrate at a zero point called the isoelectric point [51].

5.2 *Particle Size*

In 1860, nanoparticle technology was born from nanoscale or nanometer (nm) scale materials. Currently, the nanoscale concept is described as a material with a particle size of 1–1000 nm. The scope of nanotechnology lies in the manufacture of nanoscale materials with various properties and the study of these properties. While chemistry, physics, molecular biology, and materials science are related to nanotechnology, many areas of science are beginning to associate nanotechnology with the development of technological methods [52, 53]. The evolution of nanotechnology is not limited to the position of atoms in the structure of materials as they are transported from one location to another. Moreover, nanomaterials, which have a high surface area due to their nanosize, are synthesized with exclusive properties throughout their controlled size. Therefore, nanomaterials are widely used in many fields such as electronic equipment, automobiles, military, medical, manufacturing of conductive and semi-conductive materials, ceramics, surface coating materials, ink manufacturing, and so on. This technological revolution is seen by scientists as the starting point for more development over the next 10–15 years. The concept of nanotechnology in gene therapy began in 1930 with the discovery of the intracellular nanoscale structure. Today, nanotechnology approaches in gene therapy are known for developing nanoparticle systems below 200 nm. The size of the polyplex is important for functionality. The diameter limit of the polyplex is about 10 nm at the first uptake in the liver. In addition, the upper limit size of the polyplex should be less than 200 nm. The size of the polyplex is changed by changing the DNA: polymer ratio [54, 55]. In particular, factors that influence the transfection efficiency of the chitosan-DNA complex are the degree of deacetylation and the molecular weight of chitosan, pH, serum, chitosan charge ratio to DNA, and viscosity, and cell type [6, 38, 56, 57]. Transfection efficiency and DNA loading capacity increase with the degree of deacetylation, and extracellular DNA protection and intracellular DNA release increase with molecular weight. The optimum pH for transfection media is 6.8–7.0 [56].

5.3 Chitosan Modification Reactions

Chitosan is insoluble in physiological pH and has low transfection efficiency, so modification studies are needed. For this purpose, several modification reactions are performed on chitosan such as modified with PEG or glycol [58], synthesized quaternized chitosan [59], low molecular weight chitosan [60], and reducing or thiolated chitosan (Fig. 5.2) [61]. Generally, PEG, glycol, or pHPMA is used to mask the instability of nanoparticles in serum and the formation of the hydrophilic layer onto the polyplex. The introduction of grafted PEG units onto the galactosylated chitosan was investigated by Parket al. which increases stability in water and cellpermeability [62]. Mao et al. have been studied grafted methoxyPEG (mPEG) units of different molecular weights onto the trimethylchitosan [63] to produce modified chitosan such as PEG-aldehyde [64, 65], PEGcarboxylicacid [66, 67], PEG-carbonate [68], PEG-iodide [69], PEG-epoxide [70], PEG-diacrylate [71], PEG-NHS ester [72], and PEG-sulfonate [73–75]. The introduction of colloidal stabilities of polyplexes with a pHPMAlinker was studied by Luten et al. [76] which enhanced the stability in serum for in vitro transfection with low cytotoxicity. Later on, a lot of studies based on this concept have been carried out to date [72, 77–84].

This modification is preferred over the others because of improvement in transfection efficiency and solubility in water. Numerous quaternized chitosans such as *N*-(4-pyridinylmethyl)chitosans [38], *N*-trimethylated chitosan oligomers [85], methylated *N*-(4-*N*,*N*-dimethylaminobenzyl) chitosan [86], octadecyl quaternized carboxymethyl chitosans [87], PEG*graft*-quaternized chitosan [88], and other low molecular weight chitosan have been studied in non-viral gene delivery. Richardson et al. studied the effect of molecular weight of chitosan on the cytotoxicity, complexation with DNA, and relation to the protection of DNA from nuclease degradation. He found that the low molecular weight of chitosan is more effective than poly(L-lysine) for complexation with DNA, and there is no cytotoxic effect for use in gene delivery [89]. Various research articles related to the modification reactions of the low molecular weight chitosans appeared in the literature [90–102]. Reducing or thiolated chitosan is commonly used to dissociate DNA and vectors. In addition, modifications containing ester bonds [103–105] or biological macromolecules [106, 107] such as heparin and proteoglycans are used but not generally preferred by the researchers. The disulfide bonds are delivered to the gene switch through diverse strategies. One of those strategies is the cationic ligands, which can be coiled at the polymer segments with the disulfide bonds, then DNA is bonded through the electrostatic interaction, and so DNA is added into the cytosol via the dissociation of disulfide bonds from the polyplex. Alternatively, other methods are the formation of cross-linking points via disulfide bonds and the reaction with disulfide bonds via polymer segments. These bonds present in the polymer segment not only release DNA but also reduce cytotoxicity through the degradation of small molecule components [24, 25]. Lee et al. stated the thiol modification of chitosan for sustained gene transport. In this study, the thiolated chitosan/DNA nanocomplexes exhibited significantly stepped forward gene transport in vitro and in vivo via way of means of the oxidation of thiol groups to crosslink the thiolated chitosan [108]. Numerous reducible chitosan research had been stated so far [109–113].

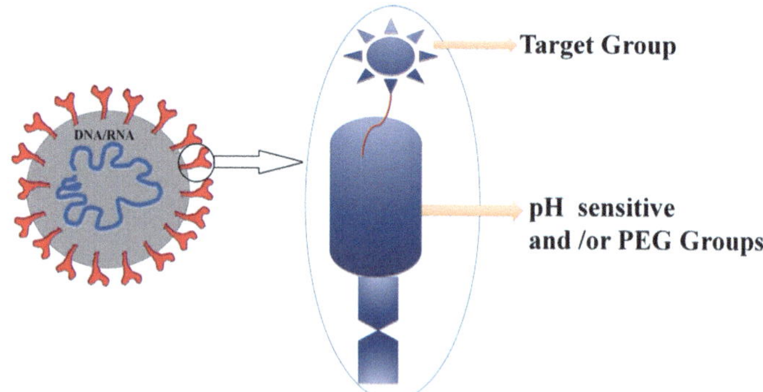

Fig. 5.2 Scheme for representation of chitosan-modified nanoparticles

6 Application of Chitosan-Based Nanocomposites in Gene Therapy

Chitosan is the most studied natural macromolecule for gene delivery. It has a positive charge, binds to negatively charged cell membranes with high affinity, and forms a complex with DNA via electrostatic interaction [22, 114]. Mumper et al. [15, 115] have introduced firstly the role of chitosan in gene therapy. Afterward, in 1996–1997, Murata et al. synthesized the chitosan gene carrier systems having galactose residues for the transportation of the DNA molecule [116, 117]. Although gene transport mechanisms of cationic nanoparticles had now no longer been absolutely understood in those years, numerous reports related to the chitosan gene carrier systems were reported in the literature. Polymeric gene transport applications are quite tough to classify. In well-known, classification is performed further to the polymer type as natural and artificial polymeric gene transport structures. However, researchers understood that this type does not explain all the gene delivery applications so, in place of making the classifications, they realized that the gene shipping mechanisms want to be understood in aggregate with polymer-DNA out of the mobile to the occurrence of the related proteins into the mobile nucleus. Despite the fact that these mechanisms have now not been completely defined yet, the modification reactions are done in step with the diagnosed mechanisms. Wong et al. described seven essential steps for the transport of a gene to the nucleus: (1) healing genes packaging; (2) entry to the cellular; (3) endolysosomal getaway; (4) the impact of DNA/provider system release; (5) progression throughout the cytoplasm and transition into the nucleus; (6) gene expression; and (7) biocompatibility [24]. Gene packaging strategies are very critical for gene delivery structures. For the prevention of the same charge impact on the cell membrane resulting from the phosphate corporations of DNA, the condensation of the cumbersome shape of DNA, and the safety of DNA from the degradation extracellularly or intracellularly, researchers stepped forward with two packaging strategies for chitosan gene transport: (1) electrostatic interplay and a couple of encapsulation [24, 118, 119]. Usually, electrostatic interaction strategies are desired for the chitosan gene transport systems. Erbacher et al. used the electrostatic interplay methods for growing chitosan/DNA complexes and they located that this complicated required some feature modifications for strong and small complexes [107]. Chitosan has amino businesses that are protonated in impartial pH and have interaction with DNA spontaneously. Due to those features of chitosan, Lee et al. studied the hydrophobically modified chitosan complexes with plasmid DNA to put together self-aggregated nanoparticles in aqueous media with adjusted pH [120]. Aside from this examination, there are various articles reported relating to those capabilities of chitosan [121–128]. Especially because the encapsulation strategies are explored, the researchers have not generally favored the electrostatic interaction techniques. The encapsulation methods are used for the protection of genes from enzymatic degradation and offer the controlled launch of DNA via the biodegradable groups of chitosan [24]. Buddy et al. classified the techniques of obtaining the encapsulated chitosan-based totally on the nanoparticle systems;

those are covalently connected nanoparticles, ionically pass-linked nanoparticles, and desolvated nanoparticles [129]. The covalently pass-related method is carried out with a chemical cross-linking agent, including glutaraldehyde, for cross-linking of chitosan [130]. The ionically cross-connected nanoparticles method or ionotropic gelation approach is the most famous technique for the use of polyanions, such as tripolyphosphate, for forming cross-connected chitosan [131–134]. In the desolvated nanoparticles approach, the desolvating agent is used for the precipitation of chitosan to extract water from chitosan polymeric chains. The complex coacervation approach is one of the desolvated strategies [129]. Bozkır and Saka studied the complicated formation of chitosan and plasmid DNA with the use of complicated coacervation and solvent evaporation techniques. In addition, they investigated the crucial parameters which include encapsulation efficiency, molecular weight, and deacetylation degree of chitosan [135]. The restrictions of the encapsulation techniques are that the polymeric provider systems are uncovered the natural solvents and high temperatures, which disrupt the genetic materials, much less encapsulation efficiency, much less DNA biocompatibility because of inadequate launch from the polymeric carrier structures, and the degradation of DNA due to the hydrolysis of ester bonds in low pH [24]. The polymer-DNA complexes as polyplexes come upon the primary barrier on the mobile, and it is far referred to as the plasma membrane. Polyplexes do not undergo passive diffusion because the transition membrane's pores and canals are very constrained dimensionally. Various strategies have been employed to overcome bodily obstacles. Endocytic uptake of the molecules, which are not handed from the cell membrane via the easy diffusion or active delivery, is passed essentially from the mobile membrane through 3 approaches: (1) phagocytosis; (2) pinocytosis; (3) receptor-mediated endocytosis. In general, the particles which are bigger than 250 nm pass via phagocytosis, and the smaller ones skip through endocytosis on the mobile membrane [136]. Polyplexes are uptaken in the mobile by means of receptor-mediated endocytosis. The focus on the agent in the provider structures is used for specific uptake of the gene in a mobile as for reaching endocytosis. The endogenous ligands, inclusive of folate and transferrin, are widely utilized in phrases of growing the biocompabilities and transfection efficiencies. However, the exogenous ligands have very constrained utilization in terms of generated immune reaction because of their overseas structures [24, 78, 114, 137–145]. The endolysosomal getaway of chitosan polyplexes is explained while after the endocytic uptake of polyplexes, they go back to the mobile floor, which is facilitated by means of lysosomes, intracell organelle, etc. This concept is expressed with the pK_a value of polycations, which is stricken by a trade-in buffer capability. For chitosan polyplexes, which have a pK_a value of approximately 6.5, the amino organizations are protonated within the cellular cytoplasm, but this function is applied for the endolysosomal break out of chitosan polyplexes [146]. In line with the effectiveness of molecular weight and degree of deacetylation, Huang et al. studied the transfection efficiency of chitosan. They said that chitosan has 2.5 times higher proton absorption ability than PLL [147]. However, Höggard et al. studied the connection between ultrapure chitosan and PEI and studied their characteristics. Consistent with their experimental consequences, the ultrapure chitosan does now not offer a sponge effect due to its primary amine groups as compared with PEI. In addition, its buffering capacity is lower than PEI on the acid endosomal pH interval of 4.5–5.5 [148]. due to these disadvantages, researchers have carried out a few modification reactions on chitosan. Moreira et al. studied to improve the transfection performance of chitosan with the aid of promoting the endosomal break out potential and buffer capability of chitosan polyplexes. For this cause, the chitosan backbone is modified with imidazole moieties with a view to escape endolysosomal degradation, much like PEI [149]. In another study, Chang et al. modified chitosan with histidine as buffering potential of histidine might help the escape of DNA inside the endosomal pH variety [150]. Comparable research was finished with the aid of others as well [151–154]. The most important steps for the transport of a gene to the nucleus are DNA/carrier device dissociation, cytosolic carrier, launch into the nucleus, and gene expression. The maximum crucial step for chitosan carrier systems is the DNA/vector dissociation among those primary steps because green gene transfer is completed with a minimum retention time of non-protective DNA in the cytosol. Efforts had been made to enhance the modification of the chitosan backbone, and charge reduction or modification of chitosan is done by modifying it with thermoresponsive groups, ester bonds, or disulfide bonds (reducible polymers). Among them, the most critical and typically desired strategies are modification with thermoresponsive groups and disulfide bonds. The thermoresponsive polymers are transformed to reversible frizz-circular form relying upon the temperature. Thus, the degree of DNA condensation is decided by the change in temperature. The frizz phase has the flexible, hydrophilic, the long-wide chain conformation, whereas the circular form has the collapse, hydrophobic, small stretched conformation. If the carrier system is circular up to the transition temperature, and frizz forms below the transition temperature, this transition temperature is called lower critical solution temperature (LCST) [155]. In fact, the thermoresponsive provider device that has an LCST value underneath the frame temperature is used for the condensation DNA with stretching form into the cell. In this regard, poly(N-isopropylacrylamide) (p-NIPAM), with an LCST value of 32 °C, is extensively utilized in transfection studies. This polymer offers excessive transfection efficiency, endosomal escape, cationic character, and hydrophobicity and NIPAM has been used for chitosan modification in many studies [156–159]. The thiolated polymers are commonly desired in gene transfer systems as a promising tool [160]. The disulfide linkages are modified to shape the gene carrier systems by the usage of various strategies; (1) electrostatic interactions, (2) reversible cross-linking, and (3) direct affiliation of the disulfide linkages at the polymer backbone. As cited above, those linkages preserve the polymer structure solid inside the cytosol, launch DNA into the cytosol, and furthermore the cytotoxicity of the carrier system is decreased with the dissociation of the carrier system to lower molecular weight components [24, 25, 160]. The thiolated chitosan carrier systems that enable gene transfection efficiency were developed by Schmitz et al. [161]. Jia et al. advanced a redox-responsive chitosan carrier system by using the PEG, PEI, and disulfide bonds for greater effective gene transfection in HeLa cells [162]. Targeted gene delivery is a significant step in chitosan carrier systems for obtaining selective and enhanced gene delivery to the

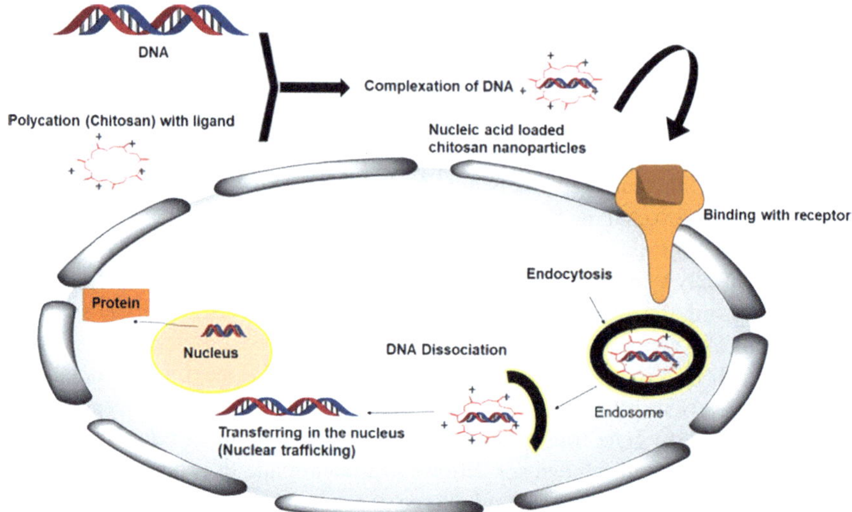

Fig. 5.3 Schematic Pathway for Chitosan nanoparticles in targeted gene delivery. Reproduced from M. Junaid Dar et. al. with permission from Elsevier 2019

target site. By using the in vivo approaches, the chitosan polyplex is administered to the body, it must be targeted to the specific sites as shown in Fig. 5.3.

Numerous reports are reported in the literature relating to targeting these specific sites, such as tumors [163], liver [164, 165], lung [166, 167], and brain [168, 169]. In most of these studies, chitosan is conjugated with the protein, transferrin, peptide, antibody, etc. [170]. A peptide functionalized chitosan-DNA nanoparticles were reported by Talvitie et al. for cellular targeting which is targeted to the required cell receptors in a specific and time-dependent manner [171]. In addition, Wang et al. synthesized the pH-sensitive gene delivery system for cancer cell-targeting which improved gene delivery by the introduction of pDNA nanocomplexes in the core and a pH-sensitive anionic polymer folic acid-modified PEG tethered carboxylated chitosan coating on the surface [172]. Numerous studies focused on targeted chitosan carrier systems have been reported in the literature [62, 173–180]. Currently, a combination of both the viral vectors and chitosan is used for efficient and permanent transfection [181]. Lameiro et al. coupled the adenovirus into the chitosan microparticle for mucosal vaccination. The main reason behind this study was to defend viruses, lower the immune response, and prolonged release. However, there are some boundaries including the difficulty of controlled release, loss of viral activity, and less loading efficiency. More studies are needed to triumph over these shortcomings [181, 182].

7 Conclusion

The idea of gene delivery was introduced in 1963, and viral vectors have been most effectively utilized in the gene therapy area. Because early 2000, these vectors were in the main abandoned because of the adverse side effects of viral-based gene therapies. Researchers have targeted the synthesis and applications of non-viral vectors, which are lipophilic or polymer-primarily based gene delivery systems. Cationic polymers are desired as a non-viral vector inside the field of gene transport. Chitosan and chitosan derivatives are commonly desired by the other cationic polymers due to their superior properties. Chitosan and chitosan derivatives are especially biodegradable and biocompatible polysaccharides. These are chemically versatile for undergoing varieties of reactions having different physicochemical properties which were tuned through modification having lower cytotoxicity, and high transfection properties. They are also called effective DNA-condensing agents and provide protection against DNAase-degradation. In connection with this fact that polymer-based gene delivery systems are yet to gain a massive presence in medical trials. Chitosan and its derivatives have been utilized in gene delivery studies after numerous modifications. A number of in vitro and in vivo studies confirm that chitosan and its derivatives are suitable and promising materials for efficient non-viral gene and DNA vaccine delivery. It is evident that chitosan and its derivatives are strong candidates to be used as the most preferred non-viral vector for gene delivery clinical trials in the future.

References

1. Thomas CE, Ehrhardt A, Kay MA (2003) Progress and problems with the use of viral vectors for gene therapy. Nat Rev Genet 4(5):346–358
2. Verma IM, Somia N (1997) Gene therapy-promises, problems and prospects. Nature 389(6648):239–242
3. Kircheis R, Wightman L, Wagner E (2001) Design and gene delivery activity of modified polyethylenimines. Adv Drug Deliv Rev 53(3):341–358
4. Boulaiz H, Marchal JA, Prados J, Melguizo C, Aranega A (2005) Non-viral and viral vectors for gene therapy. Cell Mol Biol 51(1):3–22
5. Walther W, Stein U (2000) Viral vectors for gene transfer. Drugs 60(2):249–271
6. Amiji MM (2010) Polymeric gene delivery: principles andapplications. CRC Press, Bosa Roca
7. Kwon GS (2005) Polymeric drug delivery systems. Taylor & Francis, Boca Raton
8. Wirth T, Ylä-Herttuala S (2013) History of gene therapy. Gene 525(2):162–169
9. Al-Dosari MS, Gao X (2009) Nonviral gene delivery: principle, limitations, and recent progress. AAPS J 11(4):671–681
10. Szybalska EH, Szybalski W (1962) Genetics of human celllines, IV. DNA-mediated heritable transformation of a biochemical trait. Proc Natl Acad Sci USA 48(12):2026–2034
11. Rogers S, Pfuderer P (1968) Use of viruses as carriers of added genetic information. Nature 219(5155):749–751
12. Neumann E, Schaefer-Ridder M, Wang Y, Hofschneider P (1982) Gene transfer into mouse lyoma cells by electroporation in high electric fields. EMBO J 1(7):841–845

13. Felgner PL, Gadek TR, Holm M, Roman R, Chan HW, Wenz M, Northrop JP, Ringold GM,Lipofection Danielsen M(1987) A highly efficient, lipid-mediated DNA-transfection procedure. Proc Natl Acad Sci 84(21):7413–7417
14. Boussif O, Lezoualéh F, Zanta MA, Mergny MD, Scherman D, Demeneix B, Behr J-P (1995) A versatile vector for gene and oligonucleotide transfer into cells in culture and in vivo: polyethylenimine. Proc Natl Acad Sci 92(16):7297–7301
15. Mumper R, Wang J, Claspell J, Rolland A (1995) Novel polymeric condensing carriers for gene delivery. In: Proceedings of the 22nd international symposium on controlled releasebioactive materials, controlled release society, vol 2, Seattle, Washington, July 30-August 2, pp 178–179
16. Xu F-J, Li H, Li J, Zhang Z, Kang E-T, Neoh K-G (2008) Pentablock copolymers of poly(ethylene glycol), poly((2-dimethyl amino) ethyl methacrylate) and poly(2-hydroxyethyl methacrylate) from consecutive atom transfer radical polymerizations for non-viral gene delivery. Biomaterials 29(20):3023–3033
17. Thomas M, Klibanov A (2003) Non-viral gene therapy: polycation-mediated DNA delivery. Appl Microbol Biotechnol 62(1):27–34
18. Vorhies JS, Nemunaitis JJ (2009) Synthetic vs. natural/biodegradable polymers for delivery of shRNA-based cancer therapies. In: Macromolecular drug delivery; Belting, M.,Ed.; Humana Press, Springer: New York, pp 11–29
19. Luo D, Saltzman WM (2000) Synthetic DNA delivery systems. Nat Biotechnol 18(1):33–37
20. Wu GC, Zhou F, Ge LF, Liu XM, Kong FS (2012) Novelmannan-PEG-PE modified bioadhesive PLGA nanoparticles for targeted gene delivery. J Nanomater 2012:1–9
21. Mandke R, Singh J (2012) Cationic nanomicelles for delivery of plasmids encoding interleukin-4 and interleukin-10 for prevention of autoimmune diabetes in mice. Pharm Res 29(3):883–897
22. Tiera MJ, Winnik FM, Fernandes JC (2006) Synthetic and natural polycations for gene therapy: state of the art and new perspectives. Curr Gene Ther 6(1):59–71
23. Jaeger W, Bohrisch J, Laschewsky A (2010) Syntheticpolymers with quaternary nitrogen atoms-Synthesis and structure of the most used type of cationic polyelectrolytes. Prog Polym Sci 35(5):511–577
24. Wong SY, Pelet JM, Putnam D (2007) Polymer systems for gene delivery-Past, present, and future. Prog Polym Sci 32(8):799–837
25. Osada K, Christie RJ, Kataoka K (2009) Polymeric micelles from poly(ethylene glycol)-poly(amino acid) block copolymer for drug and gene delivery. J R Soc Interface 6(Suppl 3):S325–S339
26. Lin S, Du F, Wang Y, Ji S, Liang D, Yu L, Li Z (2007) Anacid-labile block copolymer of PDMAEMA and PEGas potential carrier for intelligent gene delivery systems. Biomacromolecules 9(1):109–115
27. Hosseinkhani H, Azzam T, Tabata Y, Domb A (2004) Dextran-spermine polycation: an efficient nonviral vector for in vitro and in vivo gene transfection. Gene Ther 11(2):194–203
28. Chevallay B, Herbage D (2000) Collagen-based biomaterials as 3D scaffold for cell cultures: applications for tissue engineering and gene therapy. Med Biol Eng Comput 38(2):211–218
29. Kaul G, Amiji M (2005) Tumor-targeted gene delivery using poly(ethylene glycol)-modified gelatin nanoparticles: invitro and in vivo studies. Pharm Res 22(6):951–961
30. Kommareddy S, Amiji M (2007) Antiangiogenic gene therapy with systemically administered sFlt-1 plasmid DNA in engineered gelatin-based nanovectors. Cancer Gene Ther 14(5):488–498
31. Bsme WCS (2011) Scaleless dieting: the essential SurvivalKit for the overweight. Obese and Diabetics; Author-House, Bloomington
32. Elkins R, Hennen WJ (1996) Chitosan; Woodland Publishing, TX, USA
33. Park S, Jeong EJ, Lee J, Rhim T, Lee SK, Lee KY (2013) Preparation and characterization of non aarginine modified chitosan nanoparticles for siRNA delivery. Carbohydr Polym 92(1):57–62
34. Jeevitha D, Amarnath K (2013) Chitosan/PLA nanoparticles asa novel carrier for the delivery of anthraquinone: synthesis, characterization and in vitro cytotoxicity evaluation.Colloid Surf B 101:126–134

35. Jagani H, Rao JV, Palanimuthu VR, Hariharapura RC, Gang S (2013) A nanoformulation of siRNA and its role in cancer therapy: In vitro and in vivo evaluation. Cell Mol Biol Lett 18(1):120–136
36. Cho Y-W, Han S-S, Ko S-W (2000) PVA containing chito-oligosaccharideside chain. Polymer 41(6):2033–2039
37. Saranya N, Moorthi A, Saravanan S, Devi MP, Selvamurugan N (2011) Chitosan and its derivatives for gene delivery. Int J Biol Macromol 48(2):234–238
38. Opanasopit P, Sajomsang W, Ruktanonchai U, Mayen V, Rojanarata T, Ngawhirunpat T (2008) Methylated N-(4-pyridinylmethyl)chitosan as a novel effective safe gene carrier. Int J Pharm 364(1):127–134
39. Sajomsang W, Gonil P, Ruktanonchai UR, Petchsangsai M, Opanasopit P, Puttipipatkha-chorn S (2013) Effects of molecular weight and pyridinium moiety on water-soluble chitosan derivatives for mediated gene delivery. Carbohydr Polym 91(2):508–517
40. Alves N, Mano J (2008) Chitosan derivatives obtained by chemical modifications for biomedical and environmental applications. Int J Biol Macromol 43(5):401–414
41. Zohuriaan-Mehr MJ (2005) Advances in chitin and chitosan modification through graft copolymerization: a comprehensive review. Iran Polym J 14(3):235–265
42. Kasuya T, Kuroda SI (2009) Nanoparticles for human liver specific drug and gene delivery systems: in vitro and in vivo advances. Expert Opin Drug Deliv 6(1):39–52
43. Bhavsar MD, Amiji MM (2007) Polymeric nano-and microparticle technologies for oral gene delivery. ExpertOpin. Drug Deliv. 4(3):197–213
44. Park TG, Jeong JH, Kim SW (2006) Current status of polymeric gene delivery systems. Adv Drug Deliv Rev 58:467–486
45. Schwartz B, Ivanov M, Pitard B, Escriou V, Rangara R, Byk G, Wils P, Crouzet J, Scherman D (1999) SyntheticDNA-compacting peptides derived from human sequence enhance cationic lipid-mediated gene transfer in vitro and in vivo. Gene Ther 6(2):282–292
46. Gaspar V, Sousa F, Queiroz J, Correia I (2011) Formulationof chitosan–TPP–pDNAnanocap-sules for gene therapyapplications. Nanotechnology 22(1):1–12
47. .Plank C, Mechtler K, Szoka Jr FC, Wagner E (1996) Activation of the complement system by synthetic DNA complexes: a potential barrier for intravenous gene delivery. Hum Gene Ther 7(12):1437–1446
48. Barron LG, Gagne L, Szoka FC (1999) Lipoplex-mediated gene delivery to the lung occurs within 60 minutes of intravenous administration. Hum Gene Ther 10(10):1683–1694
49. Kunath K, Harpe A, Fischer D, Petersen H, Bickel U, Voigt K, Kissel T (2003) Low-molecular-weight polyethylenimine as a non-viral vector for DNA delivery: comparison of physicochem-ical properties, transfection efficiency andin vivo distribution with high-molecular-weight polyethylenimine. J Control Release 89(1):113–125
50. Mansouri S, Cuie Y, Winnik F, Shi Q, Lavigne P, Benderdour M, Beaumont E, Fernandes JC (2006) Characterization of folate-chitosan-DNA nanoparticles for gene therapy. Biomaterials 27(9):2060–2065
51. Hunter RJ (1981) Zeta potential in colloid science: principles and applications. Academic Press, London
52. Lundin KE, Simonson OE, Moreno PM, Zaghloul EM, Oprea II, Svahn MG, Smith CE (2009) Nanotechnologyapproaches for gene transfer. Genetica 137(1):47–56
53. Amiji MM (2006) Nanotechnology for cancer therapy. CRCPress. Taylor & Francis Groups, Boca Raton
54. . Venturoli D, Rippe B (2005) Ficoll and dextran vs. globular proteins as probes for testing glomerular perm selectivity: effects of molecular size, shape, charge, and deformability. Am J Physiol Renal Physiol 288(4):F605–F613
55. Schatzlein AG, Zinselmeyer BH, Elouzi A, Dufes C, Chim YTA, Roberts CJ, Davies MC, Munro A, Gray AI, Uchegbu IF (2005) Preferential liver gene expressionwith polypropylen-imine dendrimers. J Control Release 101(1):247–258
56. Kim T-H, Jiang H-L, Jere D, Park I-K, Cho M-H, Nah J-W, Choi Y-J, Akaike T, Cho C-S (2007) Chemicalmodification of chitosan as a gene carrier in vitro and invivo. Prog Polym Sci 32(7):726–753

57. Özbas-Turan S, Akbuga J (2011) Plasmid DNA-loaded chitosan/TPP nanoparticles for topical gene delivery. DrugDeliv. 18(3):215–222

58. Vlerken LE, Vyas TK, Amiji MM (2007) Poly(ethyleneglycol)-modified nanocarriers for tumor-targeted and intracellular delivery. Pharm Res 24(8):1405–1414

59. Li T, Shi XW, Du YM, Tang YF (2007) Quaternized chitosan/alginate nanoparticles for protein delivery. J Biomed Mater Res A 83(2):383–390

60. Xu Y, Du Y (2003) Effect of molecular structure of chitosan on protein delivery properties of chitosan nanoparticles. Int J Pharm 250(1):215–226

61. Li Z-T, Guo J, Zhang J-S, Zhao Y-P, Lv L, Ding C, Zhang X-Z (2010) Chitosan-graft-polyethylenimine with improved properties as a potential gene vector. Carbohydr Polym 80(1):254–259

62. Park IK, Kim TH, Park YH, Shin BA, Choi ES, Chowdhury EH, Akaike T, Cho CS (2001) Galactosylatedchitosan-graft-poly(ethyleneglycol) as hepatocytetargetingDNA carrier. J Control Release 76(3):349–362

63. Mao SR, Shuai XT, Unger F, Wittmar M, Xie XL, Kissel T (2005) Synthesis, characterization and cytotoxicity ofpoly(ethylene glycol)-graft-trimethyl chitosan block copolymers. Biomaterials 26(32):6343–6356

64. Wu J, Wang XF, Keum JK, Zhou HW, Gelfer M, Avila-Orta CA, Pan H, Chen WL, Chiao SM, Hsiao BS, Chu B (2007) Water soluble complexes of chitosang-MPEG and hyaluronic acid. J Biomed Mater Res A 80A(4):800–812

65. Zhang W, Pan SR, Zhang X, Luo X, Du Z (2008) Preparation of monomethyl poly(ethylene glycol)-g-chitosan copolymers with various degrees of substitution: Their ability to encapsulate and condense plasmid DNA. J Appl Polym Sci 108(5):2958–2967

66. Ghiamkazemi S, Amanzadeh A, Dinarvand R, Rafiee-Tehrani M, Amini M (2010) Synthesis, and characterization, and evaluation of cellular effects of the FOL-PEG-g-PEIGALnanoparticles as a potential non-viral vector for gene delivery. J Nanomater 2010:1–10

67. Fernandez-Megia E, Novoa-Carballal R, Quinoa E, Riguera R (2007) Conjugation of bioactive ligands to PEG-grafted chitosan at the distal end of PEG. Biomacromol 8(3):833–842

68. He H, Li Y, Jia X-R, Du J, Ying X, Lu W-L, Lou J-N, Wei Y (2011) PEGylated Poly(amidoamine) dendrimer based dual-targeting carrier for treating brain tumors. Biomaterials 32(2):478–487

69. Hu Y, Jiang H, Xu C, Wang Y, Zhu K (2005) Preparation and characterization of poly(ethylene glycol)-g-chitosan with water-and organo solubility. Carbohydr Polym 61(4):472–479

70. Casettari L, Vllasaliu D, Mantovani G, Howdle SM, Stolnik S, Illum L (2010) Effect of PEGylation on the toxicity and permeability enhancement of chitosan. Biomacromol 11(11):2854–2865

71. Prabaharan M, Mano J (2004) Chitosan-based particles as controlled drug delivery systems. Drug Deliv 12(1):41–57

72. Chan P, Kurisawa M, Chung JE, Yang Y-Y (2007) Synthesis and characterization of chitosan-g-poly(ethyleneglycol)-folate as a non-viral carrier for tumor-targeted gene delivery. Biomaterials 28(3):540–549

73. Casettari L, Vllasaliu D, Castagnino E, Stolnik S, Howdle S, Illum L (2012) PEGylated chitosan derivatives: synthesis, characterizations and pharmaceutical applications. Prog Polym Sci 37(5):659–685

74. .Deng L, Qi H, Yao C, Feng M, Dong A (2007) Investigation on the properties of methoxy poly(ethylene glycol)/chitosan graft co-polymers. J Biomater Sci Polym Ed 18(12):1575–1589

75. Amiji MM (1997) Synthesis of anionic poly(ethylene glycol)derivative for chitosan surface modification in blood-contacting applications. Carbohydr Polym 32(3):193–199

76. Luten J, Akeroyd N, Funhoff A, Lok MC, Talsma H, Hennink WE (2006) Methacrylamide polymers with hydrolysis-sensitive cationic side groups as degradable gene carriers. Bioconjug Chem 17(4):1077–1084

77. Varkouhi AK, Lammers T, Schiffelers RM, van Steenbergen MJ, Hennink WE, Storm G (2011) Gene silencing activity of siRNA polyplexes based on biodegradable polymers. Eur J Pharm Biopharm 77(3):450–457

78. Mao HQ, Roy K, Troung-Le VL, Janes KA, Lin KY, Wang Y, August JT, Leong KW (2001) Chitosan-DNAnanoparticles as gene carriers: synthesis, characterization and transfection efficiency. J Control Release 70(3):399–421

79. Park IK, Kim TH, Kim SI, Park YH, Kim WJ, Akaike T, Cho CS (2003) Visualization of transfection of hepatocytes by galactosylated chitosan-graft-poly(ethyleneglycol)/DNA complexes by confocal laser scanning microscopy. Int J Pharm 257(1):103–110

80. Yun YH, Jiang HL, Chan R, Chen WL (2005) Sustainedrelease of PEG-g-chitosan complexed DNA from poly(lactide-co-glycolide). J Biomater Sci Polym Ed 16(11):1359–1378

81. Jiang X, Dai H, Leong KW, Goh SH, Mao HQ, Yang YY (2006) Chitosan-g-PEG/DNA complexes deliver gene to the rat liver *via* intrabiliary and intraportal infusions. J Gene Med 8(4):477–487

82. Germershaus O, Mao SR, Sitterberg J, Bakowsky U, Kissel T (2008) Gene delivery using chitosan, trimethyl chitosan or polyethylenglycol-graft-trimethyl chitosan block copolymers: establishment of structure-activity relationships in vitro. J Control Release 125(2):145–154

83. Wu YD, Liu CB, Zhao XY, Xiang JN (2008) A new biodegradable polymer: PEGylated chitosan-g-PEI possessing a hydroxyl group at the PEG end. J Polym Res 15(3):181–185

84. Zhang T, Yu YY, Li D, Peng R, Li Y, Jiang Q, Dai P, Gao R (2009) Synthesis and properties of a novel methoxypoly(ethylene glycol)-modified galactosylated chitosan derivative. J Mater Sci Mater Med 20(3):673–680

85. Thanou M, Florea BI, Geldof M, Junginger HE, Borchard G (2002) Quaternized chitosan oligomers as novel gene delivery vectors in epithelial cell lines. Biomaterials 23(1):153–159

86. Rojanarata T, Petchsangsai M, Opanasopit P, Ngawhirunpat T, Ruktanonchai U, Sajomsang W, Tantayanon S (2008) Methylated N-(4-N, N-dimethylaminobenzyl)chitosan for novel effective gene carriers. Eur J Pharm Biopharm 70(1):207–214

87. Liang XF, Tian H, Luo H, Wang HJ, Chang J (2009) Novel quaternized chitosan and polymeric micelles with cross-linked ionic cores for prolonged release of minocycline. J Biomater Sci Polym Ed 20(1):115–131

88. Liang XF, Sun YM, Duan YR, Cheng YS (2012) Synthesis and characterization of PEG-graft-quaternized chitosan and cationic polymeric liposomes for drug delivery. J Appl Polym Sci 125(2):1302–1309

89. Richardson SCW, Kolbe HJV, Duncan R (1999) Potential of low molecular mass chitosan as a DNA delivery system:Biocompatibility, body distribution and ability to complex and protect DNA. Int J Pharm 178(2):231–243

90. Huang Z, Dong L, Chen JJ, Gao FB, Zhang ZP, Chen JN, Zhang JF (2012) Low-molecular weight chitosan/vascular endothelial growth factor short hairpin RNA for the treatment of hepatocellular carcinoma. Life Sci 91(23):1207–1215

91. Yang XR, Yuan XY, Cai DN, Wang SY, Zong L (2009) Low molecular weight chitosan in DNA vaccine delivery *via* mucosa. Int J Pharm 375(1):123–132

92. Heo SH, Jang MJ, Kim DG, Jeong YI, Jang MK, Nah JW (2007) Characterization and preparation of low molecular weight water soluble chitosan nanoparticle modified with cell targeting ligand for efficient gene delivery. Polym Korea 31(5):454–459

93. Ercelen S, Zhang X, Duportail G, Grandfils C, Desbrieres J, Karaeva S, Tikhonov V, Mely Y, Babak V (2006) Physicochemical properties of low molecular weight alkylated chitosans: a new class of potential nonviral vectors for gene delivery. Colloid Surf B 51(2):140–148

94. Mao SR, Shuai XT, Unger F, Simon M, Bi DZ, Kissel T (2004) The depolymerization of chitosan: effects on physicochemical and biological properties. Int J Pharm 281(1):45–54

95. Janes KA, Alonso MJ (2003) Depolymerized chitosan nanoparticles for protein delivery: preparation and characterization. J Appl Polym Sci 88(12):2769–2776

96. Yang F, Cui XQ, Yang XR (2002) Interaction of low-molecular-weight chitosan with mimic membrane studied by electrochemical methods and surface plasmon resonance. Biophys Chem 99(1):99–106

97. Lee M, Nah JW, Kwon Y, Koh JJ, Ko KS, Kim SW (2001) Water-soluble and low molecular weight chitosan-based plasmid DNA delivery. Pharm Res 18(4):427–431

98. Tripathi SK, Goyal R, Kashyap MP, Pant AB, Haq W, Kumar P, Gupta KC (2012) Depolymerized chitosansfunctionalized with bPEI as carriers of nucleic acids andtuftsin-tethered conjugate for macrophage targeting. Biomaterials 33(16):4204–4219

99. Fernandes JC, Qiu XP, Winnik FM, Benderdour M, Zhang XL, Dai KR, Shi Q (2012) Low molecular weight chitosan conjugated with folate for siRNA delivery invitro: optimization studies. Int J Nanomedicine 7:5833–5845

100. Yuan H, Lu LJ, Du YZ, Hu FQ (2011) Stearic acid-g-chitosan polymeric micelle for oral drug delivery: in vitro transport and in vivo absorption. Mol Pharm 8(1):225–238

101. Yoksan R, Akashi M (2009) Low molecular weight chitosan-g-L-phenylalanine: preparation, characterization, and complex formation with DNA. Carbohydr Polym 75(1):95–103

102. Gao SY, Chen JN, Xu XR, Ding Z, Yang YH, Hua ZC, Zhang JF (2003) Galactosylated low molecular weight chitosan as DNA carrier for hepatocyte-targeting. Int J Pharm 255(1):57–68

103. Lee K, Kwon I, Kim Y-H, Jo W, Jeong S (1998) Preparation of chitosan self-aggregates as a gene delivery system. J Control Release 51(2):213–220

104. Mintzer MA, Simanek EE (2008) Nonviral vectors for gene delivery. Chem Rev 109(2):259–302

105. Zhang S, Xu Y, Wang B, Qiao W, Liu D, Li Z (2004) Cationiccompounds used in lipoplexes and polyplexes forgene delivery. J Control Release 100(2):165–180

106. Liu Z, Jiao Y, Liu F, Zhang Z (2007) Heparin/chitosan nanoparticle carriers prepared by polyelectrolyte complexation. J Biomed Mater Res A 83(3):806–812

107. Erbacher P, Zou S, Bettinger T, Steffan A-M, Remy J-S (1998) Chitosan-based vector/DNA complexes for gene delivery: biophysical characteristics and transfection ability. Pharm Res 15(9):1332–1339

108. Lee D, Zhang W, Shirley SA, Kong X, Hellermann GR, Lockey RF, Mohapatra SS (2007) Thiolated chitosan/DNA nanocomplexes exhibit enhanced and sustained gene delivery. Pharm Res 24(1):157–167

109. Talaei F, Azizi E, Dinarvand R, Atyabi F (2011) Thiolatedchitosan nanoparticles as a delivery system for antisense therapy: evaluation against EGFR in T47D breast cancer cells. Int J Nanomed 6:1963–1975

110. Varkouhi AK, Verheul RJ, Schiffelers RM, Lammers T, Storm G, Hennink WE (2010) Gene silencing activity of siRNA polyplexes based on thiolated N, N. N-trimethylatedchitosan Bioconjug Chem 21(12):2339–2346

111. Martien R, Loretz B, Sandbichler AM, Schnurch AB (2008) Thiolated chitosan nanoparticles: transfection study in the Caco-2 differentiated cell culture. Nanotechnology 19(4):1–9

112. Martien R, Loretz B, Thaler M, Majzoob S, Bernkop-Schnuerch A (2007) Chitosan-thioglycolic acid conjugate: an alternative carrier for oral nonviral gene delivery? J Biomed Mater Res A 82A(1):1–9

113. Zhao X, Li Z, Pan H, Liu W, Lv M, Leung F, Lu WW (2013) Enhanced gene delivery by chitosan-disulfide-conjugated LMW-PEI for facilitating osteogenic differentiation. Acta Biomater 9(5):6694–6703

114. Jeong JH, Kim SW, Park TG (2007) Molecular design of functional polymers for gene therapy. Prog Polym Sci 32(11):1239–1274

115. MacLaughlin FC, Mumper RJ, Wang J, Tagliaferri JM, Gill I, Hinchcliffe M, Rolland AP (1998) Chitosan and depolymerized chitosan oligomers as condensing carriers for in vivo plasmid delivery. J Control Release 56(1):259–272

116. Murata J, Ohya Y, Ouchi T (1996) Possibility of application of quaternary chitosan having pendant galactose residues as gene delivery tool. Carbohydr Polym 29(1):69–74

117. Murata J, Ohya Y, Ouchi T (1997) Design of quaternary chitosan conjugate having antennary galactose residues asa gene delivery tool. Carbohydr Polym 32(2):105–109

118. Abdelhady HG, Allen S, Davies MC, Roberts CJ, Tendler SJ, Williams PM (2003) Direct real-time molecular scale visualisation of the degradation of condensed DNAcomplexes exposed to DNase I. Nucleic Acids Res 31(14):4001–4005

119. Schaffer DV, Lauffenburger DA (1998) Optimization of cell surface binding enhances efficiency and specificity of molecular conjugate gene delivery. J Biol Chem 273(43):28004–28009
120. Lee KY, Kwon IC, Jo WH (2005) Jeong, Complex formation between plasmid DNA and self-aggregates of deoxycholic acid-modified chitosan. Polymer 46(19):8107–8112
121. Miao J, Zhang XG, Hong Y, Rao YF, Li Q, Xie XJ, Wo JE, Li MW (2012) Inhibition on hepatitis B viruse-gene expression of 10–23 DNAzyme delivered by novel chitosan oligosaccharide-stearic acid micelles. Carbohydr Polym 87(2):1342–1347
122. Han L, Zhao J, Zhang X, Cao WP, Hu XX, Zou GZ, Duan XL, Liang XJ (2012) Enhanced siRNA delivery and silencing gold-chitosan nanosystem with surface charge-reversal polymer assembly and good biocompatibility. ACS Nano 6(8):7340–7351
123. Liao ZX, Ho YC, Chen HL, Peng SF, Hsiao CW, Sung HW (2010) Enhancement of efficiencies of the cellular uptake and gene silencing of chitosan/siRNA complexes *via* the inclusion of a negatively charged poly(gammaglutamicacid). Biomaterials 31(33):8780–8788
124. Liu LX, Song CN, Song LP, Zhang HL, Dong X, Leng XG (2009) Effects of alkylated-chitosan-DNA nanoparticles on the function of macrophages. J Mater Sci Mater Med 20(4):943–948
125. Bao JB, Song CX (2009) Chitosan chip and application to evaluate DNA loading on the surface of the metal. Biomed Mater 4(1):1–5
126. Pimpha N, Rattanonchai U, Surassmo S, Opanasopit P, Rattanarungchai C, Sunintaboon P (2008) Preparation of PMMA/acid-modified chitosan core-shell nanoparticles and their potential as gene carriers. Colloid Polym Sci 286(8–9):907–916
127. Satoh T, Kakimoto S, Kano H, Nakatani M, Shinkai S, Nagasaki T (2007) In vitro gene delivery to HepG2 cells using galactosylated 6-amino-6-deoxychitosan as a DNA carrier. Carbohydr Res 342(11):1427–1433
128. Schatz C, Bionaz A, Lucas JM, Pichot C, Viton C, Domard A, Delair T (2005) Formation of polyelectrolyte complex particles from self-complexation of N-sulfated chitosan. Biomacromol 6(3):1642–1647
129. Pal K, Behera B, Roy S, Ray SS, Thakur G (2013) Chitosan based delivery systems on a length scale: Nano to macro. Soft Mater 11(2):125–142
130. Akbuga J, Ozbas-Turan S, Erdogan N (2004) Plasmid-DNAloaded chitosan microspheres for in vitro IL-2 expression. Eur J Pharm Biopharm 58(3):501–507
131. Vimal S, Taju G, Nambi KSN, Majeed SA, Babu VS, Ravi M, Hameed ASS (2012) Synthesis and characterization of CS/TPP nanoparticles for oral delivery of gene in fi sh. Aquaculture 358:14–22
132. Jiang H, Wu H, Xu YL, Wang JZ, Zeng Y (2011) Preparationof galactosylated chitosan/tripolyphosphate nanoparticles and application as a gene carrier for targetingSMMC7721 cells. J Biosci Bioeng 111(6):719–724
133. Fabregas A, Minarro M, Garcia-Montoya E, Perez-Lozano P, Carrillo C, Sarrate R, Sanchez N, Tico JR, Sune-Negre JM (2013) Impact of physical parameters on particle size and reaction yield when using the ionic gelation method to obtain cationic polymeric chitosan-tripolyphosphatenanoparticles. Int J Pharm 446(1):199–204
134. Nasti A, Zaki NM, de Leonardis P, Ungphaiboon S, Sansongsak P, Rimoli MG, Tirelli N (2009) Chitosan/TPPand Chitosan/TPP-hyaluronic acid nanoparticles: Systematicoptimisation of the preparative process and preliminary biological evaluation. Pharm Res 26(8):1918–1930
135. Bozkır A, Saka OM (2004) Chitosan nanoparticles for plasmid DNA delivery: effect of chitosan molecular structure on formulation and release characteristics. Drug Deliv 11(2):107–112
136. Conner SD, Schmid SL (2003) Regulated portals of entry into the cell. Nature 422(6927):37–44
137. Kadiyala I, Loo YH, Roy K, Rice J, Leong KW (2010) Transport of chitosan-DNA nanoparticles in human intestinal M-cell model versus normal intestinal enterocytes. Eur J Pharm Sci 39(1):103–109

138. Manaspon C, Viravaidya-Pasuwat K, Pimpha N (2012) Preparation of folate-conjugated pluronic F127/chitosan core-shell nanoparticles encapsulating doxorubicin for breast cancer treatment. J Nanomater 2012:1–11
139. Zhu HY, Liu F, Guo J, Xue JP, Qian ZY, Gu YQ (2011) Folate-modified chitosan micelles with enhanced tumor targeting evaluated by near infrared imaging system. Carbohydr Polym 86(3):1118–1129
140. Morris VB, Pillai CKS, Sharma CP (2011) Folic acidconjugateddepolymerized quaternized chitosan as potential targeted gene delivery vector. Polym Int 60(7):1097–1106
141. Galbiati A, Tabolacci C, Della Rocca BM, Mattioli P, Beninati S, Paradossi G, Desideri A (2011) Targetingtumor cells through chitosan-folate modified microcapsules loaded with camptothecin. Bioconjug Chem 22(6):1066–1072
142. Zheng Y, Cai Z, Song XR, Yu B, Bi YQ, Chen QH, Zhao D, Xu JP, Hou SX (2009) Receptor mediated gene delivery by folate conjugated N-trimethyl chitosan invitro. Int J Pharm 382(1):262–269
143. Jiang HL, Xu CX, Kim YK, Arote R, Jere D, Lim HT, Cho MH, Cho CS (2009) The suppression of lung tumorigenesis by aerosol-delivered folate-chitosan-graftpolyethylenimine/Akt1 shRNA complexes through the Akt signaling pathway. Biomaterials 30(29):5844–5852
144. Fernandes JC, Wang HJ, Jreyssaty C, Benderdour M, Lavigne P, Qiu XP, Winnik FM, Zhang XL, Dai KR, Shi Q (2008) Bone-protective effects of nonviral gene therapy with folate-chitosan DNA nanoparticle containing interleukin-1 receptor antagonist gene in rats with adjuvant-induced arthritis. Mol Ther 16(7):1243–1251
145. Ganta S, Devalapally H, Shahiwala A, Amiji M (2008) A review of stimuli-responsive nanocarriers for drug and gene delivery. J Control Release 126(3):187–204
146. Alvarenga ES (2011) Characterization and properties of chitosan. In: Biotechnology of biopolymers; Elnashar, M., Ed.;Intech, Crotia, pp 91–108
147. Huang M, Fong CW, Khor E, Lim LY (2005) Transfectionefficiency of chitosan vectors: effect of polymer molecularweight and degree of deacetylation. J Control Release 106(3):391–406
148. Koping-Hoggard M, Tubulekas I, Guan H, Edwards K, Nilsson M, Varum KM, Artursson P (2001) Chitosan as anoviral gene delivery system. Structure-property relationshipsand characteristics compared with polyethyleniminein vitro and after lung administration in vivo. GeneTher 8(14):1108–1121
149. Moreira C, Oliveira H, Pires LR, Simoes S, Barbosa MA, Pego AP (2009) Improving chitosan-mediated genetransfer by the introduction of intracellular bufferingmoieties into the chitosan backbone. Acta Biomater 5(8):2995–3006
150. Chang KL, Higuchi Y, Kawakami S, Yamashita F, Hashida M (2010) Efficient gene transfection by histidine-modified chitosan through enhancement of endosomal escape. Bioconjug Chem 21(6):1087–1095
151. Zaki NM, Nasti A, Tirelli N (2011) Nanocarriers for cytoplasmicdelivery: cellular uptake and intracellular fate ofchitosan and hyaluronic acid-coated chitosan nanoparticlesin a phagocytic cell model. Macromol Biosci 11(12):1747–1760
152. Thibault M, Astolfi M, Tran-Khanh N, Lavertu M, Darras V, Merzouki A, Buschmann MD (2011) Excess polycationmediates efficient chitosan-based gene transfer bypromoting lysosomal release of the polyplexes. Biomaterials 32(20):4639–4646
153. Chang KL, Higuchi Y, Kawakami S, Yamashita F, Hashida M (2011) Development of lysine-histidine Dendron modified chitosan for improving transfection efficiencyin HEK293 cells. J Control Release 156(2):195–202
154. Dehousse V, Garbacki N, Colige A, Evrard B (2010) Developmentof pH-responsive nanocarriers using trimethylchitosansand methacrylic acid copolymer for siRNAdelivery. Biomaterials 31(7):1839–1849
155. HerasAlarcón C, Pennadam S, Alexander C (2005) Stimuliresponsive polymers for biomedical applications. Chem Soc Rev 34(3):276–285
156. Bao HQ, Li L, Gan LH, Ping YA, Li J, Ravi P (2010) Thermo- and pH-responsive association behavior of dualhydrophilic graft chitosan terpolymer synthesized viaatrpand click chemistry. Macromol 43(13):5679–5687

157. McMahon SS, Nikolskaya N, Choileain SN, Hennessy N, O'Brien T, Strappe PM, Gorelov A, Rochev Y (2011) Thermosensitive hydrogel for prolonged delivery of lentiviralvector expressing neurotrophin-3 in vitro. J. GeneMed. 13(11):591–601
158. Hastings CL, Kelly HM, Murphy MJ, Barry FP, O'Brien FJ, Duffy GP (2012) Development of a thermoresponsivechitosan gel combined with human mesenchymalstem cells and desferrioxamine as a multimodalpro-angiogenic therapeutic for the treatment of criticallimb ischaemia. J Control Release 161(1):73–80
159. Cheng YH, Yang SH, Lin FH (2011) Thermosensitive chitosan-gelatin-glycerol phosphate hydrogel as a controlledrelease system of ferulic acid for nucleus pulposus regeneration. Biomaterials 32(29):6953–6961
160. Loretz B, Thaler M, Bernkop-Schnurch A (2007) Role ofsulfhydryl groups in transfection? A case study with Chitosan-NAC nanoparticles. Bioconjug Chem 18(4):1028–1035
161. Schmitz T, Bravo-Osuna I, Vauthier C, Ponchel G, Loretz B, Bernkop-Schnurch A (2007) Development and invitro evaluation of a thiomer-based nanoparticulate genedelivery system. Biomaterials 28(3):524–531
162. Jia LJ, Li ZY, Zhang DR, Zhang Q, Shen JY, Guo HJ, Tian XN, Liu GP, Zheng DD, Qi LS (2013) Redox-responsive catiomer based on PEG-ss-chitosanoligosaccharide-ss-polyethylenimine copolymer for effectivegene delivery. Polym Chem 4(1):156–165
163. Mitra S, Gaur U, Ghosh P, Maitra A (2001) Tumour targeteddelivery of encapsulated dextran-doxorubicin conjugateusing chitosan nanoparticles as carrier. J Control Release 74(1):317–323
164. Tian Q, Zhang C-N, Wang X-H, Wang W, Huang W, Cha R-T, Wang C-H, Yuan Z, Liu M, Wan H-Y (2010) Glycyrrhetinic acid-modified chitosan/poly(ethyleneglycol) nanoparticles for liver-targeted delivery. Biomaterials 31(17):4748–4756
165. Dai H, Jiang X, Tan GC, Chen Y, Torbenson M, Leong KW, Mao H-Q (2006) Chitosan-DNA nanoparticlesdelivered by intrabiliary infusion enhance liver-targetedgene delivery. Int J Nanomed 1(4):507–522
166. Chen Q, Wang X, Chen F, Zhang Q, Dong B, Yang H, Liu G, Zhu Y (2011) Functionalization of upconvertedluminescent NaYF$_4$:Yb/Er nanocrystals by folic acid-chitosan conjugates for targeted lung cancer cell imaging. J Mater Chem 21(21):7661–7667
167. Issa MM, Köping-Höggård M, Tømmeraas K, Vårum KM, Christensen BE, Strand SP, Artursson P (2006) Targeted gene delivery with trisaccharide-substituted chitosanoligomers in vitro and after lung administration invivo. J Control Release 115(1):103–112
168. Wang X, Chi N, Tang X (2008) Preparation of estradiol chitosannanoparticles for improving nasal absorption and braintargeting. Eur J Pharm Biopharm 70(3):735–740
169. Aktas Y, Yemisci M, Andrieux K, Gürsoy RN, Alonso MJ, Fernandez-Megia E, Novoa-Carballal R, Quiñoá E, Riguera R, Sargon MF (2005) Development andbrain delivery of chitosan-PEG nanoparticles functionalizedwith the monoclonal antibody OX26. Bioconjug Chem 16(6):1503–1511
170. Hughes JA, Rao GA (2005) Targeted polymers for gene delivery. Expert Opin Drug Deliv 2(1):145–157
171. Talvitie E, Leppiniemi J, Mikhailov A, Hytonen VP, Kellomaki M (2012) Peptide-functionalized chitosan-DNAnanoparticles for cellular targeting. Carbohydr Polym 89(3):948–954
172. Wang M, Hu H, Sun Y, Qiu L, Zhang J, Guan G, Zhao X, Qiao M, Cheng L, Cheng L (2013) A pH-sensitivegene delivery system based on folic acid-PEG-chitosan–PAMAM-plasmid DNA complexes for cancer cell targeting. Biomaterials 34(38):10120–10132
173. Zhang H, Mardyani S, Chan WC, Kumacheva E (2006) Design of biocompatible chitosan microgels for targetedpH-mediated intracellular release of cancer therapeutics. Biomacromol 7(5):1568–1572
174. Park K, Kim J-H, Nam YS, Lee S, Nam HY, Kim K, Park JH, Kim I-S, Choi K, Kim SY (2007) Effect ofpolymer molecular weight on the tumor targeting characteristicsof self-assembled glycol chitosan nanoparticles. J Control Release 122(3):305–314

175. Raghuwanshi D, Mishra V, Das D, Kaur K, Suresh MR (2012) Dendritic cell targeted chitosan nanoparticles fornasal DNA immunization against SARS CoV nucleocapsidprotein. Mol Pharm 9(4):946–956
176. Min KH, Park K, Kim Y-S, Bae SM, Lee S, Jo HG, Park R-W, Kim I-S, Jeong SY, Kim K (2008) Hydrophobically modified glycol chitosan nanoparticles-encapsulatedcamptothecin enhance the drug stability and tumortargeting in cancer therapy. J Control Release 127(3):208–218
177. Han HD, Mangala LS, Lee JW, Shahzad MM, Kim HS, Shen D, Nam EJ, Mora EM, Stone RL, Lu C (2010) Targeted gene silencing using RGD-labeledchitosan nanoparticles. Clin Cancer Res 16(15):3910–3922
178. Dufes C, Muller J-M, Couet W, Olivier J-C, Uchegbu IF, Schätzlein AG (2004) Anticancer drug deliverywith transferrin targeted polymeric chitosan vesicles. Pharm Res 21(1):101–107
179. Sun Y, Chen Z-L, Yang X-X, Huang P, X-P Zhou, Du X-X (2009) Magnetic chitosan nanoparticles as a drugdelivery system for targeting photodynamic therapy. Nanotechnology 20(13):135102
180. Nam T, Park S, Lee S-Y, Park K, Choi K, Song IC, Han MH, Leary JJ, Yuk SA, Kwon IC (2010) Tumortargeting chitosan nanoparticles for dual-modality optical/MR cancer imaging. Bioconjug Chem 21(4):578–582
181. Wang C, Pham PT (2008) Polymers for viral gene delivery.Expert Opin. Drug Deliv 5 (4):385–401
182. Lameiro MH, Malpique R, Silva AC, Alves PM, Melo E (2006) Encapsulation of adenoviral vectors into chitosan–bile salt microparticles for mucosal vaccination. J Biotechnol 126(2):152–162
183. Taira K, Kataoka K, Niidome T (2005) Non-viral gene therapy: gene design and delivery. Springer, Tokyo

Chapter 6
Recent Advancements in the Application of Chitosan-Based Nanocomposites in Tissue Engineering and Regenerative Medicine

Tailin Rieg, Angelo Oliveira Silva, Ricardo Sousa Cunha,
Karina Luzia Andrade, Dachamir Hotza,
and Ricardo Antonio Francisco Machado

Abstract Chitosan is an abundant, non-toxic, reasonably priced polysaccharide with distinct physicochemical and biological characteristics. This biopolymer has become one of the most important in a matrix of nanocomposites because of the novel and improved characteristics promoted in nanotechnological materials. In this way, there is a great range of executable arrangements to obtain nanocomposites with specific functionalities in the knowledge of tissue engineering and regenerative medicine. The former has become an important research field regarding chitosan nanocomposites' biodegradability, biocompatibility, and bio-inertness. The latter is continuously facing new applications and obstacles and must find biocompatible materials and non-immunogenic species. Chitosan nanocomposites can promote cell adhesion and also replace some of the major components of bone, skin, and cartilaginous tissue. The natural biopolymer chitosan is hypoallergenic and does not stimulate body inflammation which makes chitosan nanocomposites an interesting type of material for these areas. In addition, chitosan nanocomposite production can be performed in several shapes, such as nanoparticles, scaffolds, fibers, and 3D printed structures, among others. Finally, this chapter presents and discusses the trend and challenges linked to the applications areas of chitosan nanocomposites fields of tissue engineering and regenerative medicine.

Keywords Chitosan · Nanocomposites · Scaffolds · Tissue engineering · Wound healing

Abbreviations

3D   Three dimensional

T. Rieg · A. O. Silva · R. S. Cunha · K. L. Andrade · D. Hotza · R. A. F. Machado (✉)
Department of Chemical Engineering and Food Engineering, Federal University of Santa Catarina (UFSC), Florianópolis, Santa Catarina, Brazil
e-mail: ricardo.machado@ufsc.br

  145
S. Gulati (ed.), *Chitosan-Based Nanocomposite Materials*,
https://doi.org/10.1007/978-981-19-5338-5_6

Ber	Berberine
BG	Active glass
CNM	Carbon nanomaterials
CNTs	Carbon nanotubes
CPT	Camptothecin
CS	Chitosan
DD	Deacetylation degree
GB	Genipin
GO	Graphene oxide
mBG	Microparticles of active glass
MWCNTs	Multiwalled carbon nanotubes
Nar	Naringin
nBG	Nanoparticles of active glass
NGCs	Neural guidance channels
nHA	Nanohydroxyapatite
NP	Nanoparticle
PLGA-NPs	Poly(lactic-co-glycolicacid) nanoparticles
PNS	Peripheral nervous system
rGO	Reduced graphene oxide rGO
SWCNTs	Single-walled carbon nanotubes

1 Introduction

In recent years, the ability to heal injuries and repair tissues, employing the concept of medical engineering, has provided a significant improvement in the quality and methods of disease therapy [51]. Tissue engineering has become one of the most important scientific therapeutic approaches, receiving much attention regarding its biomedical applications [76]. This branch of biomedical engineering is an emerging field with potential alternatives that address several disciplines and that associates stem cells and scaffolds with appropriate growth factors, cytokines, and chemokines, aiming at the improvement, replacement, or regeneration of tissues and organs [46]. In the same approach, regenerative medicine also deserves to be highlighted due to its potential for healing or displace of damaged cells and organs, besides acting in the normalization of congenital defects [55].

Tissue engineering is directed toward the study of cells and the development of scaffolds with the possibility of acting in the regeneration or replacement of damaged tissue. On the other hand, regenerative medicine associates this knowledge with other fundamentals, such as treatment involving cells, genetic administration, and immunomodulation, aiming at inducing tissue regeneration in vivo. Therefore, in a way, this medical field is a scientific branch that includes tissue engineering but also incorporates other knowledge and rolls different applications [35].

In the last decades, there has been significant development of research involving the applicability of chitosan-based (CS) materials in these concerned areas [30, 37, 41, 42, 61]. The CS (poli [β-(1 → 4)-2-amino-2-deoxy-d-glucopyranose]) is a naturally sourced copolymer formed from glucosamine and N-acetylglucosamine units linked by β-1,4-glycosidic bonds [5]. It is a natural biopolymer that originates from the polysaccharide chitin, with exceptional biodegradable, biocompatible, and antibacterial properties, which are justified by the presence of free amines in the surface layer of a polymer chain. The presence of these groups enables chemical modifications, which facilitates the increment of unique functional properties of great utility and potential for the advancement of compatible materials desirable for tissue and regeneration applications [22, 42, 80].

Therefore, the main objective of this chapter is to highlight the usage of chitosan nanocomposites as an important instrument in tissue engineering and regenerative medicine. Initially, key concepts and physicochemical and biological characteristics are presented. The main processes used to prepare chitosan nanocomposites applied in biomedical engineering are briefly discussed. In sequence, it is discussed chitosan nanocomposites application, providing a review of recent studies that illustrate the fundamental role of this biomaterial in the area of tissue engineering and regenerative medicine.

2 Chitosan Nanocomposites Definition and Obtention Applied to Tissue Engineering and Regenerative Medicine

Nanocomposites are classified as systems developed from the association of two or more distinct materials, where one of them reaches nanoscale dimensions (1–100 nm) that allow the resurgence of physical and chemical properties that differ from the primitive characteristics of their constituents. Within this class are the natural nanocomposites (bionanocomposites), where at least one of its constituents comes from a natural source (for example, biopolymers), favoring the improvement of physical, mechanical, strength, biodegradable and biocompatible properties of the composites [8, 75].

The development of natural, biodegradable, and biocompatible nanocomposites for various biomedical applications has been the aim of numerous researches in recent years. In this scenario, chitosan (CS) proves to be one of the most widespread natural polymers applied in biomedical engineering [44, 80]. Among its main applications are wound dressings, bone replacement, drug delivery systems, tissue engineering, regenerative medicine, etc. [71].

Recently, nanoparticles incorporated into CS matrices have demonstrated superior biological activity through faster cell differentiation and proliferation due to improved system interaction in the nanocomposite form [61]. In order to accelerate the healing process of injuries, Blažević et al. [15] developed CS/lecithin nanoparticles for melatonin release, observing a great impact on the migration and proliferation of keratinocytes, an essential factor for wound epithelialization. Active glass (BG) particles were incorporated into CS membranes to verify which scale provides better results to the final material. CS membranes with nanoparticles of BG (nBG) showed better mechanical properties and higher bioactivity when compared to membranes with microparticles of BG (mBG). In this study, the authors mention the potential of CS/nBG membranes in the development of bioactive composites for possible application for bone regeneration [20]. The use of CS-incorporated vancomycin has shown a promising effect and efficiency in infection prophylaxis in orthopedic surgeries. The authors mention that such an effect is justified by the high potential to modify the surface of surgical implants [63]. More detail on the various applications of CS as a composite applied in tissue engineering and regenerative medicine is given in Sect. 4.

One of the key aspects for this polymer to be designated as a potential candidate for biomedical engineering applications refers to the biological properties of chitosan. Highlights include biodegradable, cytocompatible, non-toxic properties, antimicrobial, fungicidal, antioxidant, anti-inflammatory, anti-tumor, macrophage activation, angiogenesis stimulation, mucoadhesion, granulation, scar formulation, and wound healing stimulation [4, 37, 87]. In addition, chitosan also has hemostatic effects, coagulates blood, and stimulates cell growth and tissue organization, which justifies that the material aids wound healing through the formation of fibroblasts and macrophages, the accumulation of hyaluronic acid [6].

It is also worth noting that chitosan has excellent physicochemical properties that demonstrate the potential of this natural polysaccharide in biomedical applications, but its use in regenerative medicine may be limited due to the inadequate barrier, thermal and mechanical properties [42]. On the other hand, the presence of active chemical groups in the chemical structure of CS can promote the modification of the material according to the objective application; thus, characteristics such as resistance, mechanical properties, and biodegradation time can be improved [44, 80].

There are some adequated processes for shaping nanocomposites, each of them with the advantages or drawbacks for tissue engineering and regenerative medicine. Some of the most usual processes which are commonly used to prepare chitosan nanocomposites for this field are present as follows:

(a) Solvent Casting Method: This is the most simple process in producing chitosan nanocomposites [13]. Nanocomposite is obtained by blending in the solution or coating, and this technique is limited to a plate-like film morphology. For regenerative medicine, solvent casting method is strongly applied in wound healing and scaffolds, among others.

(b) Freeze Casting: This technique consists in dispersing material in a solvent followed by freezing and lyophilization of the solvent. This technique normally produces a highly porous monolith geometry and is the main route for scaffold production with chitosan [78].

(c) Electrospinning and Electrospray: The electric charged or melted polymeric solution is submitted to a high-voltage electrostatic field surpassing its own surface tension, producing small jets, which are then stretched and accelerated in the direction of a metallic collector. Finally, the jet falls onto the collector after the solvent evaporation or melt cooling, there is the production of fibers (electrospinning) or particles (electrospray). This technique can be applied in scaffolds, nanofiltration, wound healing, and drug delivery [93].

(d) 3D printing: Three-dimensional printing is a mixing of several kinds of techniques, based on 3D mathematical modeling, which can apply in biological and non-biological systems, allowing the construction of physics objects by continuous stacking. This technique possesses a higher precision of the material geometry, corresponding to a computer-aided design (CAD). 3D printing has been a great increase in scaffold production [88].

Furthermore, biological and physicochemical characteristics of chitosan favor its use as a major component in various types of shaped composites, such as in the development of membranes, films, scaffolds, 3D printed scaffolds, fibers, nanoparticles, nanovehicles, nanocapsules, drugs, hydrogels, and also as a component in grafts [44, 61].

3 Chitosan Nanocomposites Physicochemical Properties

In recent decades, chitosan has demonstrated distinct physicochemical properties for biomedical applications [68, 94]. In this scenario, chitosan-derived nanocomposites have aroused great interest because their use can provide distinct properties to the nanocomposite formed, mainly due to their prominent physical and chemical properties, such as biodegradability, biocompatibility, and chelating, among others, which directly influence the nanocomposite formed.

In this sense, chitosan has gained support because it is an ecologically correct, low-cost, sustainable, and renewable matrix of nanocomposite [52]. In its chemical structure, the amine (NH_2) and hydroxyl (OH) groups play an important role in promoting inter- and intramolecular hydrogen bonds that assure better incorporation of nanoparticles in their matrix [21, 33].

From another perspective, according to Whyte et al. [89], the chitosan matrix has a common disadvantage among hydrogels that is low mechanical strength. Because of this, other nanomaterials can be used as a coating or applied together with chitosan to ameliorate its mechanical properties [73].

According to Vandghanooni and Eskandani [86], the application of materials with complex structures or chitosan compounds in scaffolds development has been fomented by virtue of their excellent biological, chemical, and physical properties, which when combined can result in good cytocompatibility, greater polymer crosslinking, improved mechanical properties, better process parameters, a high antibacterial effect attributed to the cationic structure of chitosan, besides that to showing stronger interactions with proteins and adhesive receptors.

4 Applications of Chitosan-Based Nanocomposites in Tissue Engineering and Regenerative Medicine

Kołodziejska et al. [44] highlighted chitosan properties that make this biopolymer a promising material for several applications in biomedicine. Additionally, the authors listed some examples of the use of chitosan, remarkably citing tissue engineering and regenerative medicine applications. To illustrate this great potential, a diagram is presented in Fig. 6.1.

Owing to the structural and chemical properties of its backbone, especially the presence of amino groups that can ionize, chitosan possesses the ability to support cell attachment and multiplication [22, 53]. Moreover, the structural similarity to glycosaminoglycans, the main constituent of extracellular matrix, stands out this biopolymer as an important material for biomedical purposes [53]. Additionally, controllable degradation rates and aptitude to be molded as scaffolds with morphomechanical properties similar to the collagen material emphasize the huge potential of chitosan in the field of tissue engineering.

As explained by Zhu et al. [98], the chemical improvement from chitosan can produce tailored chemical and biophysical properties. To illustrate this statement, they stabilized Fe_3O_4 nanoparticles into distinctive active polysaccharides (chitosan,

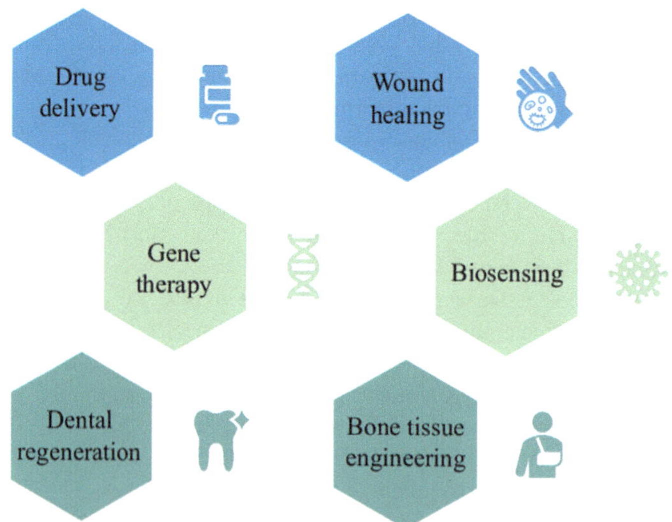

Fig. 6.1 Examples of the use of chitosan in regenerative medicine

O-carboxymethylchitosan, and N-succinyl-O-carboxymethylchitosan) aiming to increase the bioactivity, evaluate the releasement profile, and the in vitro cancer cell inhibition activity of camptothecin (CPT). Results revealed that polysaccharide character strongly influences the size distribution of CPT-loaded polysaccharide-modified Fe_3O_4 nanoparticles, adsorption efficiency, and drug release profile. Cytotoxicity test against 7721 liver cancer cells showed that CPT-loaded polysaccharide-modified Fe_3O_4 nanoparticles had better in vitro cancer cell inhibition activity than CPT-free drug.

Innovative scenarios in combining chitosan with nanostructures-derived materials have generated new porous material with elevated open porosity and large specific surface areas [43]. These features highlight the suitability of chitosan nanocomposites as a material of choice for tissue engineering and regenerative medicine.

4.1 Tissue Engineering

According to [49], some factors ought to be considered in tissue engineering: (a) cells, which are the most basic unit of organisms, (b) scaffolds, which are the framework materials specially designed to support the cellular growth and organization; (c) growth factors, which are extracellular factors that directly influence cellular activity.

Scaffolds are supported structures designed mostly to enhance cellular interaction as needed, making the genesis of new active tissues possible [89]. In this sense, for tissue engineering purposes, scaffolds must present a specific prominent pore structure with higher interconnectivity, imitating bone's pattern, and favoring cell attachment, proliferation, and growth [53]. Cellular phenomena, like migration, growth, and new tissue formation, will be facilitated in those proper porous conditions [74].

Innovative fabrication strategies have been pursued to develop scaffolds and matrices for bone tissue engineering. Figure 6.2 summarizes the main prerequisite characteristics of scaffolds regarding their application as a promising therapeutic tool.

There is a certain difficulty for chitosan and other natural-based polymeric raw materials to produce elevated porosity with controlled pore size and high pore interconnectivity. That difficulty is mainly related to the heat sensitivity of this type of material [72]. For those reasons, chitosan scaffolds and their nanoderivative are strongly aimed at freeze-drying processes or combinations even though electrospinning has already been used [86].

The association between chitosan scaffolds and some nanoparticles, such as nano-hydroxyapatite (nHA), expands the variety of process manufacturing possibilities [17, 72]. Nanoadditives are inserted in chitosan scaffold development to keep chitosan biocompatibility levels unchanged [1]. In this context, nHA, which resembles the inorganic component of natural bone material, is combined among chitosan composite scaffolds to overcome the drawbacks of a single-material shaping, improving the biological and physicochemical attributes of the designed scaffold [49].

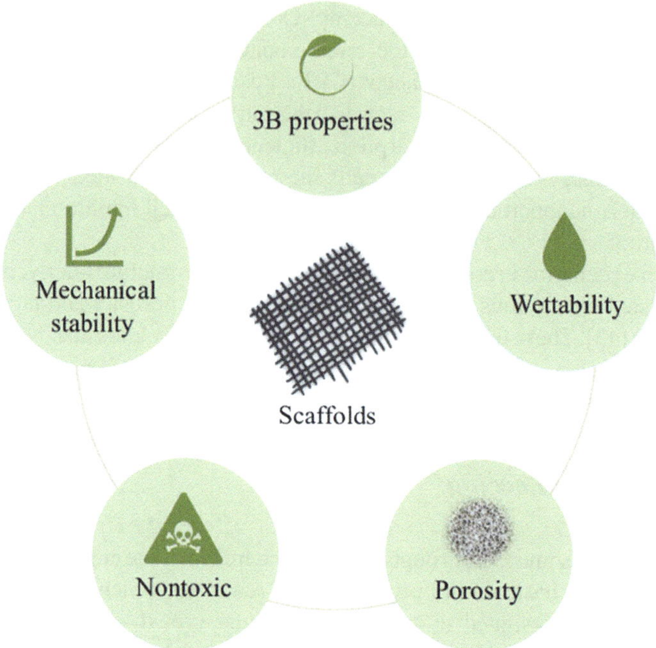

Fig. 6.2 Scaffolds' main characteristics for tissue engineering application

Improvement of mechanical properties, formation of extracellular matrix, and cytoplasmic extensions were pronounced on chitosan–nanohydroxyapatite (nHA) nanocomposite scaffold developed by Maganti et al. [53] in comparison with pure chitosan ones. According to their results, scaffolds demonstrated cytocompatibility and appropriate porosity formation irrespective of deacetylation degree (%DD) and nHA content. Although, despite the similarity in their constitution, careful examination clearly suggested the determining importance of the nanoparticles in developing improved cellular interactions of cell–nanocomposite constructs. While the ability to retain water was similar for both, characteristics such as greater indentation modulus, reduced water uptake, and slower degradation rate were better in nanocomposite scaffolds.

Targeting a potential use as bone tissue, Thein and Misra [85] synthesized a biomimetic nanocomposite by developing chitosan scaffolds with different nHA fractions. Superior mechanical, physicochemical, and biological properties compared to pure chitosan scaffolds for bone tissue engineering were achieved. Results pointed out that cell proliferation in composite scaffolds was about 1.5 times greater than pure chitosan after 7 days of culture and beyond. Water retention ability exhibited similarity between chitosan scaffolds and their nanocomposites. Greater bone-binding proprieties and biodegradation control were seen after the addition of nHa into chitosan samples.

Biodegradable scaffolds act as a critical part of tissue engineering, and three-dimensional (3D) forms provide the desired support for cells to grow and maintain their distinct functions [96]. Maganti et al. [53] explained that the scaffolds mainly act as a biodegradable template allowing their own replacement by the newly formed bone tissue.

The development of scaffolds for the 3D printing shaping technique is also possible. Bioinks, which are the precursor materials used in a 3D printer, could be specially designed for biological application purposes [70]. In the selection of bioink components, there are important parameters that must be considered, such as viscoelasticity, yield stress, cost, printability, shear-thinning, shelf life, creep, and cross-linking time [48]. Sommer et al. [79] described the development of a bioink consisting of an oil-in-water emulsion stabilized by the association of modified silica nanoparticles and chitosan. With the resulting stability of the emulsion and higher storage modulus, yield stress, and elastic recovery, the resultant ink was a better fit for 3D printing standards.

Nanomaterials have strengthened their foothold among tissue engineering strategies for treating bone and dental defects [91]. For craniofacial and dental tissue engineering, the frequently used nanomaterials include nanoparticles, nanofibers, nanotubes, and nanosheets [49]. For example, a combination of chitosan with bioactive glass nanoparticles aiming to produce a novel guided tissue and bone regeneration membrane was proposed by Mota et al. [60]. The introduction of bioactive glass nanoparticles into the chitosan increased the stiffness of the membrane, which also showed adequate extensibility in wet conditions. The composite membranes were able to promote the deposition of an apatite layer, evidence of osteoconductive potential. Results indicate that the chitosan/bioactive glass composite membrane might potentially be applied as a certain temporary tissue regeneration membrane in periodontal regeneration.

Therefore, a variety of chitosan preparation methods can be applied in combination with different types of nano-species targeting different tissues and purposes in tissue engineering. Table 6.1 summarize some works.

In recent years, researchers have been carried out focusing on chitosan as a matrix molecule in tissue engineering. Numerous studies presented in this section evidence the important improvements that can be achieved through the development of chitosan nanocomposites. Fundamental characteristics of biomedical engineerings, such as mechanical properties, conductivity, and bioactivity, were enhanced successfully by combining proper nanomaterials. A variety of methods have been employed to prepare chitosan nanocomposites, and strategies to heal different tissues have been conquered, showing promising results.

4.2 Regenerative Medicine

Chitosan has been studied in several regenerative medicine fields. In the form of nanocomposites, chitosan can be a good candidate for the process of healing wounds

Table 6.1 Chitosan nanocomposites scaffolds according to nanomaterial type, preparation method, target tissue, and the properties enhanced

Nanomaterial	Preparation method	Tissue	Properties enhanced	References
Graphene oxide	Freeze casting	Bone	Mechanical, pore formation, bioactivity,	Dinescu et al. [26]
Graphene oxide	Solvent casting	Bone	Mechanical, thermal stability	Tavakoli et al. [84]
Hydroxyapatite	3D printing	Bone	Bioactivity	Pradeep et al. [65]
MWCNTs*	Solvent casting	Bone	Bioactivity, reduced biodegradability	Bakhtiari et al. [12]
Carbon nanofiber	Freeze casting	Cardiac	Elasticity, conductivity	Martins et al. [56]
Graphene oxide	Freeze casting	Cardiac	Conductivity	Jiang et al. [38]
Gold	Freeze casting	Cardiac	Conductivity, bioactivity	Baei et al. [10]
Graphene oxide	Electrospinning	Cartilage	Mechanical	Cao et al. [18]
Graphene oxide	Freeze casting	Cartilage	Mechanical	Shamekhi et al. [74]
Single-walled carbon nanotube	Electrospinning	Neural	Mechanical, bioactivity	Shokrgozar et al. [77]
Ag	Freeze casting	Skin	Mechanical, antibacterial activity	Niu et al. [62]
Cu_2O	Electrospinning	Skin	Hydrophilicity, antibacterial activity	Zhou et al. [97]

*Multiwalled carbon nanotubes

due to their low mammalian cell toxicity and the antimicrobial properties related to the chitosan cationic structure tendency. These properties also can be improved with the presence of some nanoparticles [57, 66].

Metallic silver, gold, titanium, copper, and other metallic oxide nanoparticles have presented important therapeutic effects on wound healing. Their unique shape characteristics in combination with the general type of polymers, in the form of nanocapsules, polymersomes solid lipid nanoparticles, or polymeric nanocomplexes, are ideal transport to promote the correct action of pharmaceutical medicine (antibiotics, growth factors) for a proper wound healing [2]. Chitosan has interesting qualities in this field including the easing of fibroblast attachment and the accelerating tissue regeneration [31].

Ding et al. [27] described the test of a chitosan bilayer dressing comprised of silver nanoparticles for healing biotissues. In this study, the authors documented a decremental effect in the composites' antimicrobial response following the crosslinking of chitosan with genipin (GB). Notwithstanding, the deleterious effect was eliminated by the association of silver nanoparticles to the compound. Archana et al. [9] developed a composited chitosan nano-dressing of chitosan–pectin–titanium dioxide with increased biocompatible efficiency and antibacterial behavior. The nano-dressing presents stronger antimicrobial activity against five pathogens types and also has a good blood-compatible effect. Furthermore, chitosan–pectin–TiO_2 dressing material can also control water loss in evaporation from wound beds at one optimal level and absorb more exudates, keeping these wounds without risk of dehydration or exudate accumulation.

Focusing on skin regeneration applications, Ghaee et al. [31] developed biomimetic nanocomposite scaffolds based on surface-modified polycaprolactone nanofibers containing curcumin embedded in chitosan/gelatin. The fabricated nanofibrous-hydrogel scaffolds showed a highly interconnected porous structure, which promotes deep nutrient flow to cells and satisfies demands for application in skin tissue engineering. The cell attachment and biocompatible efficacy were confirmed under specific biological tests in the nanocomposite scaffolds obtained.

Nanocomposites have also been studied as therapeutic agent carriers. As explained by Tarhan et al. [83], because of their nanoparticle size and increased biocompatibility, nano-derivative polymeric agents can be attached to the bloodstream for a longer period of time until reaching the target site. This approach has been used to improve the clinical results of chemotherapy, for example, and the usage and advantages of magnetic nanoparticles are being explored for effective treatment without damaging healthy cells [82].

Kankala et al. [40] associated a siliceous-zinc framework loaded with doxorubicin together with a platinum (Pt) nanoparticles-enriched chitosan composite. This interesting drug delivery method contributed to tissue penetration and helped to ease the inhibition effect seen in the chemotherapeutic treatment.

Effective drug delivery and wound healing have been accomplished also through chitosan bioadhesive systems. Bioadhesive polymers are a category of polymers to adhere two surfaces where at least one of them is living tissue [34]. Chitosan, as a biopolymer with immune-stimulatory activities, biodegradability, cationic properties, and high adsorption capacity, conquers great attention in this area [23, 34]. Ameeduzzafar et al. [7] developed a bioadhesive chitosan-coated poly(lactic-co-glycolic acid) nanoparticles (PLGA-NPs) for effective treatment of bacterial conjunctivitis. An extended drug release with prolonged retention and better tolerability at the corneal site was observed in this study. The antimicrobial activity of developed nanoparticles (NPs) was superior than marketed drug compound.

Chitosan is also a promising candidate in the field of biosensing technologies, with known capabilities like specificity for many biological and chemical species [14]. Due to the relatively low conductivity of chitosan, the transfer of the electric signal to the transducer on chitosan-based biosensors is impaired. For that reason, materials such as NPs have been combined with chitosan in order to make nanocomposites addressed to enhance their conductive properties [24, 81].

A biosensor aimed at the detection of sarcosine, a biomarker for prostate cancer with increased amperometric properties, was developed by Deswal et al. [25].

This biosensor was established on covalent immobilization of an enzyme, sarcosine oxidase, over the chitosan nanocomposite bounded with grapheme nanoribbons which were electrodeposited in gold electrode, aiming to test the utility for detection of sarcosine in blood. Results showed high sensitivity, broad linear range, lower detection limit, good reproducibility, and longer stability (180 days), which are indications that the developed biosensor could be employed in the diagnosis of prostate cancer.

Biomaterials have been used for repairing lesioned ducts of soft tissues in the body, such as weasand, blood vessels, and nerve regeneration. Although synthetic polymers have been widely applied in this field, their biocompatibility and degradability limit their applications [95]. Chitosan, beyond exceptional properties such as antibacterial activity, biomechanics, biocompatibility, toxicity, biodegradability, and non-antigenicity, has seemed to stimulate Schwann cells genesis and nerve regeneration, important features for biomaterials applied in tissue engineering and regenerative medicine [50, 90, 92].

In terms of lesions of the extensive peripheral nervous system (PNS), the therapeutic employment of neural guidance channels (NGCs) has increased importantly. NGCs are polymer-derivated tubes that can be applied for linking injured nerve ends. Chitosan is, among other biomaterials, a possible candidate for use in this field [28, 64, 67]. However, even in face of promising results, researchers have been trying to amplify the NGC's healing proprieties through different strategies [19, 45].

Concerning drug delivery by NGCs, secondary drug delivery vehicles such as nanoparticles are an advantageous method related to their adequated release profile, drug delivery specificity, high encapsulation efficiency, and high surface-to-volume ratio [16, 88]. Ebrahimi et al. [29] prepared a sophisticated chitosan/alginate hydrogel composited with berberine (Ber) and naringin (Nar), both natural substances, encapsulated into CS NPs. According to their results, the authors reported 50% encapsulation efficacy and size distribution of around 500 nm with a described sustained release of almost 90% in 24 days following an initial drug delivery burst on the first days. Allegedly, in vivo results showed better anatomical healing as well as higher motor and sensory recovery when using the therapy with Ber- and Nar-loaded chitosan/alginate hydrogel. These results could indicate new uses for the designed hydrogel such as nerve damage recovery.

In order to broaden the range of examples in which chitosan nanocomposites have been applied in regenerative medicine, Table 6.2 is presented. In comparison to Table 6.1, it is also possible to notice an elevated extension of possibilities regarding chitosan nanocomposite among different target issues, promoting different properties enhancement in different application fields.

Many studies have confirmed the versatility of chitosan nanocomposite's application toward regenerative engineering. In this section, we have addressed the main trends regarding the development of nanocomposites in this biomedical field. Different strategies such as wound healing, electrical stimulation, and drug delivery approaches were considered.

Silver (Ag) emerged as an important nanomaterial for wound healing applications. Active against a vast spectrum of infectious agents, it is acknowledged and used as

Table 6.2 Chitosan nanocomposites in regenerative medicine according to nanomaterial type, application field, target issue, and properties enhanced

Nanomaterial	Application field	Target issue	Properties enhanced	Reference
Graphene oxide	Biosensing	Detection of glucose	Electrochemical, sensitivity, detection limit	Mehdizadeh et al. [58]
SWCNTs*	Biosensing	Serum leptin detection	Conductivity, biocompatibility, accuracy	Bakhtiari et al. [12]
Graphene oxide/Sulfonated graphene oxide	Bone regeneration	Bone engineering and drug delivery	Hydrophilicity, biocompatibility, drug release kinetics	Mahanta et al. [54]
Hydroxyapatite	Bone regeneration	Dental repair/regeneration	Mechanical, osteoconductivity	Dinescu et al. [26]
Graphene oxide	Drug delivery	Yet to be studied	Mechanical, biocompatibility	Zhou et al. [97]
Reduced graphene oxide (rGO)	Drug delivery	Transdermal drug delivery	Mechanical, conductivity, drug release profile	Justin and Chen [39]
Silica	Drug delivery	Cartilage regeneration	Sustained drug release, chondrocyte proliferation, mechanical	Niu et al. [62]
SWCNTs*	Drug delivery	Cancer cells	Water solubility, biocompatibility	Mo et al. [59]
Gold	Gene therapy	DNA carriers	Transfection efficiency	Abrica-gonzález et al. [3]
Silver	Wound dressing	Yet to be studied	Water vapor transmission, mechanical, antibacterial activity	Pradeep et al. [65]
Silver	Wound dressing	Thermal burns injuries	Antibacterial, bioactivity	Jiang et al. [38]
Silver-titanium	Wound dressing	Yet to be studied	Porosity, biocompatibility, water adsorption, cell growth	Liang et al. [51]

*Single-walled carbon nanotubes

an antimicrobe agent, both in its metallic and ionic forms [47]. As explained by Hasan et al. [36], many of the implant-associated problems and their deleterious health effects can be avoided, whenever such antimicrobial agents are associated with scaffolds/hydrogels preparation.

The carbon nanomaterials (CNMs) group, notedly graphene oxide (GO) and carbon nanotubes (CNTs), also appeared as an interesting constituent of chitosan nanocomposites. Alongside their sustainable physicochemical properties, superior characteristics such as elevated thermal and electrical conductivity, optical properties, and great stability are highlighted by Khorsandi et al. [43], proving this group as a notably attractive material for biomedical applications. In this sense, regenerative medicine represents a multidisciplinary field that brings innovative solutions and amplifies the healing prospect.

5 Concluding Remarks and Future Perspectives

Over the last years, chitosan has demonstrated unparalleled characteristics for tissue engineering and regenerative medicine applications. However, improved physicochemical properties are able produced with the presence of nanomaterials. The presence of surface hydroxyl and amino groups enables chemical modifications, which facilitates the increment of unique functional properties of great utility and potential for the development of nanobiomaterials that are desirable for tissue engineering and regenerative medicine.

There are many processes to design chitosan nanocomposites with the advantages or drawbacks for tissue engineering and regenerative medicine. The most common are solvent and freeze casting although there is an increment in more sophisticated techniques such as electrospinning and 3D printing.

In this chapter, we provide several examples that testify to the potential of chitosan nanocomposites in scaffolds, repair, and regeneration of a variety of human tissues. From the arguments and examples provided, chitosan nanocomposites appeared as promising tools in regenerative medicine and tissue engineering.

The future perspectives in chitosan nanocomposites in the related fields are in the study of insertion in distinct nanomaterials species and shapes, The future perspectives in chitosan nanocomposites in the related fields are in the study of insertion in distinct nanomaterials species and shapes, and the study of alternative production routes that can produces chitosan nanoparticles with sophisticated geometric features. And finally, there is still a great lack of distinct applications in tissue engineering and regenerative medicine where chitosan nanocomposite could be further applied.

References

1. Abbasian M et al (2019) Scaffolding polymeric biomaterials: are naturally occurring biological macromolecules more appropriate for tissue engineering ? Int J Biol Macromol 134:673–694
2. Abbaszadeh A et al (2019) Effects of chitosan/nano selenium biofilm on infected wound healing in rats; an experimental study. Bull Emerg Trauma 7, n. 3, p. 284–291, 2019.
3. Abrica-González P et al (2019) Gold nanoparticles with chitosan, N-acylated chitosan, and chitosan oligosaccharide as DNA carriers. Nanoscale Res Lett 14(258):1–14
4. Ahsan SM et al (2018) Chitosan as biomaterial in drug delivery and tissue engineering. Int J Biol Macromol 110:97–109
5. Ali Khan Z et al (2020) Chitosan based hybrid materials used for wound healing applications—A short review. Int J Polym Mater Polym Biomater 69(7):419–436
6. Ali SW, Bairagi S (2021) Chitosan-based bionanocomposites 7 in wound healing. In: Ahmed S, Annu (eds) Bionanocomposites in tissue engineering and regenerative medicine. Elsevier, pp 149–186
7. Ameeduzzafar et al (2020) Improvement of ocular efficacy of levofloxacin by bioadhesive Chitosan coated PLGA nanoparticles: Box-behnken design, in-vitro characterization, antibacterial evaluation and scintigraphy study. Iran J Pharm Res 19(1):292–311
8. Annu, Ahmed S (2021) Bionanocomposites: an overview. In: Ahmed S, Annu Bionanocomposites in tissue engineering and regenerative medicine. Elsevier, pp 1–6
9. Archana D, Dutta J, Dutta PK (2013) Evaluation of chitosan nano dressing for wound healing: characterization, in vitro and in vivo studies. Int J Biol Macromol 57:193–203
10. Baei P et al (2016) Electrically conductive gold nanoparticle-chitosan thermosensitive hydrogels for cardiac tissue engineering. Mater Sci Eng C 63:131–141
11. Del Bakhshayesh RA et al (2019) An overview of advanced biocompatible and biomimetic materials for creation of replacement structures in the musculoskeletal systems : focusing on cartilage tissue engineering. J Biol Eng 9
12. Bakhtiari SSE et al (2019) Chitosan/MWCNTs composite as bone substitute: physical, mechanical, bioactivity, and biodegradation evaluation. Polym Compos 40(S2):E1622–E1632
13. Bakshi PS et al (2020) Chitosan as an environment friendly biomaterial—a review on recent modifications and applications. Int J Biol Macromol 150:1072–1083
14. Baranwal A et al (2018) Chitosan: an undisputed bio-fabrication material for tissue engineering and bio-sensing applications. Int J Biol Macromol 110:110–123
15. Blažević F et al (2016) Nanoparticle-mediated interplay of chitosan and melatonin for improved wound epithelialisation. Carbohydr Polym 146:445–454
16. Boni R et al (2018) Current and novel polymeric biomaterials for neural tissue engineering. J Biomed Sci 25(90):1–21
17. Bulanov E et al (2020) Study of physicochemical properties of nanohydroxyapatite–chitosan composites. Bull Mater Sci 43(1):1–6
18. Cao L et al (2017) Fabrication of chitosan/graphene oxide polymer nanofiber and its biocompatibility for cartilage tissue engineering. Mater Sci Eng C 79:697–701
19. Cao S et al (2022) Chitosan nanoparticles, as biological macromolecule-based drug delivery systems to improve the healing potential of artificial neural guidance channels: a review. Int J Biol Macromol 201:569–579
20. Caridade SG et al (2013) Chitosan membranes containing micro or nano-size bioactive glass particles: evolution of biomineralization followed by in situ dynamic mechanical analysis. J Mech Behav Biomed Mater 20:173–183
21. Chatterjee R et al (2022) Chitosan: source, chemistry, and properties. In: Hassain MS, Beg S, Nayak AK (eds) Chitosan in drug delivery. Academic Press, pp 1–22
22. Croisier F, Jérôme C (2013) Chitosan-based biomaterials for tissue engineering. Eur Polymer J 49:780–792
23. Dash M et al (2011) Chitosan—a versatile semi-synthetic polymer in biomedical applications. Prog Polym Sci 36:981–1014

24. Dervisevic M et al (2017) Novel electrochemical xanthine biosensor based on chitosan–polypyrrole–gold nanoparticles hybrid bio-nanocomposite platform. J Food Drug Anal 25:510–519
25. Deswal R et al (2022) An improved amperometric sarcosine biosensor based on graphene nanoribbon/chitosan nanocomposite for detection of prostate cancer. Sens Int 3:100174
26. Dinescu S et al (2014) In vitro cytocompatibility evaluation of chitosan/graphene oxide 3D scaffold composites designed for bone tissue engineering. Bio-Med Mater Eng 24(6):2249–2256
27. Ding L et al (2017) Spongy bilayer dressing composed of chitosan–Ag nanoparticles and chitosan–Bletilla striata polysaccharide for wound healing applications. Carbohydr Polym 157:1538–1547
28. Du J et al (2018) Biomimetic neural scaffolds: a crucial step towards optimal peripheral nerve regeneration. Biomater Sci 6:1299–1311
29. Ebrahimi MH et al (2020) Peripheral nerve regeneration in rats by chitosan/alginate hydrogel composited with Berberine and Naringin nanoparticles: in vitro and in vivo study. J Mol Liquids 318:114226
30. Feng W, Wang Z (2022) Biomedical applications of chitosan-graphene oxide nanocomposites. iScience 25:103629
31. Ghaee A et al (2019) Biomimetic nanocomposite scaffolds based on surface modified PCL-nanofibers containing curcumin embedded in chitosan/gelatin for skin regeneration. Compos Part B Eng 177:107339
32. Greenwood HL et al (2006) Regenerative medicine and the developing world. PLoS Med 3(9):1496–1500
33. Gzyra-Jagieła K et al (2019) Physicochemical properties of chitosan and its degradation products. In: Van Den Broek LAM, Boeriu CG (eds) Chitin and chitosan: properties and applications. Wiley, pp 61–80
34. Hamedi H et al (2022) Chitosan based bioadhesives for biomedical applications: a review. Carbohydr Polym 282:119100
35. Han F et al (2020) Tissue engineering and regenerative medicine: achievements, future, and sustainability in Asia. Frontiers Bioeng Biotechnol 8:1–35
36. Hasan A et al (2018) Nano-biocomposite scaffolds of chitosan, carboxymethyl cellulose and silver nanoparticle modified cellulose nanowhiskers for bone tissue engineering applications. Int J Biol Macromol 111:923–934
37. Hossain MR, Mallik AK, Rahman MM (2020) Fundamentals of chitosan for biomedical applications. In: Gopi S, Thomas S, Pius A (2020) Handbook of chitin and chitosan. Elsevier Inc., pp 199–230
38. Jiang L et al (2019) Preparation of an electrically conductive graphene oxide/chitosan scaffold for cardiac tissue engineering. Appl Biochem Biotechnol 188(4):952–964
39. Justin R, Chen B (2014) Strong and conductive chitosan-reduced graphene oxide nanocomposites for transdermal drug delivery. J Mater Chem B 2:3759–3770
40. Kankala RK et al (2020) Ultrasmall platinum nanoparticles enable deep tumor penetration and synergistic therapeutic abilities through free radical species-assisted catalysis to combat cancer multidrug resistance. Chem Eng J 383:123138
41. Karthika V et al (2020) Chitosan overlaid Fe_3O_4/rGO nanocomposite for targeted drug delivery, imaging, and biomedical applications. Sci Rep 10:18912
42. Khan A, Wang B, Ni Y (2020) Chitosan-nanocellulose composites for regenerative medicine applications. Curr Med Chem 27:4584–4592
43. Khorsandi Z et al (2021) Carbon nanomaterials with chitosan: a winning combination for drug delivery systems. J Drug Deliv Sci Technol 66:102847
44. Kołodziejska M et al (2021) Chitosan as an underrated polymer in modern tissue engineering. Nanomaterials 11:3019
45. Lackington WA, Ryan AJ, O'Brien FJ (2016) Advances in nerve guidance conduit-based therapeutics for peripheral nerve repair. ACS Biomater Sci Eng 3(7):1221–1235
46. Langer R, Vacanti JP (1993) Tissue engineering. Science 260:920–925

47. Lara HH et al (2011) Silver nanoparticles are broad-spectrum bactericidal and virucidal compounds. J Nanobiotechnol 9(30):1–9
48. Lee JM, Yeong WY (2016) Design and printing strategies in 3D bioprinting of cell-hydrogels: a review. Adv Healthcare Mater 5(22):2856–2865
49. Li G et al (2017) Nanomaterials for craniofacial and dental tissue engineering. J Dent Res 96(7):725–732
50. Li G et al (2018) Spatially featured porous chitosan conduits with micropatterned inner wall and seamless sidewall for bridging peripheral nerve regeneration. Carbohydr Polym 194:225–235
51. Liang H et al (2022) Fabrication of tragacanthin gum-carboxymethyl chitosan bio-nanocomposite wound dressing with silver-titanium nanoparticles using freeze-drying method. Mater Chem Phys 279:125770
52. Lizardi-Mendoza J, Argüelles Monal WM, Goycoolea Valencia FM (2016) Chemical Characteristics And Functional Properties Of Chitosan. In: Bautista-Banos S, Romanazzi G, Jiménez-Aparicio A (2016) Chitosan in the preservation of agricultural commodities. Elsevier Inc., pp 3–31
53. Maganti N et al (2011) Structure-process-property relationship of biomimetic chitosan-based nanocomposite scaffolds for tissue engineering: biological, physico-chemical, and mechanical functions. Adv Eng Mater 13(3):108–122
54. Mahanta AK, Patel DK, Maiti P (2019) Nanohybrid scaffold of chitosan and functionalized graphene oxide for controlled drug delivery and bone regeneration. ACS Biomater Sci Eng 5:5139–5149
55. Mao AS, Mooney DJ (2015) Regenerative medicine: current therapies and future directions. PNAS 112(47):14452–14459
56. Martins AM et al (2014) Electrically conductive chitosan/carbon scaffolds for cardiac tissue engineering. Biomacromolecules 15(2):635–643
57. Matica MA et al (2019) Chitosan as a wound dressing starting material: antimicrobial properties and mode of action. Int J Mol Sci 20(1)
58. Mehdizadeh B et al (2020) Glucose sensing by a glassy carbon electrode modified with glucose oxidase/chitosan/graphene oxide nanofibers. Diamond Relat Mater 109:108073
59. Mo Y et al (2015) Controlled release and targeted delivery to cancer cells of doxorubicin from polysaccharide-functionalised single-walled carbon nanotubes. J Mater Chem B 3:1846–1855
60. Mota J et al (2012) Chitosan/bioactive glass nanoparticle composite membranes for periodontal regeneration. Acta Biomaterialia 8:4173–4180
61. Murugesan S, Scheibel T (2021) Chitosan-based nanocomposites for medical applications. J Polym Sci 59:1610–1642
62. Niu X et al (2020) Silver-loaded microspheres reinforced chitosan scaffolds for skin tissue engineering. Eur Polym J 134:109861
63. Pawar V et al (2019) Chitosan sponges as a sustained release carrier system for the prophylaxis of orthopedic implant-associated infections. Int J Biol Macromol 134:100–112
64. Petcu EB et al (2018) 3D printing strategies for peripheral nerve regeneration. Biofabrication 10:032001
65. Pradeep N et al (2021) 3D printing of shrimp derived chitosan with HAp as a bio-composite scaffold. Mater Manuf Process 1–10
66. Rahimi M et al (2019) A novel bioactive quaternized chitosan and its silver-containing nanocomposites as a potent antimicrobial wound dressing: structural and biological properties. Mater Sci Eng C 101:360–369
67. Raza C et al (2020) Repair strategies for injured peripheral nerve: review. Life Sci 243:117308
68. Rezaei FS et al (2021) Chitosan films and scaffolds for regenerative medicine applications: a review. Carbohydr Polym 273:118631
69. Rizeq BR et al (2019) Synthesis, bioapplications, and toxicity evaluation of chitosan-based nanoparticles. Int J Mol Sci 20:1–24
70. Sahranavard M et al (2020) A critical review on three dimensional-printed chitosan hydrogels for development of tissue engineering. Bioprinting 17(e00063):1–9

71. Sainitya R et al (2015) Scaffolds containing chitosan/carboxymethyl cellulose/mesoporous wollastonite for bone tissue engineering. Int J Biol Macromol 80:481–488
72. Sampath UGTM et al (2016) Fabrication of porous materials from natural/synthetic biopolymers and their composites. Materials 9(12):1–32
73. Semba JA, Mieloch AA, Rybka JD (2020) Bioprinting introduction to the state-of-the-art 3D bioprinting methods, design, and applications in orthopedics. Bioprinting 18(e00070):1–17
74. Shamekhi MA et al (2019) Graphene oxide containing chitosan scaffolds for cartilage tissue engineering. Int J Biol Macromol 127:396–405
75. Shchipunov Y (2012) Bionanocomposites: green sustainable materials for the near future. Pure Appl Che 84(12):2579–2607
76. Shin SR et al (2016) Graphene-based materials for tissue engineering ☆. Adv Drug Deliv Rev 105:255–274
77. Shokrgozar MA et al (2011) Fabrication of porous chitosan/poly(vinyl alcohol) reinforced single-walled carbon nanotube nanocomposites for neural tissue engineering. J Biomed Nanotechnol 7(2):276–284
78. Silva AO et al (2021) Chitosan as a matrix of nanocomposites: a review on nanostructures, processes, properties, and applications. Carbohydr Polym 272:118472
79. Sommer MR et al (2017) 3D printing of concentrated emulsions into multiphase biocompatible soft materials. Soft Matter 13(9):1794–1803
80. Sultankulov B et al (2019) Progress in the development of chitosan-based biomaterials for tissue engineering and regenerative medicine. Biomolecules 9:470
81. Susanto H et al (2013) Immobilization of glucose oxidase on chitosan-based porous composite membranes and their potential use in biosensors. Enzyme Microbial Technol 52:386–392
82. Tarhan T (2021) Synthesis and characterization of magnetic nanocomposite for in vitro evaluation of irinotecan using human cell lines. Turk J Chem 45:540–550
83. Tarhan T, Tural B, Tural S (2019) Synthesis and characterization of new branched magnetic nanocomposite for loading and release of topotecan anti-cancer drug. J Anal Sci Technol 10(30):1
84. Tavakoli M, Karbasi S, Soleymani Eil Bakhtiari S (2020) Evaluation of physical, mechanical, and biodegradation of chitosan/graphene oxide composite as bone substitutes. Polym-Plast Technol Mater 59(4):430–440
85. Thein-Han WW, Misra RDK (2009) Biomimetic chitosan-nanohydroxyapatite composite scaffolds for bone tissue engineering. Acta Biomater 5:1182–1197
86. Vandghanooni S, Eskandani M (2019) Electrically conductive biomaterials based on natural polysaccharides: challenges and applications in tissue engineering. Int J Biol Macromol 141:636–662
87. Vunain E, Mishra AK, Mamba BB (2017) Fundamentals of chitosan for biomedical applications, vol 1. Elsevier
88. Wang G et al (2021) Prospects and challenges of anticancer agents' delivery via chitosan-based drug carriers to combat breast cancer: a review. Carbohydr Polym 268:118192
89. Whyte DJ et al (2019) Bioprinting a review on the challenges of 3D printing of organic powders. Bioprinting 16(e00057):1–10
90. Yang Y et al (2011) Nerve conduits based on immobilization of nerve growth factor onto modified chitosan by using genipin as a crosslinking agent. Eur J Pharm Biopharm 79:519–525
91. Yazdanian M et al (2020) Current and advanced nanomaterials in dentistry as regeneration agents: an update. Mini-Rev Med Chem 21:899–918
92. Yuan Y et al (2004) The interaction of Schwann cells with chitosan membranes and fibers in vitro. Biomaterials 25(18):4273–4278
93. Xue et al (2019) Electrospinning and electrospun nanofibers: methods, materials, and applications. Chem Rev 119:5298–5415
94. Zhang Z et al (2021) Research progress of chitosan-based biomimetic materials. Marine Drugs 19(7):372
95. Zhao CQ et al (2020) Chitosan ducts fabricated by extrusion-based 3D printing for soft-tissue engineering. Carbohydr Polym 236:116058

96. Zhao L, Chang J (2004) Preparation and characterization of macroporous chitosan/wollastonite composite scaffolds for tissue engineering. J Mater Sci Mater Med 15:625–629
97. Zhou X et al (2021) In situ deposition of nano Cu_2O on electrospun chitosan nanofibrous scaffolds and their antimicrobial properties. Int J Biol Macromol 191:600–607
98. Zhu A et al (2009) Polysaccharide surface modified Fe_3O_4 nanoparticles for camptothecin loading and release. Acta Biomaterialia 5:1489–1498

Chapter 7
Emerging Applications
of Chitosan-Based Nanocomposites
in Multifarious Cancer Diagnosis
and Therapeutics

Nandini Sharma, Shikha Gulati, and Jeevika Bhat

Abstract Cancer, a debilitating disease by uncontrolled cell differentiation in the body, has been described to have over 200 different characteristic manifestations and clinical types. Conventional treatment strategies like chemotherapy, surgery, and radiotherapy (applying radiations) have been employed to treat the majority of the malignancies but acute side effects like hair loss, anaemia, oedema, bruising, fatigue, etc., have compelled scientists all around the world to look for alternate treatment regimes. Recent developments in nanoscience have revealed it to be highly effective in the detection and cure of cancers. One such class of nanomaterials that possesses a lot of potential in the biomedical domain is nanocomposites which can be roughly defined as a combination of nanoscale substances having no less than one dimension in the nanoscale range that are arranged in terms of a polymeric matrix, with the materials being in various combinations of organic and inorganic origin. Their rise in demand and research is because of their unusual properties and flexible nature that is relevant in the biomedical landscape. When associated with other biomaterials, they become even more functionally advantageous and the most promising one in cancer diagnostics, and treatment is chitosan. Chitosan, being a biopolymer, is produced by deacetylating chitin, a widely found polymeric form of N-acetylglucosamine which contains active functional groups that are highly susceptible to chemical reactions. This results in many unique properties like biocompatibility, biodegradability, non-toxicity, antimicrobial activity, etc. Thus, this chapter focusses on the up-gradation of nanocomposite properties when introduced with chitosan along with highlighting the multifarious uses of these chitosan-based nanocomposites in the domain of biomedicine, with special emphasis on cancer diagnostics and treatment. The chapter also aims to put into perspective, the recent developments in these biomaterials and discusses their functionalities and attributes whilst describing their applications in cancer healthcare concerning future advancements.

N. Sharma · J. Bhat
Department of Biological Science, Sri Venkateswara College, University of Delhi, Delhi 110021, India

S. Gulati (✉)
Department of Chemistry, Sri Venkateswara College, University of Delhi, Delhi 110021, India
e-mail: shikha2gulati@gmail.com

S. Gulati (ed.), *Chitosan-Based Nanocomposite Materials*,
https://doi.org/10.1007/978-981-19-5338-5_7

Keywords Nanotechnology · Bionanocomposites · Chitosan · Chitosan-based nanocomposites · Cancer · Biopolymer · Oncotherapy

Abbreviations

A549	Adenocarcinomic human alveolar basal epithelial cell line
API	Active pharmaceutical ingredient
CHI	Chitosan
CHI/GO	Chitosan conjugated graphene oxide
HCT-15	Human colorectal carcinoma cell line
HeLa cells	Henrietta lacks cell line
HepG2	Liver hepatocellular carcinoma
HT-29	Human colorectal adenocarcinoma cell line
IARC	International agency for research on cancer
ISO	International organization for standardization
MCF-7	Michigan cancer foundation
NP	Nanoparticle
PLGA	Poly(lactic-co-glycolic) acid
ROS	Reactive oxygen species
XRD	X-ray diffraction

1 Introduction

Currently, cancer is one of the most debilitating diseases and is amongst the leading causes of death across the globe, after the spectrum of diseases like cardiovascular problems. According to the well-researched data by the International Agency for Research on Cancer (IARC), approximately 13.1, a million casualties associated with cancer are estimated by 2030. Some reasons behind the occurrence of metastasis may be external, such as urbanization, lifestyle, and overgrowing population, or internal, such as hormonal, poor immune system, or even genetic [1]. Evidence suggests that carcinogenesis is primarily caused due to DNA damages that lead to genomic instability, a phenomenon in which chronic mutations are induced in the genetic material and can be initiated by agents like ionizing radiation, heavy metals, etc. This instability can be passed on to the next generation and is responsible for increasing the risk of carcinogenesis and may even lead to secondary cancers in patients undergoing chemotherapy or radiotherapy. Traditional methods of treating cancer include surgery, chemotherapy, and radiotherapy, with chemotherapy being the one that is used most widely. But it has its detrimental effects such as sub-par bioavailability, multi-drug resistance, deleterious physiological manifestations in the form of hair loss, anaemia, etc., due to poor targeting and low-cost efficacy.

In the process of finding novel methods of treating cancer, scientists have started exploring nanotechnology for overcoming the limitations of conventional treatment procedures.

Ever since Richard Feynman gave the definition of nanoparticles, science has progressed a long way in the arena of nanotechnology and is making continuous strides in increasing the applications of this field. In a myriad of fields, nanoformulations are being applied because of their distinctive properties which are exemplified by their high-surface area, remarkable mobility in their free states, reactivity, stable carrier structures, quantum effects, etc. Such attributes are particularly useful in the field of biomedicine where these are put to use in a multitude of ways, ranging from diagnostics to therapy. There are a large variety of nanoscale formulations possible like nanoparticles of numerous origins (metallic, organic, inorganic, etc.), nanotubes, dendrimers, liposomes, polymeric nanoparticles, nanocomposites, and many more. Out of these, one class that has garnered large-scale attention is nanocomposites especially the ones of polymeric origin.

To recapitulate, nanocomposites are combinations of nanoscale substances with at least one dimension in the nanorange that are arranged in terms of a polymeric matrix, with the materials being a combination of organic/organic, organic/inorganic, or even inorganic/inorganic [2]. The textbook definition of nanocomposites has experienced many changes with the ongoing research and presently, is referred to as those systems with combinations of varied dimensional materials, mixed with amorphous materials at the nanoscale [3]. They showcase a multitude of beneficial properties that are applicable in industries like food packaging, healthcare, purification, etc., but they still possess the problem of biodegradability which is solved by another sub-class of these nanomaterials called "bionanocomposites" [4]. These are hybrid substances that comprise inorganic solids and biopolymers with the most important advantage being their biodegradability, that is, these green and eco-friendly formulations are subject to degradation by the action of living organisms in the biosphere, making them safe for the environment. They are also abundant and renewable, making them an ideal option over the other existing alternatives. Amidst the magnitude of biopolymers available, one of the first, most researched, and widely applicable biopolymer is "chitosan", obtained from the second most abundant polysaccharide. This compound is derived after limited or incomplete deacetylation of chitin (Fig. 1), and some of its multifarious properties are excellent biodegradability, biocompatibility, non-toxicity, biodistribution, along with significant anti-tumour activity, and such unique and highly advantageous attributes have a huge scope in oncotherapy, which is explored in its depth throughout the course of this chapter.

2 Emerging Role of Nanotechnology in Oncotherapy

Throughout the history of scientific research, experts have been looking for the correct answer to treat one of the deadliest diseases to ever plague mankind, which is cancer. Extensive research has been going on for years, in search of newer approaches

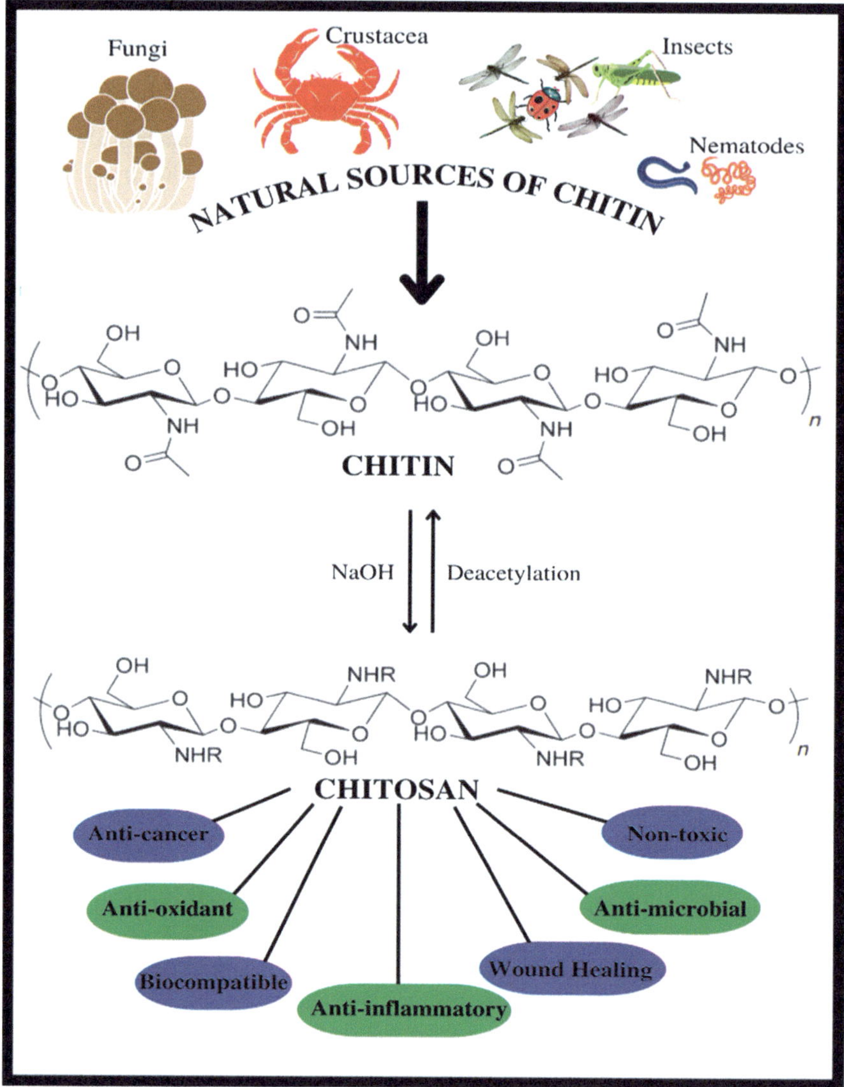

Fig. 1 Chitosan isolation and applications in the biomedical domain

for cancer therapy that do not have the disadvantages of conventional methods and are safer for public use. A promising answer has been discovered in the form of nanotechnology. Nanoformulations are extremely short in size, their dimensions ranging between 10 and 1000 nm. Its most utilized form is a "nanoparticle", which can be organic or inorganic and can be created from a variety of materials that provides it with specific and high-yielding properties, that are utilized in cancer therapy. Nanotechnology can rightly be referred to as the "Science of today", owing to its tremendous applicability in every field with far better results than the conventional solutions, and consequently, the area of cancer detection and treatment has seen vast development in the landscape of nanoscience. Cancer nanotechnology provides a distinguished approach where it provides a holistic solution and covers areas of prevention, timely diagnosis, effective therapy, and even precision medicine.

2.1 Nanoparticles: What Are They?

Discovered almost 50 years before by scientist Richard Feynman, who is "The Father of Nanotechnology", nanoparticles are reported to be sub-microscopic, ultrafine units that have the ability to be measured in dimensions of nanometres (nm; 1 nm = 10^{-9} m). Its definition was initially proposed in 2008 by the union, International Organization for Standardization (ISO) as an individual nano-object having all the 3 Cartesian dimensions to not be more than 100 nm. Following this, in the year 2011, the Commission of the European Union gave its approval to a more-technical and all-encompassing definition: *a natural, incidental, or manufactured material containing particles, in an unbound state or as an aggregate or as an agglomerate and where for 50% or more of the particles in the number size distribution, one or more external dimensions are in the size range 1* nm–*100* nm [5]. Based on this expression, a nanoscale object is required to have a minimum of at least one dimension in the range 1–100 nm so that it can be classified as a nanostructure, even if its remaining two dimensions are not incorporated in the range specified above.

Nanoparticles have been segregated into numerous types based on a multitude of reasons like- size, shape, properties of a material, source, etc. A few classifications bifurcate nanoparticles into nanoparticles of organic and inorganic origin, the former includes metallic oxide nanoparticles, metallic nanoparticles (gold, silver, silicon, zinc, copper, etc.), quantum dots, etc., and the latter accounts for nanoparticles made of polymers like carbon and carbon-based, fullerenes, ceramics, liposomes, micelles, quantum dots, chitosan, dendrimers, etc. Nanoparticles are able to depict several distinctive effects and unique properties in different compositions due to their three major attributes:

1. Possess a high level of mobility when present in the free state.
2. Own exceptionally massive specific areas.
3. Showcase unique quantum effects in a variety of compositions and conditions.

The catalytic, mechanical, and thermal properties of nanoparticles have the tendency to become altered either by increasing or by decreasing the surface area/volume ratio, and the new as well as individual properties generated by such changes allow applications of nanoformulations in a variety of biological fields like cancer therapy, drug administration, wound healing, diagnostic tests, biosensors, etc. The nanoparticles of chemically manufactured structures like polysaccharides, metals, and other polymers incorporated with active drugs and anti-cancer pharmaceuticals like Abraxane, Doxil, DaunoXome, Onyvide, etc., pose a highly efficient, safe, and targeted treatment for cancer [6, 7].

3 Advantages of Nanoformulations in Cancer Therapy

Nanotechnology is an extremely rapidly emerging field of science with groundbreaking applications in various areas, especially in medicine. Its most prominent application is found to be the treatment of cancer mainly because of the following characteristics of nanoformulations,

(i) Extremely minute size enables them to cross physiological barriers easily and reach the target size properly,

(ii) Enhanced stability inside anatomic sites of action which in turn increases the time of action of therapeutic agents adsorbed on the surface of these nanoformulations,

(iii) Safer as compared to other alternatives as they can reduce the toxicity of chemotherapeutic drugs when combined with active pharmaceutical ingredients (APIs)

(iv) Enables targeted mode of action and precision in working due to its bioavailability, etc. [8],

(v) Interaction of nanoparticles does not cause any harm or necrosis to the physiological constituents of protein and cellulose,

(vi) The presence of targeting ligands on the surface prevents the scattering of the therapeutic agent in other premises of the body and payloads it on the required site,

(vii) Biodegradable composition of nanoformulations can be achieved so that complete removal can be achieved after its action is accomplished

These advantages of nanoformulations can be summarized in the following Fig. 2.

4 What Are Nanocomposites?

A nanocomposite is a multiphase solid substance with one, two, or three proportions of less than 100 nm in one of the phases, or even structures with nanoscale distances between the distinct phases that make up the material, nanocomposites are utilized as building blocks to construct materials with exceptional flexibility and enhancement in physical properties. In a broad sense, this term can refer to porous media, colloids, gels, and copolymers, although it is most usually associated with the solid combination of a bulk matrix and a nano-dimensional phase (s) with properties that differ due to structural and chemical similarities [9].

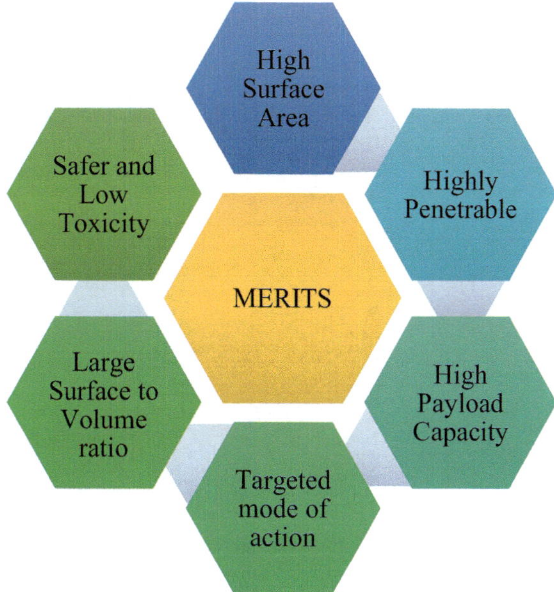

Fig. 2 Major advantages of nanoformulations when used in cancer therapy

Abalone shells and bone structures are examples of nanocomposites found in natural environments [10]. The application of nanoparticle-rich materials predates our understanding of their physicochemical properties. Jose-Yacaman et al. looked into the origins of Maya blue paint's colour depth and resilience to acids and bio-corrosion, putting it down to a nanoparticle mechanism [11]. Nanoscale organo-clays have been employed to control the flow of polymer solutions (e.g. as paint viscosi-fiers) and the compositions of gels since the mid-1950s (e.g. as a thickening substance in cosmetics, keeping the preparations in homogeneous form). Ceramic matrix nanocomposites, metal matrix nanocomposites, polymer matrix nanocomposites, and magnetic matrix nanocomposites are examples of nanocomposites (Fig. 3).

Because of the extremely high-surface-to-volume ratio of the reinforcing phase, nanocomposites differ from traditional composite materials in terms of mechan-ical structure. Particles (e.g. minerals), sheets (e.g. exfoliated clay stacks), or fibres can be used as reinforcing materials (e.g. carbon nanotubes or electrospun fibres). In comparison with conventional composite materials, the area of the interphase between the matrix and reinforcement phase(s) is often an order of magnitude larger. In the area of reinforcement, the matrix material's characteristics are considerably influenced [12]. Due to the vast amount of surface area available for reinforcement, even a little amount of nanoscale reinforcement can have a significant impact on the macroscale properties of the composite. Carbon nanotubes, for example, improve both electrical and thermal conductivity. Other types of nanoparticulate can improve optical properties, dielectric properties, heat resistance, or mechanical properties like stiffness, strength, and wear and damage resistance. During processing, the rein-forcement is generally dispersed into matrices. Because of the low-filler percolation

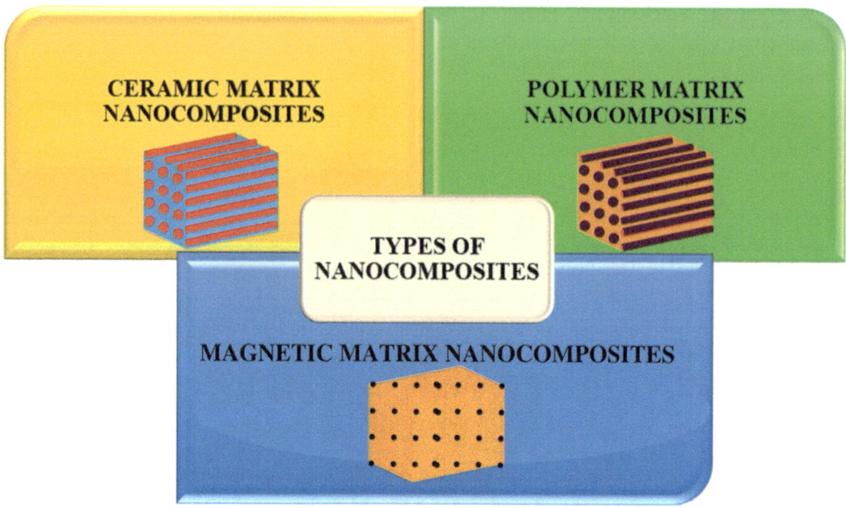

Fig. 3 Types of nanocomposites and their basic structure

threshold, the mass fraction of nanoparticles introduced can remain very low (about 0.5–5%), especially for the most commonly used non-spherical, high-aspect ratio filler (e.g. nanometre-thin platelets, like clays, or nanometre-diameter cylinders, like carbon nanotubes). The effectiveness of thermal conductivity of nanocomposites is highly influenced by the orientation and arrangement of asymmetrical nanoparticles, thermal characteristics that do not match at the interface, interface density per unit volume at the nanocomposite, and nanoparticle polydispersity.

5 Chitosan: A Versatile Super Polymer in Treating Cancer

Chitosan, being a linear polysaccharide, is polycationic in nature (having a positive charge at multiple sites) with a composition of β-(1-4)-linked-D-glucosamine in association with N-acetyl-D-glucosamine, and additionally is a derivative of "chitin" [13]. This biopolymer is extremely beneficial in the biomedical domain owing to its remarkable properties which arise as a result of its amino groups, conferring it anti-cancerous attributes, namely efficient cell permeability, biocompatibility, anti-angiogenesis, immune enhancement, anti-oxidation, and even apoptotic tendencies [14]. In general, chitosan exhibits a myriad of unique characteristics that make them ideal in numerous aspects like hydrophilicity, tunable molecular weight, antimicrobial nature, wound healing, anti-virulence, analgesic, haemostatic, and mucoadhesive behaviour that render them superior to other similar compounds [15–18]. Chitosan can induce apoptosis in malignant outgrowths by activating specific pathways of caspase-3 and caspase-8 which results in cell cycle dysfunction and inhibition of

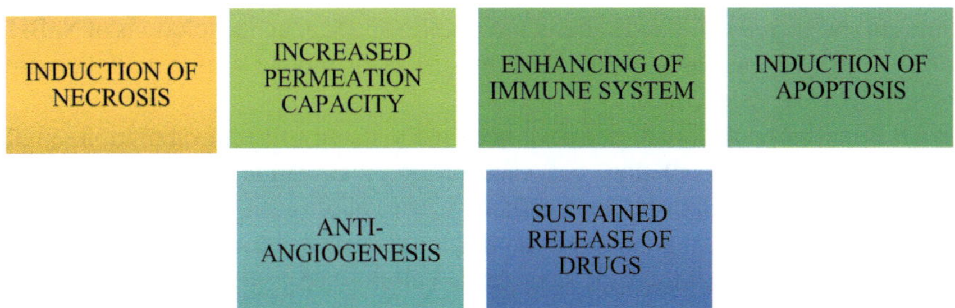

Fig. 4 Advantages of biopolymer chitosan in cancer theranostics

transcription along with a translation, including disruption of hormonal cascades [19]. This multifunctional biopolymer also has reported improvement in immunity by arresting the dividing cells in G_1/S phases that in turn downregulate the expression of caspase-3, resulting in apoptosis (controlled cell death). As mentioned previously, the amine group of chitosan plays a highly essential role in exhibiting anti-tumour activity as it expresses cytotoxicity amongst varying malignancies by interacting directly, for example, owing to its low-molecular weight and high degrees of electrostatic interactions, chitosan adheres to the membranes of cancer cells to aid in their endocytosis and therefore, stands to be a remarkable delivery vehicle for anti-cancer drug formulations [20]. The advantageous traits of chitosan, with respect to cancer therapy, are showcased in Fig. 4.

The functional groups present on chitosan, which are the amine and hydroxyl groups have the ability to become functionalized and with a variety of chemicals and drugs that lead to chemical alterations resulting in increased solubility of chitosan-based carriers and even enhanced drug loading capacity. Its unique properties allow improved binding capacity towards mucosal membranes, resulting in successful transmucosal delivery, intrapulmonary, and intranasal delivery of therapeutic drugs against metastasis, and sometimes even acting as an immunoadjuvant for immunotherapy against cancer [21].

6 Synthesis and Attributes of Chitosan-Based Bionanocomposites

Because of the poor solubility of chitosan, a solution of 2 mg/ml in 1% acetic acid was prepared first. The mixture was gyrated to completely dissolve and kept overnight before being filtered through 0.22 m Millipore syringe filters to remove any impurities. Metal–chitosan nanocomposites are easy to make; in general, metal nanoparticles are obtained via chemical reduction of metal salts to yield zero-valent metal nanoparticles with $NaBH_4$. The concentration of $NaBH_4$ has to be 10 times higher than that of the metal salt to enable complete reduction. A 50 l, 20 mM $HAuCl_4$, $AgNO_3$, H_2PtCl_6, or Na_2PdCl_4 aqueous solution were mixed with 3 ml, 2.0 mg/ml chitosan and stirred for 30 min, after which freshly made aqueous solutions of $NaBH_4$ (50 1, 0.2 M) were added to the mixture and stirred for another 90 min until the entire metal salts were reduced [22].

Chitosan-based nanoparticles have been made using solvent evaporation, emulsion, diffusion, ionic gelation, coacervation or precipitation, spray dyeing, self-assembly, and cross-linking. Chitosan/nanometric cellulose composites combine the properties of chitosan (e.g. biodegradability, antibacterial and antimicrobial activity, transparency, etc.) with those of nanometric cellulose (e.g. high-surface area, very good barrier as mechanical properties) to produce composite materials that can be used in the packaging industry (e.g. food film, paper coatings), the chemical industry (e.g. catalysts, adsorbents) (e.g. carrier of active substances, filaments) [23, 24].

7 Biomedical Relevance of Chitosan-Based Nanoformulations

Pharmaceutical formulation and drug delivery systems such as antibiotics, anti-inflammatory drugs, vaccines, proteins, and peptides are amongst the possible biomedical applications of chitosan. Antimicrobial applications, gene delivery, gene therapy, wound healing and burns, regenerative medicine, tissue engineering on bone, ligament, cartilage, tendon, liver, neural and skin regeneration, cancer applications (treatment, therapy, diagnostic strategy), dermatology, ophthalmology, dentistry, biosensors, and many other applications including bioimaging (for example, magnetic resonance imaging), supporting immobilized enzymes, and veterinary medicine.

7.1 Antimicrobial Outcomes

Despite brilliant progress in the development of antimicrobial mediums, many infectious diseases are still difficult to treat, due to a lot of reasons such as the rise and spread of resistant clones, the lack of the antimicrobial bodies, and inadequate pharmaceutical properties of existent antimicrobial bodies, that sometimes are difficult to reach active concentrations inside bacterial strains or in some body areas. Chitosan is a promising biocompatible and biodegradable biopolymer that shows potential antimicrobial activity [25]. Unlike chitosan, the nanocomposites of chitosan have been found to exhibit a broad spectrum of antimicrobial activity against various pathogens (both gram-positive and gram-negative bacteria) [15].

Lavorgna et al. [16] created silver montmorillonite antibacterial nanocomposites with a chitosan matrix by substituting the natural montmorillonite's (MMT) Na+ ions with silver ions. They were able to improve mechanical performance, but more importantly, they demonstrated that after 24 h, pseudomonas aeruginosa acquired a considerable delay [26]. Lim et al. also looked into how rGO and CHI nanocomposites affected pseudomonas aeruginosa development. The results demonstrated that bacterial growth was not reliant on rGO concentration or size, but that a modest concentration of rGO in the chitosan solution may totally suppress bacterial growth, resulting in a 100% viability loss [27].

7.2 Anti-Cancer Drugs

Despite incredible advances in medical research, cancer remains one of the leading causes of mortality worldwide. The foundation of successful cancer management is accurate diagnosis and tailored treatment strategies. Chitosan and its derivatives are ideal for cancer diagnosis because their chemical features make them simple to produce into gels, sponges, membranes, beads, and scaffolds.

The chitosan oligosaccharide-arachidonic acid conjugate has been successfully produced and used to develop self-assembled nanoparticles for the administration of doxorubicin [28]. Grafting the targeting function via the thiolation procedure, for example, RGD (arginine-glycine-aspartic acid) improves selective intratumoural transport of siRNA packaged in PGD-chitosan nanoparticles (RGD-CH-NPS) and measures antitumoural activity [29].

A biodegradable polymer-drug combination of doxorubicin conjugated with stearic-acid-grafted chitosan nanosized oligosaccharide recently demonstrated excellent effectiveness for cellular absorption and tumour growth.

7.3 Drug Delivery Frameworks

Because of non-specific cell targeting and tissue biodispersion, as well as their quick metabolism and excretion, standard medications are becoming increasingly limited, resulting in a high need for high-performance solutions [30, 31]. Nanocarriers have emerged as one of the most promising drug delivery methods due to their ability to interface with the cell membranes and enter via endocytosis, departing for the endosomal compartment and releasing medication in cytosolic compartments [32]. Biodegradability and biocompatibility, which are often offered by natural polymers such as CHI, are the most important prerequisites for creating these systems.

It is less expensive to apply novel distribution methods than it is to develop new medications. Chitosan nanoparticles were found to be well adjusted towards nanocarriers used as an antimalarial drug delivery system [33], as a promising drug delivery system to improve antiviral drugs for the treatment of HIV infection [34], transformed by folic acid towards targeted drug delivery [35], developed responsive hybrid nanogels by poly(methacrylic acid) for pH-responsive drug release, and magnetic chitosan nanoparticles used as multifunctional nanocarriers were loaded with bleomycin to function as a nanocarrier proved to be effective towards targeting system [36].

7.4 Tissue Engineering Applications

Tissue engineering is the study of how to use structural and functional relationships in normal and infected tissues to generate biological substitutes that restore and improve biofunction. Different forms of bone grafts, such as autografts, allografts, and synthetic bone grafts, were employed for fracture repair in bone tissue engineering [37]. Autografting is the gold standard for bone repair, but it has several drawbacks, including limited availability, donor site morbidity, and danger of disease transmission from donor to recipient [38]. Allografts have some drawbacks, such as a lack of osteoinduction, the risk of disease transmission, and poor mechanical qualities. As a result, the development of synthetic bone grafts that overcome these disadvantages represents a huge demand for bone regeneration, especially through biomimetic devices with osteoconductive properties.

Biocompatible polysaccharides, such as chitosan-based materials, increase cell adhesion, proliferation, and differentiation and have been utilized extensively in orthopaedic tissue engineering [39]. Bone tissue engineering materials [40], cartilage regeneration [41], and liver and nerve tissue engineering [42] have all been described using chitosan-based hybrid nanocomposites.

7.5 Wound Healing Applications

Wound healing is a complex series of biologically regulated processes linked to tissue growth and regeneration. It goes through a variety of stages in which numerous cellular and matrix components work together to repair the damage and restore the tissues that have been lost [43]. Chitosan-based nanomaterials have been employed in a variety of wound healing applications, including composite scaffolds, chitosan-based sponges, immobilized scaffolds, and drug-loaded scaffolds [44].

Sulfated chitosan has the ability to disrupt the coagulation process. There are numerous medical applications for this. This chitosan derivative has been proven to have high-anticoagulant action when compared to heparin. Sulfated chitosan, unlike heparin, is not known to have antiplatelet activity, which can lead to excessive bleeding in individuals. Pure CHI-based hydrogels with high toughness may be obtained using, for example, double-network strategies; however, using nanocomposites based on CHI matrices is much more general. Considering these features, Lu et al. reported the use of CHI-PVA/graphene nanofibers, produced by electrospinning, for wound healing applications [45]. The potential of these membranes was tested on mouse and rabbit skin wounds and found that after five days, the wound area significantly decreased, and at the end of 10 days, the skin was completely recovered, whereas, for membranes lacking graphene, these wound areas still existed. Aguzzi and co-workers explored the use of CHI/MMT nanocomposites combined with silver sulfadiazine for the same purpose. They demonstrated a successful loading of the silver sulfadiazine in the nanocomposite structure, as XRD tests have shown no free drug in the composite matrix, which revealed that the intercalated nanocomposite was made by insertion of drug and/or polymer molecules, having a homogeneous dispersion in the nanocomposite structures [46].

7.6 Gene Therapy and Bioimaging Applications

The transfer of nucleic acids by the cell is at the heart of gene therapy, which offers the potential to heal a wide range of currently incurable diseases. Chitosan possesses a number of characteristics that make it an ideal gene delivery system. Chitosan formulation is simple, as the positive amine group reacts with negatively charged phosphate groups on the DNA, resulting in increased stability and improved gene transfer capabilities [47]. Because of its biocompatibility, bioimaging applications of chitosan nanoparticles are also gaining traction. Imaging agents, such as Fe_3O_4 for MRI, were incorporated into self-assembled nanoparticles to target tumour imaging [48]. Nanoparticles containing imaging agents were investigated for radio-pharmacological and MRI applications [49, 50].

7.7 Biosensing

Biosensing is the detection of target molecules using principles similar to those utilized by a living system such as the immune system. The main factors to consider when performing biosensing are detection specificity and sensitivity [51]. As shown by Singh and co-workers, CHI/GO nanocomposites have shown the ability to detect DNA for rapid and sensitive detection of typhoid, using a *Salmonella typhi* specific 5′-amine labelled single strand (ss) DNA (5′NH2 -ssDNA), covalently bound through CHI/GO by glutaraldehyde. The developed bioelectrode was able to distinguish between complementary and non-complementary sequences, which could be owing in part to the basic properties listed above, but to CHI compatibility, which enhances the DNA immobility and facilitates electron transfer between DNA and electrode surface [52].

Song et al. developed a glucose biosensor with cytochrome c and glucose oxidase entrapped on Au NPs and CHI over a glassy carbon electrode, demonstrating that the deposition of CHI/Au NPs increased its roughness to 9.5 0.1 nm, which was found to be critical for achieving a high-surface-to-volume ratio. Furthermore, higher glucose sensitivity and a lower detection limit were found [53].

Zhang et al. announced the development of a haemoglobin/Au NPs/CHI/graphene biosensor for hydrogen peroxide detection based on a glassy carbon electrode. The electron transport parameters of the biosensor were investigated using electrochemical impedance spectroscopy. The data showed that using haemoglobin, Au NPs, and graphene improves the electron transfer, reducing the transfer resistance supplied by CHI. In addition, a low-detection limit (0.35 μM), good stability (94%) for over one month, and a high sensitivity (347.1 mA/cm^2 M) were found for these biosensors [54].

8   Chitosan-Based Nanocomposites: An Upcoming Domain in Cancer Diagnosis and Therapy

Bionanocomposites are distinctive, hybrid nanoformulations that possess inherent properties like biodegradability, biocompatibility, non-toxicity, etc., accompanied by special structural and functional qualities, owing to their unique composition of inorganic solids and natural polymers, including the cumulative impact of green/natural nanofillers that leads to increased sustainability, making them a green choice. Amongst these, chitosan-based nanocomposites have garnered significant attention in the arena of biomedical applications because of enhanced chemical, biological, and physical characteristics in a cost-efficient and eco-friendly manner. As discussed previously, chitosan has promising functional groups that possess the ability to interact with a variety of compounds, namely gold, silver, zinc oxide, copper, zinc oxide, etc., and therefore are attainable as films, hydrogels, beads, membranes, pervious frameworks, fibrous meshes, and even powdered form. Such variable chitosan nanocomposites hold a great value in the domain of cancer treatment and diagnosis owing to their intensified surface area-to-volume ratio, bioavailability, solubility, active, and specific targeting, reduced systemic damages, a high index of therapy, and even a greater circulation time within the biological setting [55]. Consequently, chitosan-based bionanocomposites are employed in oncotherapy in multiple formats like drug delivery systems, induction of apoptosis as well as necrosis, tumour imaging, biosensing of malignant tissues, photodynamic therapy, etc.

Some of the required abilities of efficient nanocarriers in promising drug delivery systems are susceptibility towards the light, pH, enzymes, temperature, and even magnetism that are portrayed by chitosan bionanocomposites after conjugation with a drug-induced pH-sensitive linker, allowing malignancy disruption. Resultantly, such a difference between the pH of malignant and typical cells allows the nanocarriers to become even more profitable as compared to other conventional drug delivery systems [21]. In the subsequent sections, the applications of chitosan-based bionanocomposites are described in various models for cancer theranostics.

1.   **Chitosan-based Gold Nanocomposites**

These bionanocomposites can be developed for utilization in photothermal therapy which was created by Zhang et al. [56]. After amplified laser ablation efficiency, such nanocomposites displayed a higher degree of accumulation on the surface of cancerous growths, when compared to normal healthy cells.

Additionally, these nanocomposites possessed an increased tendency of selectivity towards malignancies than the normal cells, during photothermal ablation, which, in this case, involved cell lines of human dermal fibroblast cells/HDF as well as hepatocellular carcinoma/HepG2. After adherence to the cancer cells, these components generate a significant amount of heat that is able to destroy cancerous cells but not healthy ones; therefore, this study proved chitosan-based gold nanocomposites to have advantages like faster ablation with low-power near-infrared radiation, accompanied with low dosage.

Apart from this, scientists recently developed gold-chitosan nanoparticle-based films for rapidly and accurately detecting prostate cancer with the help of a special prostate-specific antigen, known as a "bio-marker". This advancement of an electrochemical immunosensor that resembles a sandwich model reportedly showcases increased stability, promising biocompatibility, more current response tendency along with higher electrocatalytic activity [57].

Another example is the development of a localized combinatorial therapy for delivering anti-cancer drugs with the help of injectable chitosan-hydrogel incorporated gold nanocomposites or porous silica nanocomposites, which exhibited the advantage of a sustained release capacity for doxorubicin hydrochloride under acidic environment, near the malignant area [58].

2. **Chitosan-based Copper Nanocomposites**

The ability to respond to biomarkers is a promising ability with respect to precisely diagnosing cancer, and this can be done in nanocomposites via functionalization with special molecules that are sensitive to these biomarkers, like copper. For instance, the diagnosis of prostate cancer is tedious as it requires the detection of sarcosine biomarkers which is expensive, complex, and time-taking. To combat this, chitosan-copper nanoparticles conjugated with carbon nanotubes were probed and characterized to become immobilized by the help of sarcosine oxidase and their analysis suggested that such biosensors portray a superior analytical performance with the ability to examine them in real-time [59]. A similar bioinspired fabrication was applied to chitosan-copper oxide bionanocomposites that were reported with bioflavonoid rutin. In this case, the *in vitro* examinations propounded the concentration-dependent activity regarding anti-proliferation along with apoptotic induction in the human cancerous cell line of A549 cells [60].

3. **Chitosan-based Silver Nanocomposites**

A vital example of this category is chitosan conjugated with silver and phyco-erythrin nanocomposites, abbreviated as "CS-Ag-PE-NCs" which have given a promising result of significant apoptotic activity induction in breast cancer cells of triple-negative nature. Owing to their internalization by cancerous cells due to an amplification in the reactive oxygen species (ROS), subsequent activation of caspase cascades was observed, resulting in the subsequent release of apoptotic factors that caused mitochondrial-moderated apoptosis in malignancies [61].

Another instance of chitosan stabilized by Ag nanoparticles was studied to give vital evidence in inducing apoptosis in cancer cells of HT-29 cell line, along with increased production of ROS intracellularly, which stimulated the required caspase signalling cascades needed for apoptotic induction [62].

Comparably, their usage against A549 lung carcinoma cell line and *Salmonella* sp. also shows similar action of ROS production to induce apoptosis with additional cytotoxic activity against A549 cells as a result of green synthesis along with inhibition of associated pathogen, *Mycobacterium tuberculosis* [63]. After morphogenic analysis of A549 carcinomic cells, following treatment with this nanocomposite indicated a concentration-dependent downfall in the capacity of adherence in tumour cells when compared with normal cells, accompanying a shape of round cells in carcinoma tissue in contrast with a normal polygonal occurrence in WI 38 cells.

4. Chitosan-based Iron Oxide Nanocomposites

The anti-cancerous qualities of iron oxide chitosan bionanocomposites were originally examined *in vitro* against HepG2 cell line, as a result of enhanced cytotoxicity in tumour cells because of iron oxide incorporation [64]. Their incorporation also reported the activity of chitosan as a chelating agent towards metal ions causing a rise in the occurrence of hydroxyl and amino groups that were speculated to be involved in anti-cancerous outcomes due to free radical scavenging. Additionally, these iron oxide-based chitosan nanocomposites have given evidence of their activity against malignancies as well as multi-drug resistance in pathogens [65]. Another indication of apoptosis in nanocomposite treated A549 carcinoma cell line was observed to be fragmentation in nuclear material accompanied by nuclear condensation.

5. Chitosan-based Graphene Oxide Nanocomposites

The nanocomposites functionalized with graphene oxide have reported a good yield of delivery of the anti-cancerous drug, doxorubicin hydrochloride, due to their internalization inside the MCF-7 cells causing the successive release of doxorubicin to kill tumorigenesis [66]. Moreover, the conjugation of iron oxide with chitosan nanocomposites also renders them more soluble in the biological environment and also reduces the binding of proteins of non-specific nature in the physiological setting.

The modelling of 5-fluorouracil-loaded chitosan-graphene oxide bionanocomposites along with curcumin was employed in treating HT-29 colon cancer cell lines [67]. Besides this, polyelectrolytic chitosan nanocomposites conjugated with graphene oxide were developed recently for achieving targeted drug delivery of doxorubicin and its proper loading in the polymeric matrix along with controlled release was examined in HeLa cells [68].

6. Chitosan-based Zinc Oxide Nanocomposites

A potent anti-cancer and antimicrobial activity has been studied in polypyrrole-grafted ZnO/chitosan nanocomposites that were synthesized *in situ*. The created structure was observed to have the ability to induce apoptosis in the cell line of cervical cancer, studied in HeLa cells, and even breast cancer which was studied in MCF-7 cells [69].

7. Chitosan-based Magnetic Nanocomposites

The synthesis of folic acid-based magnetic nanocomposites conjugated with chitosan has proven to show targeted tumour imaging [70]. Selective toxicity against tumorigenesis was exhibited by these drug-loaded nanocomposites which were coupled with controlled drug release and extremely sensitive selection of cancerous cells in acidic pH, making them suitable for breast cancer treatment. Besides this, magnetic resonance imaging studies done *in vivo* and *in vitro* displayed an enhanced negative signal in the malignancy, confirming the MRI contrast performance.

8. Chitosan-based Carbon Nanocomposites

Multiwalled carbon nanotubes coated with chitosan/silver and enveloped with 5-fluorouracil were formulated and analyzed to exhibit cytotoxicity against the cancerous cell line of MCF-7 with reported sustained release of anti-cancerous drug formulations against the same cell line [71].

9. Other Chitosan-based Nanocomposites

Photothermal ablation of tumorigenesis via photoacoustic imaging was seen in polypyrrole-based chitosan bionanocomposites with the results showing complete healing of mice tumour with dual treatment of nanocomposites and NIR 808 nm laser [72]. Moreover, the viability of breast carcinoma cells, namely MDA-MB-231, was lessened due to the photothermal effect of these nanocomposites. Apart from this, *in vivo* photodynamic therapy and targeted drug release done by involving near-infrared radiation achieved by graphene oxide nanocomposites have also been reported [73].

Colon cancer treatment has been reported in celecoxib-loaded chitosan nanocomposites based on hydroxyapatite, with their *in vitro* analysis showing a time-dependent take-up of nanocomposite in HT-29 and HCT-15 cancerous cell line succeeded by apoptosis and anti-proliferation, observed during *in vivo* studies in mice [74].

Some major applications of chitosan-based bionanocomposites are listed in Table 1.

Table 1 Oncotherapeutic bionanocomposites based on chitosan with their specific applications

S.No	Base of nanocomposite	Type of additive	Biomedical effect	Cell line studied	Form of malignancy	Reference
1	Chitosan	Silver (Ag)	Tumour cytotoxicity	A549	Lung carcinoma	[75]
2	Chitosan	Gold (Au)	Photoablation	HepG2	Multiple cancers	[56]
3	Chitosan	Copper (Cu)	Induction of apoptosis	A549	Lung cancer	[60]
4	Chitosan	Polypyrrole/ ZnO	Apoptosis	HeLa	Cervical carcinoma	[69]
5	Chitosan	ZnO/polypyrrole	Apoptosis	MCF-7	Breast carcinoma	[69]
6	Chitosan	Tamoxifen	Apoptosis	MCF-7	Breast cancer	[76]
7	Chitosan	Celecoxib-loaded hydroxyapatite	Apoptosis and anti-proliferation	HCT-15, HT-29	Colon carcinoma	[74]
8	Chitosan	Herceptin-conjugated with gemcitabine	Apoptosis	Mia PaCa-2 and PANC-1	Pancreatic cancer	[77]
9	Chitosan	Polyelectrolyte/graphene oxide	Cytotoxicity	HeLa	Colon cancer	[68]
10	Chitosan	Gold (Au)	Electro-chemical immune sensor	–	Prostate cancer	[57]

9 Concluding Remarks and Future Perspective

Cancer is the second-largest contributor to the global disease burden and novel studies, and research on domains like nanocomposites and chitosan is the need of the hour so that this enfeebling malady can be cured of its core. The unique and outstanding properties of chitosan, namely biocompatibility, chemical receptivity, and anti-inflammatory and non-toxic nature, are especially beneficial in exhibiting anti-malignant behaviour. Chitosan-based nanocomposites derived from combining organic/inorganic nanoformulations with chitosan biopolymers have exhibited their enormous potential in numerous fields ranging from food packaging, material science, and water purification to mainstream biomedicine, especially oncotherapy.

The anti-tumour efficacy shown by chitosan-based nanocomposites shows precise diagnostic methods, preventive measures, treatment strategies, and personalized medicine through efficient drug delivery systems, biosensors, gene delivery systems, phototherapy agents, and other theragnostic instruments that have proven their remarkable potential in cancer diagnosis and treatment. The most promising outcome of chitosan bionanocomposites is their ability to provide target-specific drug deliveries that abstain from negatively influencing the usual healthy tissues around the target site.

However, there are still many challenges faced by developed bionanocomposites that have clinical applications in cancer with respect to toxicity-related issues, the specific underlying mechanism of treatment, and other complications. Thus, there is an urgent need to critically examine and study these revolutionary structures in the field of oncotherapy along with toxicological surveys, molecular medicine, clinical trials, and antibody generation so as to transform the clinical landscape in the future of cancer therapeutics as well as diagnostics.

References

1. Gulati S, Kumar S, Singh P, Diwan A, Mongia A (2020) Biocompatible Chitosan-coated gold nanoparticles: novel, efficient, and promising nanosystems for cancer treatment. In: Handbook of polymer and ceramic nanotechnology. Springer International Publishing, pp 1–29
2. Arora B, Bhatia R, Attri P (2018) In: Bionanocomposites: green materials for a sustainable future. Elsevier Inc.
3. Kamel S (2007) Nanotechnology and its applications in lignocellulosic composites, a mini review. Express Polym Lett 1(9):546–575. https://doi.org/10.3144/expresspolymlett.2007.78
4. Gulati S et al (2022) Starch based bio-nanocomposites: modern and benign materials in food packaging industry. In: Handbook of consumer nanoproducts, pp 881–909. https://doi.org/10.1007/978-981-16-8698-6_96
5. Wang EC, Wang AZ (2014) Nanoparticles and their applications in cell and molecular biology. Integr Biol (United Kingdom) 6(1):9–26. https://doi.org/10.1039/c3ib40165k
6. Kumar S, Mongia A, Gulati S, Singh P, Diwan A, Shukla S (2020) Emerging theranostic gold nanostructures to combat cancer: novel probes for combinatorial immunotherapy and photothermal therapy. Cancer Treat Res Commun 25:100258. https://doi.org/10.1016/J.CTARC.2020.100258

7. van der Meel R, Lammers T, Hennink WE (2017) Cancer nanomedicines: oversold or under-appreciated? Expert Opinion Drug Delivery 14(1):1–5, 20 Jan 2017. Taylor and Francis Ltd. https://doi.org/10.1080/17425247.2017.1262346

8. Grodzinski P, Kircher M, Goldberg M, Gabizon A (2019) Integrating nanotechnology into cancer care. ACS Nano 13(7):7370–7376. https://doi.org/10.1021/acsnano.9b04266

9. Chikwendu Okpala C The benefits and applications of nanocomposites. Int J Adv Eng Technol

10. Yuan J, Zhang X, Qian H (2010) A novel approach to fabrication of superparamagnetite hollow silica/magnetic composite spheres. J Magn Magn Mater 15(322):2172–2176. https://doi.org/10.1016/J.JMMM.2010.02.004

11. José-Yacamán M, Rendón L, Arenas J, Serra Puche MC (1996) Maya blue paint: an ancient nanostructured material. Science (80) 273(5272):223–225. https://doi.org/10.1126/SCIENCE.273.5272.223

12. Tian Z, Hu H, Sun Y (2013) A molecular dynamics study of effective thermal conductivity in nanocomposites. Int J Heat Mass Transf 61(1):577–582. https://doi.org/10.1016/J.IJHEATMASSTRANSFER.2013.02.023

13. Cheung RCF, Ng TB, Wong JH, Chan WY (2015) Chitosan: an update on potential biomedical and pharmaceutical applications. Mar Drugs 13(8):5156. https://doi.org/10.3390/MD13085156

14. Adhikari HS, Yadav PN (2018) Anticancer activity of chitosan, chitosan derivatives, and their mechanism of action. Int J Biomater 2018. https://doi.org/10.1155/2018/2952085

15. Benhabiles MS, Salah R, Lounici H, Drouiche N, Goosen MFA, Mameri N (2012) Antibacterial activity of chitin, chitosan and its oligomers prepared from shrimp shell waste. Food Hydrocoll 29(1):48–56. https://doi.org/10.1016/J.FOODHYD.2012.02.013

16. Sandri G, Rossi S, Bonferoni MC, Ferrari F, Mori M, Caramella C (2012) The role of chitosan as a mucoadhesive agent in mucosal drug delivery. J Drug Deliv Sci Technol 22(4):275–284. https://doi.org/10.1016/S1773-2247(12)50046-8

17. Pogorielov MV, Sikora VZ (2015) Chitosan as a hemostatic agent: current state. Eur J Med Ser B 2(1):24–33. https://doi.org/10.13187/EJM.S.B.2015.2.24

18. Hu Z et al (2018) Investigation of the effects of molecular parameters on the hemostatic properties of chitosan. Molecules 23(12). https://doi.org/10.3390/MOLECULES23123147

19. Hasegawa M, Yagi K, Iwakawa S, Hirai M (2001) Chitosan induces apoptosis via caspase-3 activation in bladder tumor cells. Jpn J Cancer Res 92(4):459–466. https://doi.org/10.1111/J.1349-7006.2001.TB01116.X

20. Wang W et al (2020) Chitosan derivatives and their application in biomedicine. Int J Mol Sci 21(2). https://doi.org/10.3390/IJMS21020487

21. Babu A, Ramesh R (2017) Multifaceted applications of chitosan in cancer drug delivery and therapy. Mar Drugs 15(4). https://doi.org/10.3390/MD15040096

22. Huang H, Yuan Q, Yang X (2004) Preparation and characterization of metal–chitosan nanocomposites. Colloids Surf B Biointerfaces 39(1–2):31–37. https://doi.org/10.1016/J.COLSURFB.2004.08.014

23. Thakur VK, Thakur MK (2014) Recent advances in graft copolymerization and applications of chitosan: a review. ACS Sustain Chem Eng 2(12):2637–2652. https://doi.org/10.1021/SC500634P/ASSET/IMAGES/MEDIUM/SC-2014-00634P_0011.GIF

24. Fernandes SCM, Freire CSR, Silvestre AJD, Pascoal Neto C, Gandini A (2011) Novel materials based on chitosan and cellulose. Polym Int 60(6):875–882. https://doi.org/10.1002/PI.3024

25. Zheng LY, Zhu JF (2003) Study on antimicrobial activity of chitosan with different molecular weights. Carbohydr Polym 4(54):527–530. https://doi.org/10.1016/J.CARBPOL.2003.07.009

26. Lavorgna M et al (2014) MMT-supported Ag nanoparticles for chitosan nanocomposites: structural properties and antibacterial activity. Carbohydr Polym 102(1):385–392. https://doi.org/10.1016/J.CARBPOL.2013.11.026

27. Lim HN, Huang NM, Loo CH (2012) Facile preparation of graphene-based chitosan films: enhanced thermal, mechanical and antibacterial properties. J Non Cryst Solids 3(358):525–530. https://doi.org/10.1016/J.JNONCRYSOL.2011.11.007

28. Termsarasab U et al (2013) Chitosan oligosaccharide–arachidic acid-based nanoparticles for anti-cancer drug delivery. Int J Pharm 441(1–2):373–380. https://doi.org/10.1016/J.IJPHARM.2012.11.018

29. Han HD et al (2010) Targeted gene silencing using RGD-labeled chitosan nanoparticles. Clin Cancer Res 16(15):3910–3922. https://doi.org/10.1158/1078-0432.CCR-10-0005
30. Mura S, Nicolas J, Couvreur P (2013) Stimuli-responsive nanocarriers for drug delivery. Nat Mater 2013 1211 12(11):991–1003. https://doi.org/10.1038/nmat3776
31. Lima AC, Sher P, Mano JF (2012) Production methodologies of polymeric and hydrogel particles for drug delivery applications. 9(2):231–248. https://doi.org/10.1517/17425247.2012.652614
32. Kavaz A et al (2010) Bleomycin loaded magnetic chitosan nanoparticles as multifunctional nanocarriers. https://doi.org/10.1177/0883911509360735
33. Tripathy S, Das S, Chakraborty SP, Sahu SK, Pramanik P, Roy S (2012) Synthesis, characterization of chitosan-tripolyphosphate conjugated chloroquine nanoparticle and its in vivo anti-malarial efficacy against rodent parasite: a dose and duration dependent approach. Int J Pharm 434(1–2):292–305. https://doi.org/10.1016/J.IJPHARM.2012.05.064
34. Wu D et al (2016) Zinc-stabilized chitosan-chondroitin sulfate nanocomplexes for HIV-1 infection inhibition application. Mol Pharm 13(9):3279–3291. https://doi.org/10.1021/ACS.MOL PHARMACEUT.6B00568/SUPPL_FILE/MP6B00568_SI_001.PDF
35. Yuan Q, Hein S, Misra RDK (2010) New generation of chitosan-encapsulated ZnO quantum dots loaded with drug: synthesis, characterization and in vitro drug delivery response. Acta Biomater 6(7):2732–2739. https://doi.org/10.1016/J.ACTBIO.2010.01.025
36. Wu W, Shen J, Banerjee P, Zhou S (2010) Chitosan-based responsive hybrid nanogels for integration of optical pH-sensing, tumor cell imaging and controlled drug delivery. Biomaterials 31(32):8371–8381. https://doi.org/10.1016/J.BIOMATERIALS.2010.07.061
37. James R et al (2011) Nanocomposites and bone regeneration. Front Mater Sci 5(4):342–357. https://doi.org/10.1007/S11706-011-0151-3
38. Duan B, Wang M, Zhou WY, Cheung WL, Li ZY, Lu WW (2010) Three-dimensional nanocomposite scaffolds fabricated via selective laser sintering for bone tissue engineering. Acta Biomater 6(12):4495–4505. https://doi.org/10.1016/j.actbio.2010.06.024
39. Di Martino A, Sittinger M, Risbud MV (2005) Chitosan: a versatile biopolymer for orthopaedic tissue-engineering. Biomaterials 26(30):5983–5990. https://doi.org/10.1016/J.BIOMATERIALS.2005.03.016
40. Bhowmick A et al (2016) Multifunctional zirconium oxide doped chitosan based hybrid nanocomposites as bone tissue engineering materials. Carbohydr Polym 151:879–888. https://doi.org/10.1016/J.CARBPOL.2016.06.034
41. In KS et al (2009) Chitosan nano-/microfibrous double-layered membrane with rolled-up three-dimensional structures for chondrocyte cultivation. J Biomed Mater Res A 90(2):595–602. https://doi.org/10.1002/JBM.A.32109
42. Mottaghitalab F, Farokhi M, Mottaghitalab V, Ziabari M, Divsalar A, Shokrgozar MA (2011) Enhancement of neural cell lines proliferation using nano-structured chitosan/poly(vinyl alcohol) scaffolds conjugated with nerve growth factor. Carbohydr Polym 86(2):526–535. https://doi.org/10.1016/J.CARBPOL.2011.04.066
43. Boateng JS, Matthews KH, Stevens HNE, Eccleston GM (2008) Wound healing dressings and drug delivery systems: a review. J Pharm Sci 97(8):2892–2923. https://doi.org/10.1002/JPS.21210
44. Ali A, Ahmed S (2018) A review on chitosan and its nanocomposites in drug delivery. Int J Biol Macromol 109:273–286. https://doi.org/10.1016/J.IJBIOMAC.2017.12.078
45. Lu B et al (2012) Graphene-based composite materials beneficial to wound healing. Nanoscale 4(9):2978–2982. https://doi.org/10.1039/C2NR11958G
46. Aguzzi C et al (2014) Solid state characterisation of silver sulfadiazine loaded on montmorillonite/chitosan nanocomposite for wound healing. Colloids Surf B Biointerfaces 113:152–157. https://doi.org/10.1016/J.COLSURFB.2013.08.043
47. Malmo J, Vårum KM, Strand SP (2011) Effect of chitosan chain architecture on gene delivery: comparison of self-branched and linear chitosans. Biomacromol 12(3):721–729. https://doi.org/10.1021/BM1013525

48. Lee CM et al (2011) Oleyl-chitosan nanoparticles based on a dual probe for optical/MR imaging in vivo. Bioconjug Chem 22(2):186–192. https://doi.org/10.1021/BC100241A
49. Kumar MNVR, Muzzarelli RAA, Muzzarelli C, Sashiwa H, Domb AJ (2004) Chitosan chemistry and pharmaceutical perspectives. Chem Rev 104(12):6017–6084. https://doi.org/10.1021/CR030441B/ASSET/CR030441B.FP.PNG_V03
50. Dash M, Chiellini F, Ottenbrite RM, Chiellini E (2011) Chitosan—a versatile semi-synthetic polymer in biomedical applications. Prog Polym Sci 36:981–1014. https://doi.org/10.1016/j.progpolymsci.2011.02.001
51. Yasuga H, Shoji K, Koiwai K, Kawano R (2021) New sensing technologies: Microtas/NEMS/MEMS. Ref Modul Biomed Sci. https://doi.org/10.1016/B978-0-12-822548-6.00046-7
52. Singh A et al (2013) Graphene oxide-chitosan nanocomposite based electrochemical DNA biosensor for detection of typhoid. Sens Actuators, B Chem 185:675–684. https://doi.org/10.1016/J.SNB.2013.05.014
53. Qin FX, Jia SY, Wang FF, Wu SH, Song J, Liu Y (2013) Hemin@metal-organic framework with peroxidase-like activity and its application to glucose detection. Catal Sci Technol 3(10):2761–2768. https://doi.org/10.1039/c3cy00268c
54. Zhang L, Han G, Liu Y, Tang J, Tang W (2014) Immobilizing haemoglobin on gold/graphene–chitosan nanocomposite as efficient hydrogen peroxide biosensor. Sens Actuators B Chem (197):164–171. https://doi.org/10.1016/J.SNB.2014.02.077
55. Din FU et al (2017) Effective use of nanocarriers as drug delivery systems for the treatment of selected tumors. Int J Nanomed 12:7291. https://doi.org/10.2147/IJN.S146315
56. Zhang G, Sun X, Jasinski J, Patel D, Gobin AM (2012) Gold/chitosan nanocomposites with specific near infrared absorption for photothermal therapy applications. J Nanomater 2012. https://doi.org/10.1155/2012/853416
57. Suresh L, Brahman PK, Reddy KR, Bondili JS (2018) Development of an electrochemical immunosensor based on gold nanoparticles incorporated chitosan biopolymer nanocomposite film for the detection of prostate cancer using PSA as biomarker. Enzyme Microb Technol 112:43–51. https://doi.org/10.1016/J.ENZMICTEC.2017.10.009
58. Xia B, Zhang W, Tong H, Li J, Chen Z, Shi J (2019) Multifunctional Chitosan/Porous Silicon@Au nanocomposite hydrogels for long-term and repeatedly localized combinatorial therapy of cancer via a single injection. ACS Biomater Sci Eng 5(4):1857–1867. https://doi.org/10.1021/ACSBIOMATERIALS.8B01533/SUPPL_FILE/AB8B01533_SI_001.PDF
59. Narwal V, Kumar P, Joon P, Pundir CS (2018) Fabrication of an amperometric sarcosine biosensor based on sarcosine oxidase/chitosan/CuNPs/c-MWCNT/Au electrode for detection of prostate cancer. Enzyme Microb Technol 113:44–51. https://doi.org/10.1016/J.ENZMICTEC.2018.02.010
60. Bharathi D, Ranjithkumar R, Chandarshekar B, Bhuvaneshwari V (2019) Bio-inspired synthesis of chitosan/copper oxide nanocomposite using rutin and their anti-proliferative activity in human lung cancer cells. Int J Biol Macromol 141:476–483. https://doi.org/10.1016/J.IJBIOMAC.2019.08.235
61. Thangam R et al (2015) Theranostic potentials of multifunctional chitosan-silver-phycoerythrin nanocomposites against triple negative breast cancer cells. RSC Adv 5(16):12209–12223. https://doi.org/10.1039/C4RA14043E
62. Sanpui P, Chattopadhyay A, Ghosh SS (2011) Induction of apoptosis in cancer cells at low silver nanoparticle concentrations using chitosan nanocarrier. ACS Appl Mater Interfaces 3(2):218–228. https://doi.org/10.1021/AM100840C/SUPPL_FILE/AM100840C_SI_001.PDF
63. Arjunan N, Kumari HLJ, Singaravelu CM, Kandasamy R, Kandasamy J (2016) Physicochemical investigations of biogenic chitosan-silver nanocomposite as antimicrobial and anticancer agent. Int J Biol Macromol 92:77–87. https://doi.org/10.1016/J.IJBIOMAC.2016.07.003
64. Dawy Badry M, Ahmed Wahba M, Khaled R, Moawad Ali M, Ali Farghali A (2017) Synthesis, characterization, and in-vitro anticancer evaluation of iron oxide/chitosan nanocomposites. Inorganic Nano-Metal Chem 47(3):405–411. https://doi.org/10.1080/15533174.2016.1186064

65. Rabel AM, Namasivayam SKR, Prasanna M, Bharani RSA (2019) A green chemistry to produce iron oxide—chitosan nanocomposite (CS-IONC) for the upgraded bio-restorative and pharmacotherapeutic activities—supra molecular nanoformulation against drug-resistant pathogens and malignant growth. Int J Biol Macromol 138:1109–1129. https://doi.org/10.1016/J.IJBIOMAC.2019.07.158
66. Lei H, Xie M, Zhao Y, Zhang F, Xu Y, Xie J (2016) Chitosan/sodium alginate modified graphene oxide-based nanocomposite as a carrier for drug delivery. Ceram Int 42(15):17798–17805. https://doi.org/10.1016/J.CERAMINT.2016.08.108
67. Dhanavel S et al (2020) 5-Fluorouracil and curcumin co-encapsulated chitosan/reduced graphene oxide nanocomposites against human colon cancer cell lines. Polym Bull 77(1):213–233. https://doi.org/10.1007/S00289-019-02734-X
68. Anirudhan TS, Chithra Sekhar V, Athira VS (2020) Graphene oxide based functionalized chitosan polyelectrolyte nanocomposite for targeted and pH responsive drug delivery. Int J Biol Macromol 150:468–479. https://doi.org/10.1016/J.IJBIOMAC.2020.02.053
69. Ahmad N, Sultana S, Faisal SM, Ahmed A, Sabir S, Khan MZ (2019) Zinc oxide-decorated polypyrrole/chitosan bionanocomposites with enhanced photocatalytic, antibacterial and anticancer performance. RSC Adv 9(70):41135–41150. https://doi.org/10.1039/C9RA06493A
70. Nejadshafiee V et al (2019) Magnetic bio-metal–organic framework nanocomposites decorated with folic acid conjugated chitosan as a promising biocompatible targeted theranostic system for cancer treatment. Mater Sci Eng C 99:805–815. https://doi.org/10.1016/J.MSEC.2019.02.017
71. Nivethaa EAK, Dhanavel S, Rebekah A, Narayanan V, Stephen A (2016) A comparative study of 5-Fluorouracil release from chitosan/silver and chitosan/silver/MWCNT nanocomposites and their cytotoxicity towards MCF-7. Mater Sci Eng C Mater Biol Appl 66:244–250. https://doi.org/10.1016/J.MSEC.2016.04.080
72. Manivasagan P et al (2017) Multifunctional biocompatible chitosan-polypyrrole nanocomposites as novel agents for photoacoustic imaging-guided photothermal ablation of cancer. Sci Rep 7. https://doi.org/10.1038/SREP43593
73. Sharma H, Mondal S (2020) Functionalized graphene oxide for chemotherapeutic drug delivery and cancer treatment: a promising material in nanomedicine. Int J Mol Sci 21(17):1–42. https://doi.org/10.3390/IJMS21176280
74. Venkatesan P et al (2011) The potential of celecoxib-loaded hydroxyapatite-chitosan nanocomposite for the treatment of colon cancer. Biomaterials 32(15):3794–3806. https://doi.org/10.1016/J.BIOMATERIALS.2011.01.027
75. Abdel-Aziz MM, Elella MHA, Mohamed RR (2020) Green synthesis of quaternized chitosan/silver nanocomposites for targeting mycobacterium tuberculosis and lung carcinoma cells (A-549). Int J Biol Macromol 142:244–253. https://doi.org/10.1016/J.IJBIOMAC.2019.09.096
76. Vivek R, Nipun Babu V, Thangam R, Subramanian KS, Kannan S (2013) PH-responsive drug delivery of chitosan nanoparticles as Tamoxifen carriers for effective anti-tumor activity in breast cancer cells. Colloids Surf B Biointerfaces 111:117–123. https://doi.org/10.1016/J.COLSURFB.2013.05.018
77. Arya G, Vandana M, Acharya S, Sahoo SK (2011) Enhanced antiproliferative activity of Herceptin (HER2)-conjugated gemcitabine-loaded chitosan nanoparticle in pancreatic cancer therapy. Nanomed Nanotechnol Biol Med 7(6):859–870. https://doi.org/10.1016/J.NANO.2011.03.009

Important Websites

78. https://www.azom.com/article.aspx?ArticleID=17329
79. https://www.mddionline.com/news/enhancing-medical-device-performance-nanocomposite-polymers

80. https://royalsocietypublishing.org/doi/10.1098/rsta.2017.0040
81. https://worldwidescience.org/topicpages/b/bio-nanocomposites+biopolymer+matrix.html
82. https://www.britannica.com/technology/nanotechnology
83. https://www.polymerexpert.biz/industries/173-composites
84. https://www.drugs.com/npp/chitosan.html

Chapter 8
An Overview of the Application of Chitosan-Based Nanocomposites in Bioimaging

Ishita Chakraborty, Sharmila Sajankila Nadumane, Rajib Biswas, and Nirmal Mazumder

Abstract Bioimaging methods are used to visualize biological samples and processes in real-time non-invasively. Chitosan is conjugated with other materials to form chitosan composites and is used in bioimaging due to its outstanding biodegradability and non-toxic nature. Chitosan-based nanocomposites are applied for cancer and tumour diagnosis through various imaging techniques based on fluorescence, ultrasound, photoacoustic, magnetic resonance imaging etc. The chapter discusses the applications, strengths, and limitations of chitosan-based nanocomposites in bioimaging techniques for diagnosis of various types of cancer.

Keywords Chitosan nanocomposites · Bioimaging · Optical imaging · Ultrasound imaging · Magnetic resonance imaging · Computer tomography

Abbreviations

BMP-2	Bone morphogenetic protein 2
CA	5-Cholanic acid
CNP	Chitosan nanoparticles
Echo-CNPs	Echogenic CNPs
GC-CA	Glycol chitosan-5-cholanic acid conjugates
MRI	Magnetic resonance imaging
PAI	Photoacoustic imaging

I. Chakraborty · S. S. Nadumane · N. Mazumder (✉)
Department of Biophysics, Manipal School of Life Sciences, Manipal Academy of Higher Education, Manipal, Karnataka 576104, India
e-mail: nirmaluva@gmail.com

R. Biswas (✉)
Applied Optics and Photonics Lab, Department of Physics, Tezpur University, Napaam, Tezpur, Assam 784028, India
e-mail: rajivb27@gmail.com

S. Gulati (ed.), *Chitosan-Based Nanocomposite Materials*,
https://doi.org/10.1007/978-981-19-5338-5_8

PTT	Photothermal therapy
QNC	Quercetin nanocomposites
ROI	Region of interest

1 Introduction

Chitosan is comprised of β-(1 → 4)-linked d-glucosamine (deacetylated unit) and N-acetyl-d-glucosamine (acetylated unit) after being deacetylated from chitin [5]. Chitosan possesses chelation properties and can selectively bind to metal ions and biomolecules like proteins, cholesterol, and tumour cells [14]. Due to this property, it is extensively used in various fields of biomedical research. It works as a tumour inhibitor, antimicrobial agent, wound-healing agent, and immunostimulant. Nanobiocomposites are hybrid systems composed of a biopolymer strengthened by introducing nanostructures [6]. Recently, chitosan-based nanocomposites have surfaced, which enhance specific properties such as mechanical, physical, and thermal stability of chitosan. Chitosan nanocomposites are usually formed between chitosan and metal, carbon, and other polymers via physical or chemical interactions [19] and are being majorly used in drug delivery and bioimaging. Several techniques exist for in vitro and in vivo imaging biological specimens [17] including computer tomography and magnetic resonance imaging. However, all imaging techniques have their pros and cons. The imaging technique should be selected based on the properties of the biological sample, and sensitivity resolution, complexity, and data acquisition of the imaging technique. The present chapter discusses the implementation of chitosan nanocomposites in the field of bioimaging.

2 Chitosan Nanoparticles in Bioimaging

2.1 Optical Imaging

Biological optical imaging works on forming an image by controlling the excitation and emitting signals originating from the interaction of light and the sample. However, recent challenges in cancer therapy and diagnosis include the lack of imaging agents with specific targets to tumour sites. Recently, 5-cholanic acid (CA) was used to chemically modify chitosan and synthesize amphiphilic glycol chitosan-5-cholanic acid conjugates (GC-CA) with Cy5.5 dye for optical imaging. It was revealed that the chitosan nanocomposites were easily taken up, localized, and distributed in the target cell upon administration into the mice tumour. The developed nanocomposite systems proved to be potential contrast agents for optical and non-optical imaging methods [16].

Further, chitosan nanoparticles (CNPs) were synthesized using glycol and exhibited the ability to specifically target tumours. CNPs labelled with Cy5.5 were used for in vivo imaging. It was revealed that CNPs had an excellent affinity for brain and liver tumours, and metastasis tumour models [15]. Glutaraldehyde was used as a crosslinking agent for magnetic nanoparticles and quantum dots conjugated to the chitosan matrix. The developed nanocomposites exhibited both fluorescent and magnetic properties along with in vitro good stability. Therefore, they have the potential to function as optical and magnetic resonance imaging agents [3]. Atherosclerotic lesions were imaged using probes synthesized by combining atherosclerotic peptide and modified glycol chitosan. The synthesized nanocomposites were again labelled with Cy5.5. The designed nanocomposites were utilized to illuminate pathophysiological changes in the atherosclerotic endothelium. Similarly, to image cells, a fluorescent and magnetic nanoparticle was coated with modified chitosan which was labelled with a fluorescent dye. Fluorescence and electron microscopy revealed that the magnetic particles were lodged in cell (SMMC-7721) interior and exterior [18].

2.2 Ultrasound Imaging

Nanosized contrast agents are instrumental to enhance the identification, sensitivity, and quantification through ultrasound imaging. It is based on linear imaging, where an ultrasonic pulse passing through the tissues produces nonlinear propagation and nonlinear vibration to give rise to a nonlinear acoustic signal [23]. Echogenic CNPs (Echo-CNPs) were synthesized and targeted in non-invasive ultrasound imaging for cancer diagnosis in vivo. Echo-CNPs exhibited a strong ultrasound signal that persisted for 120 min, with an echogenicity half-life of 49 min. Tumour-targeted ultrasound imaging was studied by intravenously injecting the Echo-CNPs into SCC 7 tumour-bearing mice. Post injecting, a bright echo signal was observed in ultrasound imaging, suggesting an accumulation of echo-CNPs in the tumour [13].

2.3 Magnetic Resonance Imaging (MRI)

MRI utilizes a magnetic field and radio waves to create comprehensive images of biological samples. Contrast agents are essential for MRI and paramagnetic materials offer boosted contrast by modifying the magnetic environment. These paramagnetic contrasting agents belong to the lanthanide and transition metal series. Recently, gold-coated Fe_3O_4 nanoparticles stabilized by chitosan were synthesized and used as MRI contrast agents. In several cases, superparamagnetic iron oxide nanocrystals have also been broadly utilized in MRI [1]. Further, nanoparticles cated with chitosan derivatives were also used as contrast agents for in vivo imaging of BALB/c mouse model through fluorescent imaging and MRI. The nanoparticles were taken up by the livers, which enhanced the MRI contrast [8].

The results indicate these chitosan nanocomposites may be potentially used for imaging various organs. It was reported that synthesized (carboxymethyl) chitosan (CMCS)-coated nanocomposites were used to visualize human mesenchymal stem cells (hMSCs) using MRI. The carboxymethylation reaction increases the water solubility of chitosan resulting in better dispersion of the nanoconjugates in aqueous media. The cells labelled with CMCS-coated nanocomposites exhibited high contrast in an aqueous agarose medium [20]. Further, iron oxide nanoparticles stabilized by a chitosan coating were synthesized by the co-precipitation method and characterized by MRI (Fig. 1). The relaxation times of the nanoparticles were measured by MRI and it revealed that the relaxation rate (R) and relaxivity (r) were: $= 0.713 \ \mathrm{mM}^{-1} \ \mathrm{s}^{-1}$, $= 238.16 \ \mathrm{mM}^{-1} \ \mathrm{s}^{-1}$, and $= 276.1 \ \mathrm{mM}^{-1} \ \mathrm{s}^{-1}$. An obtained high r2/r1 ratio (334) implied that nanocomposites exhibited a substantial prevailing effect on the transversal relaxation time in contrast to the longitudinal relaxation time [9]. Quercetin is a hydrophobic bioflavonoid that exhibits anti-cancer properties. Recently, in a study, quercetin nanocomposites (QNC) were synthesized from iron-gold nanoparticles, coated with oleyl chitosan, conjugated with folic acid and then loaded with quercetin. The synthesized nanocomposites were used as dual contrast agents for computer tomography (CT) and MRI. Fluorescence images of QNC-treated cancer cell lines are displayed in Fig. 2, indicating the efficacy of QNC as contrast agents for biomedical imaging techniques such as MRI and CT, respectively, [4].

2.4 Computed Tomographic Imaging

Computed tomography is a 3-dimensional (3D) non-destructive imaging technique that uses X-rays to scan a biological sample slice by slice. The X-rays are passed through the sample are used to record a 2-D projection image. Several 2-dimensional images are stacked up to create a final 3D-image [11]. Chitosan-coated tungsten trioxide nanoparticles were synthesized to be used as an efficient contrasting agent for X-ray tomography. Chitosan coating to tungsten trioxide nanoparticles reduced the toxicity and exhibited no significant reduction in cell survival.

In a study conducted, -N,6-O-sulfated chitosan/calcium phosphate nanocomposite hydrogels were loaded with bone morphogenetic protein 2 (BMP-2) intended for bone defect regeneration in rabbits with improved vascularisation. Figure 3a (i) shows micro-computed tomography images of new bone (NB) and was used to examine the blood vessel formation [2]. Jiang et al., synthesized nano-hydroxyapatite integrated into chitosan and cellulose intended for bone regeneration study (Fig. 3b). In vivo experiments with New Zealand white rabbits demonstrated enhanced infiltration of chitosan based nanocomposites into bone tissues. The optical images of hematoxylin and eosin-stained histological tissue sections verified that the nanocomposite is suitable for imaging bone regeneration [7].

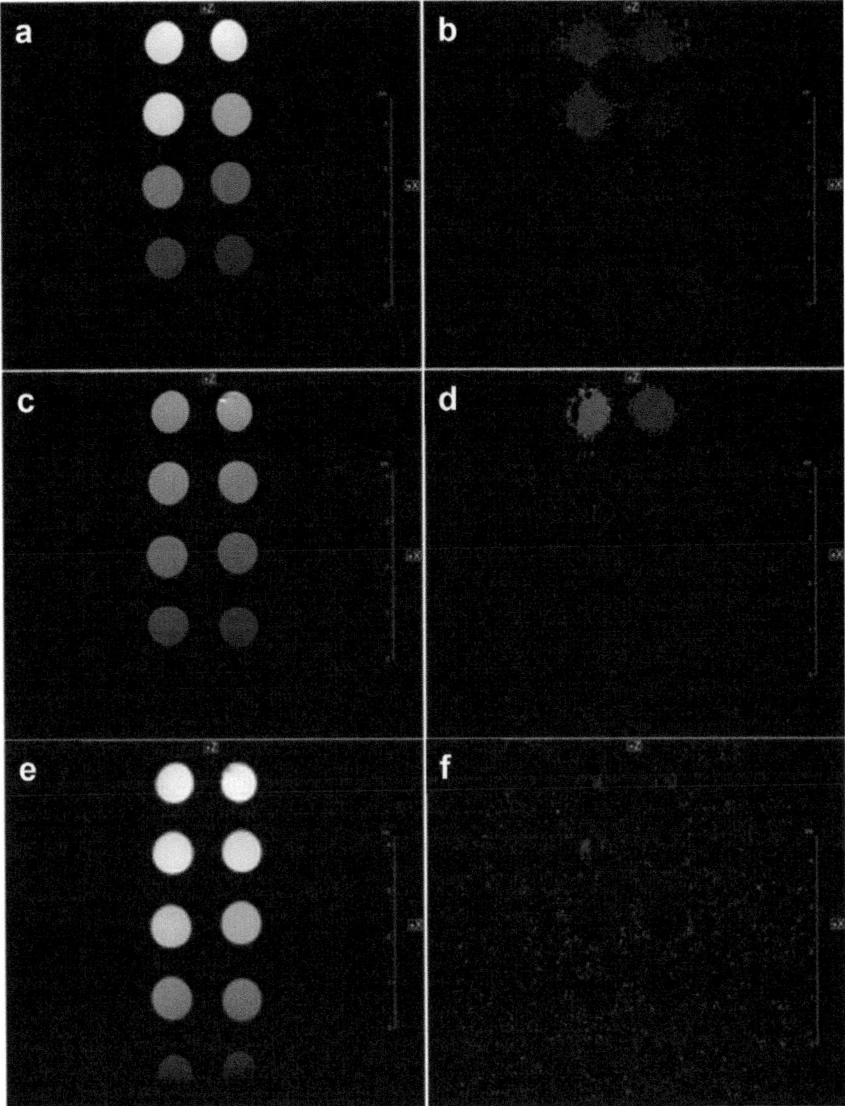

Fig. 1 Longitudinal (**a, b**) and Transversal (**c–f**) relaxation time mapping of chitosan-stabilized magnetite nanoparticles with T1, T2 and T2*—mapping MGE pulse sequence. **a** Signal intensity, **b** Relaxation time T1 mapm, **c** Signal intensity, and **d** Relaxation time T2 map acquired with T2-mapping MSME pulse sequence, **e** Signal intensity, and **f** Relaxation time T2 *map acquired with T2 * -mapping MSME pulse sequence. [9] Reproduced from Khmara et al. with permission from Elsevier

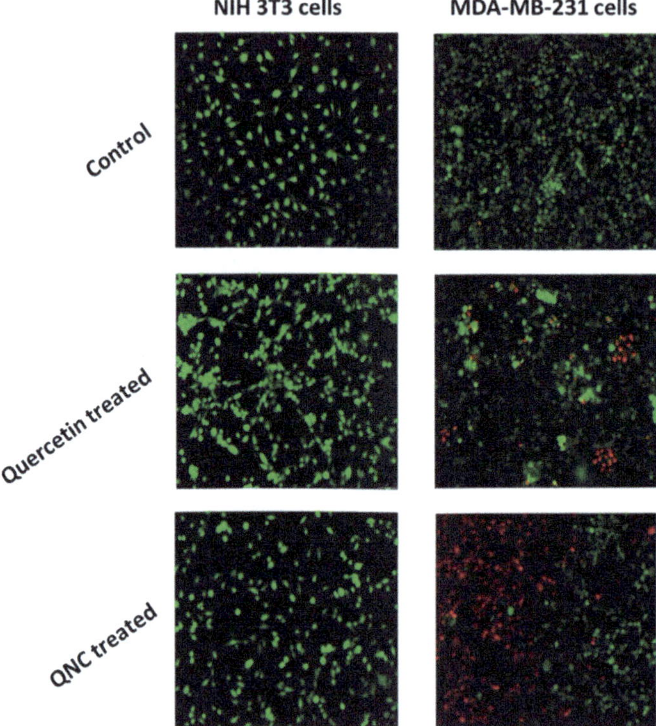

Fig. 2 Fluorescence imaging of cancer cell lines NIH 3T3, and MDA-MB-231 using calcein. Reproduced from [22] with permission from Elsevier

2.5   *Photoacoustic Imaging (PAI)*

PAI is also a non-invasive biomedical imaging technique, where ultrasonic waves are used to irradiate the biological sample with a pulsed laser and the image is reconstructed based on light energy absorption and distribution in the sample [21]. Photothermal therapy (PTT) and PAI were combined for the treatment of breast cancer. This technology is preferred since it is invasive, does not damage non-targeted areas, and at the same time can be used for breast cancer imaging [10]. A biocompatible chitosan-polypyrrole nanocomposite was developed for PAI-guided photothermal ablation of cancer owing to its various properties including strong near-infrared (NIR) absorbance. The nanocomposites exhibited great biocompatibility and can be utilized in PTT of cancer cells under 808-nm NIR laser irradiation [12].

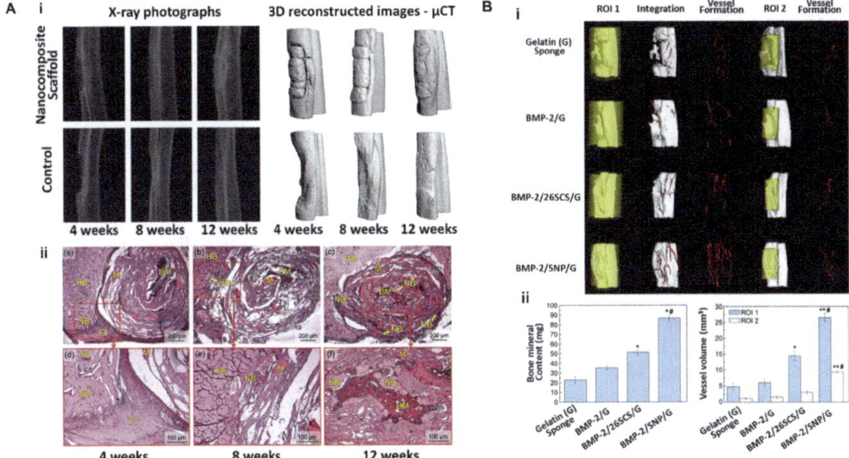

Fig. 3 **A** (i) X-ray radiographic and corresponding 3-D reconstructed micro-computed tomographic (μCT) images of spiral-cylindrical chitosan/cellulose-based scaffolds placed in a concave defect in rabbits post implantation. (ii) H & E stained histopathological images of nanocomposite scaffolds implanted in rabbits exhibiting bone marrow formation. HB: host bone, BM: bone marrow, NB: new bone, FT: fibrotic tissue, and M: scaffold. Reproduced from [7] with permission from American Chemical Society. **B** (i) 3-D μCT images of two different regions of interest (ROI) in rabbits implanted with various 2-N, 6-O-sulfated chitosan-based nanocomposite sponges. (ii) Quantification and comparison of bone mineral content and blood vessel volume. Reproduced from [2] with permission from Elsevier

3   Conclusion

Nanocomposites of chitosan are unique materials proving their as a potent contrast agent in several advanced bioimaging techniques including magnetic resonance imaging, ultrasound imaging, computer tomography, and photoacoustic imaging. The amino and hydroxyl groups present in chitosan provide a way for reaction with organic functional molecules. Use of chitosan based nanocomposites are non toxic and improve the stability of the nanocomposites themselves.

Acknowledgements Nirmal Mazumder thanks the Science and Engineering Research Board, Department of Science and Technology (DST), Government of India, India (project number: SERB/MTR/2020/000058) for financial support. Rajib Biswas thanks the Biotechnology Industry Research Assistance Council (BIRAC), Department of Biotechnology, Government of India, India (Proposal number: BIRAC/KIIT01165/BIG-17/20) for the financial support. Ms. Ishita Chakraborty thanks the Department of Science and Technology (DST), Govt. of Karnataka, India for Ph.D. fellowship (Award no: DST/KSTePS/Ph.D. Fellowship/LIF-12:2021-22/1024). Ms. Sharmila Sajankila Nadumane thanks Manipal Academy of Higher Education (MAHE), Manipal, Karnataka, India for the Dr. T.M.A. Pai Ph.D. fellowship.

References

1. Ahmad M, Manzoor K, Ikram S (2018) Chapter 2: chitosan based nanocomposites for drug, gene delivery, and bioimaging applications. In: Inamuddin, Asiri AM, Mohammad A (eds) Applications of nanocomposite materials in drug delivery. Woodhead Publishing
2. Cao L, Wang J, Hou J, Xing W, Liu C (2014) Vascularization and bone regeneration in a critical sized defect using 2-N, 6-O-sulfated chitosan nanoparticles incorporating BMP-2. Biomaterials 35(2):684–698
3. Ding Y, Yin H, Shen S, Sun K, Liu F (2017) Chitosan-based magnetic/fluorescent nanocomposites for cell labelling and controlled drug release. New J Chem 41(4):1736–1743
4. Hemalatha T, Prabu P, Gunadharini DN, Gowthaman MK (2018) Fabrication and characterization of dual acting oleyl chitosan functionalised iron oxide/gold hybrid nanoparticles for MRI and CT imaging. Int J Biol Macromol 112:250–257
5. Islam S, Bhuiyan MA, Islam MN (2017) Chitin and chitosan: structure, properties and applications in biomedical engineering. J Polym Environ 25(3):854–866
6. Jhaveri J, Raichura Z, Khan T, Momin M, Omri A (2021) Chitosan nanoparticles-insight into properties, functionalization and applications in drug delivery and theranostics. Molecules 26(2):272
7. Jiang H, Zuo Y, Zou Q, Wang H, Du J, Li Y, Yang X (2013) Biomimetic spiral-cylindrical scaffold based on hybrid chitosan/cellulose/nano-hydroxyapatite membrane for bone regeneration. ACS Appl Mater Interfaces 5(22):12036–12044
8. Kania G, Sternak M, Jasztal A, Chlopicki S, Błażejczyk A, Nasulewicz-Goldeman A, Wietrzyk J, Jasiński K, Skórka T, Zapotoczny S, Nowakowska M (2018) Uptake and bioreactivity of charged chitosan-coated superparamagnetic nanoparticles as promising contrast agents for magnetic resonance imaging. Nanomed: Nanotechnol Biol Med 14(1):131–140
9. Khmara I, Strbak O, Zavisova V, Koneracka M, Kubovcikova M, Antal I, Kopcansky P (2019) Chitosan-stabilized iron oxide nanoparticles for magnetic resonance imaging. J Magn Magn Mater 474:319–325
10. Liu Y, Bhattarai P, Dai Z, Chen X (2019) Photothermal therapy and photoacoustic imaging via nanotheranostics in fighting cancer. Chem Soc Rev 48(7):2053–2108
11. Magwaza LS, Opara UL (2014) Investigating non-destructive quantification and characterization of pomegranate fruit internal structure using X-ray computed tomography. Postharvest Biol Technol 95:1–6
12. Manivasagan P, Quang Bui N, Bharathiraja S, Santha Moorthy M, Oh YO, Song K, Oh J (2017) Multifunctional biocompatible chitosan-polypyrrole nanocomposites as novel agents for photoacoustic imaging-guided photothermal ablation of cancer. Sci Rep 7(1):1–14
13. Min HS, You DG, Son S, Jeon S, Park JH, Lee S, Kim K (2015) Echogenic glycol chitosan nanoparticles for ultrasound-triggered cancer theranostics. Theranostics 5(12):1402
14. Mohammad F, Al-Lohedan HA, Al-Haque HN (2017) Chitosan-mediated fabrication of metal nanocomposites for enhanced biomedical applications. Adv Mater Lett 8(2):89–100
15. Na JH, Koo H, Lee S, Min KH, Park K, Yoo H, Lee SH, Park JH, Kwon IC, Jeong SY, Kim K (2011) Real-time and non-invasive optical imaging of tumor-targeting glycol chitosan nanoparticles in various tumor models. Biomaterials 32(22):5252–5261
16. Nam T, Park S, Lee SY, Park K, Choi K, Song IC, Han MH, Leary JJ, Yuk SA, Kwon IC, Kim K (2010) Tumor targeting chitosan nanoparticles for dual-modality optical/MR cancer imaging. Bioconjug Chem 21(4):578–582
17. Nune SK, Gunda P, Thallapally PK, Lin YY, Laird Forrest M, Berkland CJ (2009) Nanoparticles for biomedical imaging. Expert Opin Drug Deliv 6(11):1175–1194
18. Park K, Hong HY, Moon HJ, Lee BH, Kim IS, Kwon IC, Rhee K (2008) A new atherosclerotic lesion probe based on hydrophobically modified chitosan nanoparticles functionalized by the atherosclerotic plaque targeted peptides. J Control Release 128(3):217–223
19. Salehi E, Khajavian M, Sahebjamee N, Mahmoudi M, Drioli E, Matsuura T (2022) Advances in nanocomposite and nanostructured chitosan membrane adsorbents for environmental remediation: a review. Desalination 527:115565

20. Shi Z, Neoh KG, Kang ET, Shuter B, Wang SC, Poh C, Wang W (2009) (Carboxymethyl) chitosan-modified superparamagnetic iron oxide nanoparticles for magnetic resonance imaging of stem cells. ACS Appl Mater Interfaces 1(2):328–335
21. Thomas A, Paul S, Mitra J, Singh MS (2021) Enhancement of photoacoustic signal strength with continuous wave optical pre-illumination: a non-invasive technique. Sensors 21(4):1190
22. Wang K, Zhuang J, Liu Y, Xu M, Zhuang J, Chen Z, Wei Y, Zhang Y (2018) PEGylated chitosan nanoparticles with embedded bismuth sulfide for dual-wavelength fluorescent imaging and photothermal therapy. Carbohyd Polym 184:445–452
23. Zeng F, Du M, Chen Z (2021) Nanosized contrast agents in ultrasound molecular imaging. Front Bioeng Biotechnol 9:758084

Chapter 9
Chitosan-Based Nanocomposites as Remarkably Effectual Wound Healing Agents

Sneha Vijayan, Shikha Gulati, Tanu Sahu, Meenakshi, and Sanjay Kumar

Abstract In recent years, chitosan (CS), the second-most abundant biopolymer on earth, obtained by deacetylation of chitin, has garnered significant attention for its immense potential in accelerating the wound healing process. Some of its outstanding virtues such as biocompatibility, easy degradation, ability to promote collagen deposition, low cost, and non-toxic nature, combined with its antimicrobial and anti-inflammatory properties when coupled with nanomaterials like magnetic nanoparticles, render chitosan-based nanocomposites as effectual candidates in the process of wound healing. In this chapter, we briefly describe the processing techniques of chitosan-based nanocomposites, and their application as wound healing agents has been expounded. Further, significant emphasis has been given to the properties of these nanocomposites favoring the wound repair process as well as the mechanism and effects on wounds.

Keywords Chitosan · Chitosan-based nanocomposites · Wound healing · Wound repair

Abbreviations

CAV	Chitosan Aloe vera
CS	Chitosan
DNA	Deoxyribonucleic acid
ECM	Extracellular matrix
EDTA	Ethylenediaminetetraacetic Acid
IFN-γ	Interferon gamma
IL-1	Interleukin 1
IL-10	Interleukin 10
IL-13	Interleukin 13

S. Vijayan · S. Gulati (✉) · T. Sahu · Meenakshi · S. Kumar
Department of Chemistry, Sri Venkateswara College, University of Delhi, Delhi 110021, India
e-mail: shikha2gulati@gmail.com

S. Gulati (ed.), *Chitosan-Based Nanocomposite Materials*,
https://doi.org/10.1007/978-981-19-5338-5_9

199

IL-4	Interleukin 4
IL-6	Interleukin 6
L-b-L	Layer-by-layer
MIC	Minimum Inhibitory Concentrations
MNPs	Metal nanoparticles
MRSA	Methicillin-resistant Staphylococcus aureus
NPs	Nanoparticles
OCNP	Drug Interaction Oral Contraceptive pill
PAA	Polyacrylic acid
PECs	Polyelectrolyte complexes
PLGA	Poly D,L-lactic-co-glycolic acid
PVA	Poly Vinyl Alcohol
PVA	Polyvinyl alcohol
RNA	Ribonucleic acid
ROS	Reactive Oxygen Species
SEM	Scanning Electron Microscope
TGF-β1	Transforming growth factor beta
TNF-α	Tumour Necrosis Factor alpha
VEGF	Vascular endothelial growth factor

1 Introduction

The largest exterior organ of the human body is the skin and it protects the body from mechanical harm and microbial invasion. Skin wounds are a common complication of soreness, incision, scolding, grazing, and additional types of wounding [1]. The healing of the skin is a physiological process that is influenced by a variety of circumstances which can be sped up by using the right wound treatment. A skin wound can be deep or shallow, depending on the severity of the injury [2]. Natural or synthetic materials are used to create traditional dressings which are used to treat wounds. Previously, the major purpose of dressings was to keep the wound dry by allowing fluids to evaporate besides averting hazardous bacteria from entering the wound. Nevertheless, with the passage of time, this viewpoint has evolved. Today, an ideal dressing should aim to provide optimum hydration, speed up the healing procedure, absorb a lot of exudates, and avert a second injury that could be caused by infiltration of tissue surrounding the wound.

An ideal dressing should have the following features: (1) be non-toxic and non-irritating, (2) biode in vivo can dim, (3) outstanding antibacterial qualities to prevent wound infection, (4) fine humidity, (5) enough mechanical resistance to avoid wrinkles [3, 4]. Polymers that are derived from nature, namely, polysaccharide polymers (such as in the form of hyaluronic acid, hydrogels, saccharin, cellulose, chitosan) and polypeptides derived from nature (for example, keratin, gelatin, silk fibroin, and collagen), have been extensively used as cure medium.

This chapter focuses on chitosan (CS), the world's second-most abundant polysaccharide present in nature. Chitosan is a polyaminosaccharide that is made by deproteinizing and deacetylating chitin in a basic media. It is the world's second-largest biopolymer, obtained from biotic systems such as plants, insects, and crustaceans [5]. Chitosan, among other biomaterials, has attracted much attention due to its availability, and adaptability, along with special properties such as curative effects, intrinsic antibacterial properties, non-toxicity, and biodegradability. In addition, once chitosan is broken down, it will eventually be absorbed by human tissues. Thus, many of these characteristics make it an excellent candidate for a wide range of biomedical uses, including surgical procedures, wound healing, drug delivery vehicles, tissue engineering, etc. [6]. Derived from natural sources, chitin, the antecedent of the chitosan, has restricted applications because of its least dissolvability in water and nonaqueous solvents, while chitosan has dissolvability in the majority of the natural solvents which upgrades the usefulness and demonstrates better than chitin. Therefore, chitosan is broadly utilized in the clinical field because of its high biocompatibility, biodegradability, adaptability, and microbial action.

2 Chemical Transformation of Chitin to Chitosan

Chitin lacks suitability for practical applications due to many reasons, including insolubility in different organic and aqueous solvents, impediments in selective substitution, and use of harsh reaction conditions which often result in degradation of the products [7]. These limitations are overcome by employing chitosan, which provides an effortless modification of C2 carbon (Fig. 9.1). The transformed forms of chitosan are generally in use more than the unmodified form. Chitosan can be chemically modified by the introduction of new selective groups or polymers to the chitosan backbone. The reacting sites on the chitosan backbone are the amino group and the two hydroxyl groups. Chitosan can function as a nucleophile in reaction with aldehydes to form imines [8]. This is because the non-bonding pair of electrons on the amino group acts as electron-acceptors. Various chemical modifications such as nitration, sulphonation, carboxylation, and phosphorylation can be carried out to obtain chitosan derivatives, intended for different applications.

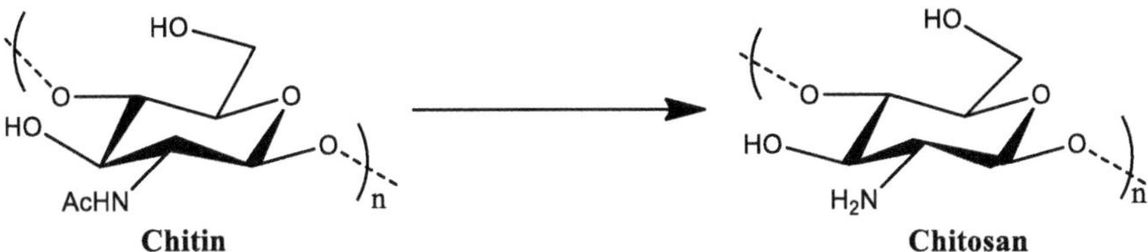

Fig. 9.1 Chemical conversion of chitin to chitosan

3 Processing Techniques for Synthesis of Chitosan-Based Nanocomposites

The two main features responsible for the effective production of nanocomposites are the large specific interfacial area and well-controlled stress transmission across the contact. The first feature is important when it comes to the properties of nanocomposites.

Chitosan is a notable naturally occurring polymer that has a wide range of uses in wound dressings, medication delivery, adhesives, antioxidants, biosensing, food packaging, films, and pharmaceuticals due to qualities like antibacterial and biocompatible, biodegradable, easy to use, non-toxic, and environmentally friendly. Because of its antimicrobial, antioxidant, and antibacterial characteristics, CS is ideal for wound treatment and tissue engineering. CS nanocomposites can be made in a variety of ways [9–11].

3.1 Mechanical Stirring

Physically mixing two or more polymers prior to scaffold fabrication is a straightforward way of generating polymeric composite scaffolds. Due to a paucity of co-solvents that can accept both polymers, forming a single homogenous phase containing chitosan and artificial polymer like polyester is problematic. In solution to this problem, Mechanical stirring can be used. A magnetic stirrer, also known as a magnetic mixer, is a laboratory instrument that employs a rotating magnetic field to rapidly rotate a stirrer immersed in a liquid, causing the rod to stir. A revolving magnet or a fixed pair of electromagnets installed below the liquid reservoir can be used to generate the rotary field. In one investigation, this technique was used to produce a homogeneous suspension of crushed chitosan microparticles in PLGA solution in methylene chloride. Then, using solvent evaporation, composite chitosan–PLGA microspheres were created, and the microspheres were fused to form scaffolds [10, 12].

3.2 Solvent Casting

It is the simplest technique used to prepare films using nanocomposites. The applicability of this technology is due to the lack of expensive equipment such as freeze dryers or excessive processing temperatures. In this technique, there are three steps to synthesize polymer nanocomposites. Mechanical stirring or ultrasonic waves are used to dissolve or disperse the polymer in a suitable volatile solvent. Then, using a petri dish or aluminum foil as a casting surface, pour the mixture into the mold. Finally, the solvent is allowed to evaporate naturally or in an air-drying oven, resulting in dry films. After drying, the film can be removed off flat surfaces. The material with the best mechanical performance will come from this gradual casting/evaporation process. This is due to Brownian motion in the suspension or solution, which allows the nanoparticles to rearrange themselves during slow water evaporation. As a result, they have more time to interact and link with one another, producing a percolating network that supports their self-reinforcing impact. Many researchers have used this low-cost, easy-to-use method to create chitosan nanocomposite films and membranes with different nanomaterials [10, 13, 14].

3.3 Freeze-Drying

Freeze drying also referred to as lyophilization or cryodesiccation, is a dehydration technique in which the product's water is frozen and then vaporized. The freeze-drying method has three steps: freezing, primary drying, and secondary drying. In the freezing phase, the temperature of the sample is lowered until the crystallization of the ice crystals begins, after which the ice crystal increases. As a result, most of the water (in the form of ice crystals) is separated from the dissolved phase, which is concentrated with a small amount of water. Sublimation removes crystalline ice formed during the first step. As a result, the chamber pressure drops much below the ice vapor pressure and the shelf temperature rises to compensate for the heat lost due to ice sublimation. At the end of the primary drying step, the sample may still contain about 20% non-frozen water in the concentrated solute step, which is desorbed in the secondary drying step, usually at high temperature and low pressure, to get the desired minimum moisture content of the final product. In freeze-drying methods, the freezing step is important. Many investigations have shown that the freezing stage has a significant impact on the final freeze-dried product's quality and morphology. In the context of chitosan-based composites, non-woven fibrous materials, such as sponges and foams can be prepared using the same freeze-drying mechanism. By forming a complex of polyelectrolytes (PEC), a stable dilute dispersion of the chitosan-based material in a solution can be obtained, which is then frozen at a high rate of cooling by immersion in liquid nitrogen (or cryogenic freezer), followed by drying. Therefore, the formation of PEC dispersions in solution is essential for the preparation of non-woven materials of nanofiber or microfiber based on chitosan, which promotes the absorption of exudates and prevents the penetration of bacteria to promote wound healing [15].

3.4 Layer-By-Layer Assembly

Layer-by-layer (L-b-L) assembly is a one-of-a-kind process for producing multilayer thin films with nanoscale precision and polymeric nanocomposites such as Chitosan. In this method, polycations and polyanions are alternately adsorbed on the substrate, and the surface charge is reversed after each layer is precipitated. Catalysis, drug delivery, wound healing, electrochemical detection, and biodetection are just some of the applications of L-b-L films. Electrostatic contact between polyelectrolytes is a major driving factor, complemented by hydrogen bonding, hydrophobic interactions, and charge exchange. Each adsorption phase involves a charge reversal, and the excess charge remains at the surface, favoring additional adsorption in subsequent steps. The key advantages of the L-b-L approach are its simplicity, independence from substrate size and form, low polymer requirement, and ease of production. Catalysis, biosensing, light-emitting diodes, wound healing agents, selective membranes, and drug delivery are just a few of the fields where it can be used because of its relative ease of production and control over many substrates [16].

3.5 Electrospinning

Nanofiber technology has recently gotten a lot of attention from researchers all over the world. Various polymer nanoparticles can have a wide range of properties, which can be further enhanced by immersing the nanofibers in a solution of additional nano-materials or by directly binding the nanomaterials to the solution of a base polymer. Compared to other processes to date, electrospinning is the most sophisticated and simple method to generate and propagate homogeneous nanofibers in the nm-μm length range. It is possible to create thin, dense lines, specially designed by certain polymeric fibers directly from the solution in the presence of large electric fields. Furthermore, the end product's shape, porosity, and content may be adjusted with comparatively simple equipment. Normally, between the polymer solution and the collector, an electric field is formed which creates an internal pushing force that causes the generation of nanoparticles in the polymer solution. The most prevalent electrospinning processes are dry electrospinning, wet electrospinning, and coaxial electrospinning. During dry spinning, a volatile solvent is used to dissolve the treated polymer and evaporate as the fiber is spun over the collector. In wet spinning, a non-volatile solvent is used to dissolve the polymer, while the second solvent is used as a collector. This second collector solvent removes the non-volatile solvent from the polymer, resulting in pure nanofibers. Coaxial spinning is not the same as the other two processes. The production of core–shell-based nanofibers is done using this method. At the same time, two distinct materials are spun. The fabrication of nanocomposites for wound disinfection, medical implants, scaffolds, and cleaning of environmental pollutants has been extensively explored using this simple and cost-effective technology. However, producing high yields and high-quality fibers from a discrete CS solution is a difficult task. This is mainly due to the relatively rigid structure of the CS chain, which does not support the tension needed to create the Taylor cone, leading to the formation of nanofibers. This problem can be solved by adding neutral salts or by using other plasticizers that reduce the electrostatic bond between the polyelectrolyte molecules, which leads to a lower critical concentration for electrospinning [10].

Moreover, electrospinning produces non-woven nanofibers with large surface areas and tiny pores for wound healing. These qualities promote wound healing by absorbing exudate and preventing the invasion of bacteria.

4 Effects and Action of Chitosan on Skin Wound

The process of wound healing comprises four major phases, namely, hemostasis, inflammation, proliferation, and skin remodeling. In the initial stage, the coagulation system gets activated following the constriction of blood vessels and aggregation of platelets. Fibrinogen is converted into insoluble fibrin which ceases bleeding by forming clots. Herein, chitosan induces the aggregation of platelets and red blood cells (erythrocytes), thereby stopping the loss of blood. Further, chitosan prevents the breakdown or dissolution of fibrin by plasmin, i.e., fibrinolysis. The inflammatory cells augmented by chitosan help clear bacteria and necrotic tissue from the wound in the inflammation phase. In the subsequent stage, chitosan quickens the proliferation of epithelial cells to produce the epithelial tissue which covers the wound. The fundamental role of chitosan in this stage is to stimulate the growth of granulation tissue, which fills the gaps in the epithelial tissue. In the ultimate stage of skin remodeling, chitosan doesn't play any significant role. This stage defines the completion of the wound repair process and is marked by the regeneration of the new epidermis and dermis.

5 Properties of Chitosan

5.1 Antibacterial Property

Chitosan is a potent antibacterial agent which can be ascribed to its cationic nature and its activity was observed against various other microorganisms such as fungi and algae [17]. An antimicrobial agent destroys microorganisms or prevents them from growing. Edible films can be obtained from chitosan as it is a non-toxic antimicrobial biofilm. These antibacterial effects are regulated by intrinsic factors, such as chitosan variety, degree of chitosan polymerization, host, natural nutritional composition or chemical nutritional composition of the substrate, and environmental conditions.

5.1.1 Antibacterial Mechanism

Coagulation, hemostasis, inflammation, granulation tissue development, angiogenesis, and re-epithelialization are all complex healing mechanisms that occur in wound skin. The wound's moist, the nutrient-rich environment creates ideal circumstances for bacterial development. As a result, wounds have antibacterial qualities. Dressings should be taken into consideration. Chitosan is a kind of chitin that is commonly used. Because of its strong antibacterial properties, it is used to treat wounds [18].

- Cell Wall and Cell Disruption

According to the gram staining results, cell wall bacteria can be classified into gram-positive bacteria and gram-negative bacteria. Gram-negative bacteria have a peptidoglycan layer along with the outer membrane in their cell wall (Fig. 9.2A). Two asymmetric monolayers make up the outer membrane. Only phospholipids are present in the inner layer, while the outer layer contains both phospholipids and lipopolysaccharides. The gram-negative bacteria are negatively charged because of the pyrophosphate and phosphate groups of lipopolysaccharides in the outer layer. Teichoic acids and peptidoglycans make up the cell wall of gram-positive bacteria (Fig. 9.2B). Due to the presence of phosphate and carboxyl groups of teichoic acids, the surface of gram-positive bacteria is positively charged. The NH_2 groups are protonated to $-NH_3^+$ cations [19], when chitosan of high molecular weight is dissolved in acidic aqueous solutions. Figure 9.2A(a) shows the electrostatic interaction between lipopolysaccharides and NH_3^+ on the cell membrane of gram-negative bacteria, while Fig. 9.2B(a) shows the electrostatic interaction between lipopolysaccharides and teichoic acid on the cell membrane of gram-positive bacteria, which causes an unequal distribution of negative charge for bacteria. As a result, there is a discrepancy between cell wall synthesis and cell wall degradation. Under unsustainable osmotic pressure, the bacterial cell membrane deforms and ruptures, resulting in cell content leakage and, in the end, cell lysis. The leaking indicates that the cell membrane has been disrupted.

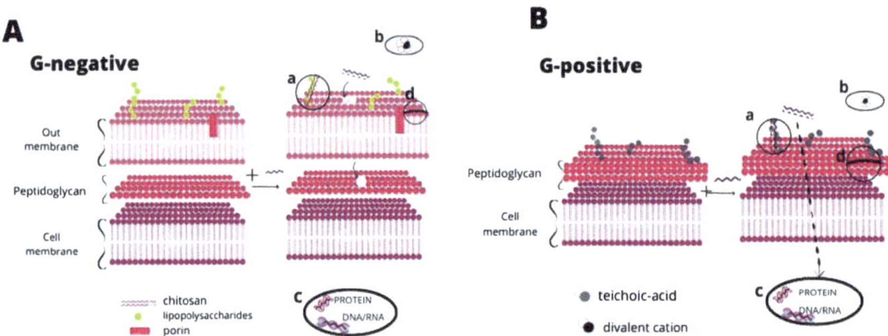

Fig. 9.2 Chitosan's antibacterial mechanism opposes gram-negative (**A**) and gram-positive bacteria (**B**) **a** The cell membrane is disrupted by electrostatic interaction between chitosan and lipopolysaccharides (or teichoic acid), allowing chitosan to permeate further into the cell. **b** Chitosan chelates divalent cations, reducing the outer membrane's stability. **c** Chitosan inside the cell can hinder DNA/RNA and protein production as well as the hinder proliferation of gram-negative bacteria. **d** Gram-negative bacteria's metabolism is hampered by the deposition of the high molecular weight of chitosan on their surfaces

Cell layer disturbance can be concluded by the spillage of particles like K^+, PO_4^{3-} and macromolecules like DNA also, RNA that are identified by solid UV retention at a frequency of 260 nm. After 5 min of contact between oleoyl chitosan and *Staphylococcus aureus* (S. aureus), an absorption peak was observed at 260 nm [20]. The after-effects of SEM perception uncovered that after 30 min of contact, chitosan atoms adhered to the outer layer of *Escherichia coli* (E. coli) and S. aureus, and cell content spillage was noticed for both bacterial strains [18].

- Chelating metallic cations

Bacteria's membrane structure can be stabilized by divalent cations. The GNB-OM model was designed by Clifton [21] to examine the stabilizing effect of divalent cations. The findings revealed that the negatively charged cations' salt bridges such as lipopolysaccharide and the divalent cations' salt bridges are essential for the outer cytoplasmic membrane's structural integrity. The anionic charge of the lipopolysaccharide molecule is neutralized by cation and hydrogen bonds, building an impenetrable network impermeable to macromolecules and hydrophobic molecules. Divalent cations such as Mg^{2+} and Ca^{2+} are used by the gram-positive bacteria's phosphate group present in the cell wall, to maintain the cytoplasmic membrane's stability.

At the point when the pH esteem of the framework drops underneath the isoelectric point of chitosan and its subordinates, protonated $-NH_3^+$ in the atomic chains are drawn in by negative charges while chitosan and its subordinates chelate the divalent cations on the cell layer surface (Fig. 9.2A(b), B(b)), prompting an unevenness in the surface potential and a shared shock among the adversely charged atoms, lastly the cell membrane burst [18].

- Interacting with Intracellular Targets

It's a process that involves interconnecting with intracellular targets. Chitosan of molecular mass <5000 D can pass through the bacterial cell wall and form complexes with the bacteria. DNA and RNA polymerase's function is jeopardized, thereby inhibiting replication and DNA and RNA transcription (Fig. 9.2A(c), B(c)), which inhibits the growth of bacteria [22, 23]. OCNPs' ability to bind to DNA was investigated by Xing et al. [23]. By determining the mobility of the bacterial genome, DNA/RNA can be discovered on an agarose gel. It is seen that with increasing OCNP concentration, the of brightness electrophoresis band decreases. The movement of DNA and RNA from *E. coli* is completely suppressed by 1000 mg/L OCNP, which may be because of electrostatic interaction between negatively charged phosphate groups and positively charged amino acids OCNP.

Furthermore, low-molecular-weight chitosan reduces microorganism protein synthesis (Fig. 9.2A(c), B(c)]. A group of over 4600 gene deletions mutations of Saccharomyces cerevisiae (S. cerevisiae) was examined with the help of an array of yeast gene deletions. Researchers could investigate interactions between chemicals and genes, such as between S. cerevisiae and chitosan. It was found that out of ten mutants, 31% were most susceptible to chitosan protein synthesis deletions of genes [18].

- Applying bacterial combination to bacterial surfaces

When dissolved in an acidic aqueous solution, chitosan with high molecular weight can form a thick polymer layer on bacterial surface that impedes nutrient delivery leading to metabolic disorders and ultimately, death of the bacteria (Fig. 9.2A(d), B(d)). SEM (scanning electron microscope) inspection revealed vesicle-like structures on the surface. *E. coli* and Salmonella outer membranes were treated with chitosan [24].

5.2 Antioxidant Property

Chitosan and its derivatives have antioxidant activity in terms of free radical scavenging. Many key biological macromolecules are made up of free radicals. DNA, proteins, and lipids are examples of macromolecules. Free radicals cause cellular harm. Radicals have been linked to a variety of illnesses, including cardiovascular disease, vascular disease, inflammatory disease, aging, diabetes mellitus, neurodegenerative disease, and cancer. Free radicals in humans are generated as superoxide anion ($O2\bullet2$), hydrogen peroxide (H_2O_2), and hydroxyl ($\bullet OH$) roots. The term "reactive oxygen species" (ROS) refers to all of these substances. ROS plays a vital role in wound healing which necessitates a delicate balance of molecules' beneficial and detrimental effects. ROS having a balanced reaction will disinfect the tissue and stimulate healthy turnover; a state of infection will occur if ROS is inhibited while an increase in ROS destroys buffer tissue [25].

6 Chitosan Nanocomposite Scaffolds Toward Wound Dressings

6.1 Chitosan Sponge for Wound Healing

Sponges are high porosity foam. Because of their microporosity, these solid structures can absorb a large quantity of fluid (>20 folds their dry weight) They are often soft and flexible while still allowing for optimal cell contact. Chitosan sponges can absorb wound exudates while also helping to heal the wound. Chitosan sponges are also employed in bone tissue engineering as a filler material. Histological outcomes were noted when the SD rat's dorsal skin wound was treated with three different types of wound dressings [26]. The lesions treated with gauze revealed immature granulation tissue, clogged arteries, and a large number of inflammatory cells. On the 12th day, the dermis was still undergoing remodeling. In the C2A2-curcumin sponge and C2A2 sponge, the collagen alignment and the granulation tissue were more advanced as compared to gauze-treated wounds. The wounds treated with these two types of sponges had collagen that is compact and coordinated, which creates a more favorable healing environment. Full-thickness wounds were produced on the backs of each rat for wound healing tests where C2A2 sponges were used. For the therapy, dressings of sponge made of chitosan glutamate (high molecular weight) and sericin were created. For the treatment of persistent skin ulcers, sponge-like dressings based on sericin and chitosan glutamate were created. The concentration of sericin in the optimal dressing is sufficient to protect human fibroblasts from oxidative damage. Furthermore, the improved dressing promotes wound healing by increasing fibroblast proliferation. Absorbable and non-absorbable dressings were made with the help of a grafted derivative of chitosan containing 2-hydroxyethyl acrylate, it was converted to a sponge via a heat-induced separation process. Results show that Levo-loaded sponges provide a promising solution for wound infection control, even at low concentrations, are generally susceptible and resistant to wound pathogens, and encourage improved antibacterial protection [27].

6.2 Immobilized Chitosan Extract Scaffolds for Wound Healing

The hydrogel film is very delicate and does not cause any problem when applied to the dressing. Plant extract is a natural microorganism that has limited compatibility in polymer solutions of hydrogel but its efficiency in hydrogel films is weaker than its original form. Film hydrogels coated with Salix alba, Juglena Regia, and gentamycin sulfate leaf extracts show antifungal as well as antibacterial activities in the disk broadcast process and in the MIC. Wounds are often infected with a variety of microorganisms, thus infection is always probable. This will encourage the development of innovative dressings that have antibacterial effects by combining or not combining antibacterial agents. The therapeutic characteristics of aloe vera can be very helpful in the creation of beneficial dressing materials. Chitosan/aloe vera-based membranes have been developed as dressing materials. The antibacterial ability of the resultant membranes is found to be enhanced by the mixing of aloe vera with chitosan. When compared to other formulations, CAV1 (CS:AV: 1:1, v/w) had the highest antibacterial activity, demonstrating that CAV1 is a more effective bactericide. It's possible that this is owing to a higher concentration of AV in the membrane, which boosts its inhibitory strength. Despite the CAV membrane's strong proliferation and cell adhesion, the greater AV content of the CAV (caveolin) membrane hampered cell adherence and proliferation. With CAV1, the anticipated increase in cell spreading and proliferation was not seen. Membranes made of chitosan and aloe vera can be considered viable wound dressing materials [27].

6.3 Natural/synthetic Polymer Blend Scaffolds for Wound Healing

In a study by Tanodekaew et al. the goal was to make a hydrogel containing chitin grafted polyacrylic acid (PAA) that may be used as a wound dressing [28]. After 14 days, the shape and behavior of the cells on the chitin-PAA film were confirmed to be usual. Fibroblast cells were used in the study by Ribeiro et al. where they created a hydrogel of chitosan that can be used for wound healing. Cells obtained from rat skin were used to test the hydrogel's cytotoxicity. The findings revealed that chitosan hydrogel was effective. As per the cell viability, hydrogel and its breakdown by-products are toxic to cells. Studies show that chitosan hydrogel could help restore skin architecture.

As a wound treatment agent, the material was found to be noncytotoxic to L929 cell growth in indirect cytotoxicity testing of films using mouse fibroblasts (L929), had good biocompatibility in vitro, and had the potential to be used as a bandage for wounds. Chitosan and its derivatives were made into porous microspheres. Antigens must be delivered in a controlled manner. The mucoadhesivity of chitosan and its cationic derivatives has been recognized and shown to improve drug adsorption, particularly at neutral pH. It was found that negatively charged cell membranes interacted with N-trimethyl chitosan chloride, inhibiting microorganism growth. The surface properties of the chitosan film can be tuned by heterogeneous chemical modification. Gonzalez et al. reported PVA/clove extract, PVA/Ag nanoparticles, PVA/bentonite, and PVA/cellulose nanocomposite hydrogels. PVA/clove hydrogels lacked homogeneity, whereas PVA/bentonite and PVA/Ag nanocomposite hydrogels showed substantial antimicrobial activity against *E. coli* growth, as well as good water vapor transmissibility. With the use of Ag nanoparticles and clay as filler, the fusion rate and water-absorbing capacity are suitable. For a wound-healing accelerator that is both effective and water-soluble [27].

7 Application of Chitosan-Based Metallic Nanocomposites in Wound Healing

7.1 Chitosan/nAu

As a commonly used natural polymer, chitosan (CS) has gotten a lot of interest in skin tissue engineering [8]. It has the ability to increase the proliferation and differentiation of many wound-healing cells, including keratinocytes, vascular endothelial cells, and fibroblasts. In addition, CS can be used directly as a reducing and stabilizing agent to produce a variety of metal nanoparticles. Despite its benefits, many studies have shown in clinical practice that the low mechanical strength of CS adversely affects the support and protection of wound surfaces. Although the exact mechanism of antimicrobial action is unknown, there are many widely accepted theories involving surface charge, mRNA inhibition, and oxygen pathway blockade.

However, a number of parameters, including the pH of the environment, inhibit this antibacterial action. Because chitosan has a low solubility at high pH, it can only act as an antibacterial in an acidic environment. Also, the pH of the film must be neutral during production. As a result, several studies have shown that the chitosan solution has antibacterial properties, but the chitosan film does not. Fortunately, antibacterial material can be used in the alteration to successfully remedy these issues. With the rapid development of nanotechnology, the use of metal nanoparticles has drawn a lot of interest and given a unique strategy for the restoration of skin defects. Nanoparticles can be useful delivery vehicles as they are small in size and have huge surface areas. They also have high antioxidant and antibacterial activity, as well as anti-inflammatory and anti-angiogenic properties for healing internal wounds. Metal nanoparticles have been shown to damage bacterial cells by releasing metal ions that are not affected by the polymer matrix. As a result, metal nanoparticles can improve the bacteriostatic properties of the CS matrix. AuNPs are much less cytotoxic than other metal nanoparticles, such as Ag, Cu, and ZnO. The antibacterial mechanism of gold nanoparticles, on the other hand, is still unknown. However, it is debatable whether this is a direct link or the emission of gold ions. AuNPs disrupt the integrity of the cell membrane by electrostatic interactions in the form of metallic gold on the surface of the bacterial cell, releasing the contents of the cells and releasing reactive oxygen species (ROS). The resulting gold ions tend to bind to the sulfhydryl groups of cellular lipoproteins and disrupt the respiratory chain and trigger ROS in the pathway through the gold ions. As a result, the AuNPs contained in the CS matrix effectively prevent bacterial infection of the wound. In addition, spherical AuNPs with a diameter of 20–40 nm induces significant anti-inflammatory effects by reducing the levels of IL-6 (interleukin 6), IL-12 (interleukin 12), and TNF-α (alpha tumor necrosis factor). Furthermore, studies have shown that AuNPs increase angiogenesis and alter collagen deposition in the extracellular matrix (ECM), both of which are beneficial for wound healing. The endogenous collagen matrix promotes angiogenesis, favoring tissue regeneration and increasing the proliferation of fibroblasts.

AuNPs, on the other hand, showed dose-dependent cytotoxicity, especially the smaller ones (less than 2 nm). Fortunately, when AuNPs are coated with CS, the cytotoxicity is greatly reduced. The particle size of AuNPs rises when they are coated with chitosan to prevent cell absorption, which reduces cytotoxicity to some extent. Furthermore, when colloidal gold nanoparticles are lowered, they tend to cluster together in the dispersions, reducing their bactericidal efficacy significantly. Fortunately, the use of CS for stabilizing gold nanoparticles has recently gained traction, preventing AuNPs from aggregating together [30].

In one study, CS/AuNPs were prepared by mixing a solution of CS and colloidal gold. Measurements (TEM) were performed to monitor the size distribution of CS/AuNP. The antibacterial and cytotoxic properties of the composites were also studied. In addition, the CS/AuNP compound had a positive effect on wound healing in mice with defective full-thickness skin [30].

Colloidal gold was self-assembled onto chitosan film in another investigation. Nanoparticles with a size of 25 nm were employed to make the film, with a molar ratio of 0.50 for the precursors, trichlorogold tetrahydrate hydrochloride to trisodium citrate. In vitro, the chitosan control had superior keratinocyte attachment in the 6th hour, whereas the three-dimensional film of chitosan gold had better keratinocyte attachment at later time points. The scaffold's three-dimensional structure may also inhibit fibroblast growth, minimizing the mixing of the two cells. The presence of heterochromatin, chondriosomes, and desmosomes in vitro keratinocyte culture studies was significant in skin regeneration [31].

7.2 Chitosan/nCu

Copper compounds are traditionally used in agriculture to protect crops from bacterial and fungal diseases and are also used as preventive feed additives and growth promoters on animal and poultry farms. Copper has antimicrobial properties through a variety of mechanisms, including cell membrane damage, protein chelation, and enzyme inactivation. Biocompatible composites with high bioavailability and antimicrobial activity of copper have the potential to become a new class of antimicrobial agents. These chitosan-metal nanocomposites replace antibiotics, potentially reducing drug resistance. Of particular interest are the use of copper-based biocomposites for the local treatment of wounds and the use of feed additives with high doses of traditional antibiotics as a prophylactic strategy to limit infection. Chitosan is a biocompatible polysaccharide substrate that has received a lot of attention in recent years due to its ability to administer drugs. Consequently, copper chitosan compounds have been studied in antibacterial, antitumor, and plant protection studies for their synergistic effects. Several methods have been developed to add copper to the chitosan matrix in effective doses to prevent toxicity to normal cells and tissues and to produce specific effects [32].

The copper-CS composite was found to be lethal to gram-negative (*E. coli*) and gram-positive bacteria when the copper level was 100 g/mL (1.57 mM) (S. aureus). Pure CS and pure copper have no bactericidal action at the same concentration. Copper nanoparticles' antibacterial mechanism is believed to include hydrolysis of the peptidoglycan layer, which destroys bacteria's cellular structure. It's also possible that this is owing to copper's thiophilic behavior, which makes it competitive enough to kill cytoplasmic iron-sulfur enzymes. Exploration of the underlying mechanism would lead to better bacteriostatic performance optimization. Copper nanoparticles can be added to CS to improve their physical and chemical properties. Copper can directly or indirectly increase cytokines or growth factors that reduce inflammation and promote wound healing by promoting fibroblast proliferation, angiogenesis, and collagen deposition. The antagonism of proinflammatory (eg IL-1, i.e. interleukin 1, TNF-α, i.e. tumor necrosis factor-alpha and IL-6) and anti-inflammatory agents (eg

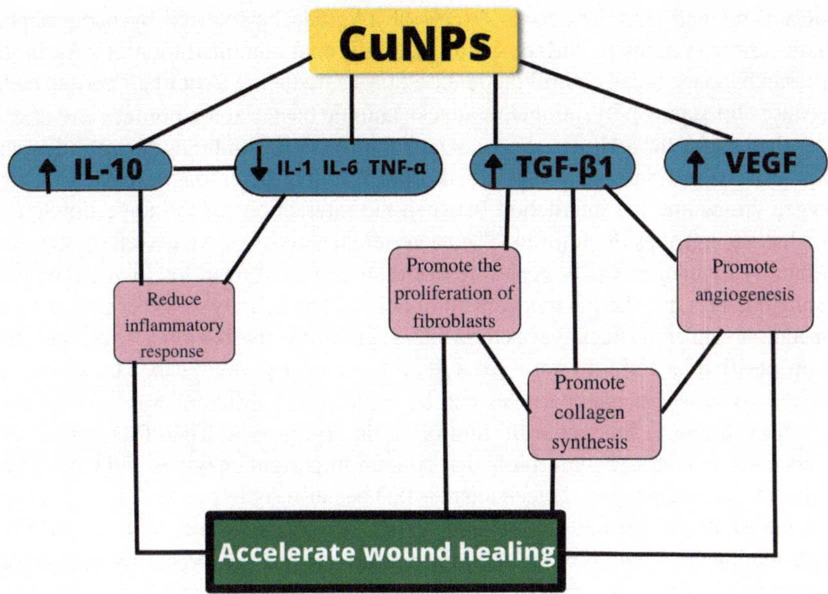

Fig. 9.3 Effect of copper on different cytokines associated with wound healing

IL-4, IL-10, and IL-13) causes an inflammatory reaction. Despite the high biocompatibility of CS, a previous study found a significantly higher expression of TNF-α in CS-treated mice at an early stage of recovery. Low levels of TNF-α have been observed in mice given CS/CuNP, suggesting that copper inhibits CS-induced inflammatory responses. Meanwhile, a rise in anti-inflammatory mediators like IL-10 (Interleukin 10) boosted the anti-inflammatory capacity even more. Vascular endothelial growth factor VEGF and Transforming growth factor-beta (TGF-β1) have been associated with angiogenesis and collagen deposition in previous studies, both of which showed high expression in the presence of CS/CuNP. The effect of copper on the various cytokines involved in wound healing is shown in Fig. 9.3. CS / CuNP accelerates wound healing by modulating cytokines and growth factors [33].

7.3 *Chitosan/nAg*

In ancient times, silver in the form of silver nitrate was used locally for burns and wounds. Nag metallic silver is widely used as a preservative in cosmetics, gels, sponges, antibacterial agents, films, and nanofiber substrates. These nAg are mainly used to treat some pathogenic bacteria that are resistant to various antibiotics. For example, nAg has led to a variety of biomedical applications, from silver-based dressings to silver-plated medical devices. nAg penetrates damaged skin more easily than undamaged skin. The toxic effects of nAg can be reduced by using appropriate carrier systems or surface coatings. Additional administration of nAg in the chitosan bandage helps control the release of nAg, reducing toxicity to normal cells. Because chitosan is polycationic, it attacks anionic bacterial membranes and aggravates damage by the addition of nAg, which inhibits cell membrane permeability and respiration. Additional advantages are the interaction of silver ions with electron-rich oxygen atoms and the interaction between the interactions of the ionic dipole and the glucoside groups of chitosan. The bactericidal activity of Ag is well known, and the ability of nAg to easily penetrate bacterial cells and form low-density regions among bacteria may be the mechanism of action. The activity of silver ions is more concentrated in gram-negative bacteria than in gram-positive bacteria. This is because Gram-positive bacteria do not have a thick layer of peptidoglycan. The antibacterial activity of silver nanoparticles can be explained in different ways. Silver ions have been found to interact with thiol or sulfhydryl groups. Therefore, when nAg enters cells, it interacts with thiols that contain important enzymes and causes cell death. The second proposed mechanism is that because Ag is a weak acid, it interacts with the weak acid group of genetic material, such as phosphate, to destroy DNA / RNA and interfere with protein translation. nAg is mainly adsorbed by the bacterial membrane to form pores in the membrane. Silver ions are known to penetrate bacterial cell membranes and trigger the synthesis of reactive oxygen species and free radicals, which cause structural and integrative damage to bacterial cells, promoting wound healing [31, 33].

The characteristics of chitosan/sago starch film impregnated with NAg and an antibiotic, gentamicin, were tested in vivo using an open excision wound model by Arockianathan et al. Films are typically employed when the wound exudate is minimal. They assist debride damaged skin and protecting it from compression. As a result, stronger mechanical qualities are frequently a deciding factor for such dressings. It would be fascinating to see if such dressings can prevent microbial diseases from spreading.

In a thermal injury model, nAg had a faster wound healing rate of 26.5 0.93 days, whereas control and silver sulfadiazine had a wound healing rate of 35–37 days on average. The thermal injury model developed the infection within 7 days, after which nAg therapy did not show an additional increase in bacterial proliferation. After treatment with nAg, inflammatory cytokines such as IL-6, TGF-b1, IL-10, VEGF, and IFN-γ have been shown to regulate wounds [31].

7.4 Chitosan/nTiO$_2$

Titanium is known to irritate the skin. In healthy skin, nTiO$_2$ does not penetrate beyond the stratum corneum layer, but in damaged skin, the penetration of nTiO$_2$ greatly increases systemic toxicity, so it is necessary to coat or encapsulate the nanoparticles in a substrate to prevent the explosion of nTiO$_2$. After chronic inflammation or mitochondrial dysfunction, nTiO$_2$ captures electrons that form ROS through Fenton chemistry, leading to the formation of superoxide radicals. Under the influence of light and humidity, $nTiO_2$ absorbs electrons from nearby atoms to form hydroxyl and superoxide anion radicals. This can damage the lipid and peptidoglycan layers of bacteria.

In addition, $nTiO_2$ may affect ROS-induced toxicity in DNA sequencing, DNA damage, activation of p53-mediated DNA damage control signals, G2M cell cycle termination, mutagenesis, reactive nitrogen production, oxidative stress, and apoptosis. Due to its favorable biological characteristics, $nTiO_2$ has found a useful role in wound healing. Photocatalysis is used on $nTiO_2$. Prolonged-release of titanium ions has been shown to suppress microbial growth and promote wound healing.

In an excision wound model, Origanum vulgare engineered $nTiO_2$ demonstrated better-wound closure. Wound contraction began on day 4 and reached nearly 94% by day 12. On TiO_2 treatment, histopathological analyzes of wound tissue revealed a higher number of macrophages, fibroblasts, and deposition of collagen. The antimicrobial activity of $nTiO_2$ was found to be higher in vitro with electrospun graphene against gram-negative *E. coli*. This was done by generating reactive oxygen species (ROS) in microorganisms.

In vitro antimicrobial activity against *E. coli, S. Aureus, P. aeruginosa, B. subtilis,* and *A. niger* were evaluated using triple chitosan-pectin bandages with $nTiO_2$ (0.001%). A polyelectrolyte complex was formed between the amine group of chitosan and the negatively charged carboxyl group of pectin. The bandage has high tensile strength and a low swelling factor. Open cut wounds treated with a triple chitosan-pectin bandage containing $nTiO_2$ heal faster than controls. The epidermis and developed dermis of the skin was found to be intact histologically. The chitosan control group wound closure rate was 95% on day 16 and the wounds had fewer scars than in the control group. This lends credence to chitosan's anti-scarring properties. It also has a matrix of chitosan pectin, which can be easily removed without damaging the skin [31].

7.5 Chitosan/nZnO

Nanocomposite mixtures of chitosan and zinc oxide (ZnO) NPs are another trend in wound-healing [34]. Nano zinc oxide, among other metal oxides, is frequently used in the manufacture of wound treatment bandages due to its antibacterial versatility. ZnO NPs, like the other MNPs (Au and TiO_2), can release Zn ions to provide antibacterial activity. Another important antimicrobial mechanism available for ZnO nanoparticles is the photocatalytic production of hydrogen peroxide. Many studies have shown that electron–hole pairs are formed when ZnO NP semiconductors emit light with wavelengths corresponding to energies >3.2 eV. Water and air on the surface surrounding positively charged holes and negatively charged electrons can generate reactive oxygen species such as superoxide (O2•−) radical anions, hydroxyl (•OH) radicals, and hydrogen peroxide (H_2O_2) and nucleic acid lesions leading to bacterial oxidation, lipid peroxidation [33]. Figure 9.4 shows the synthesis of reactive

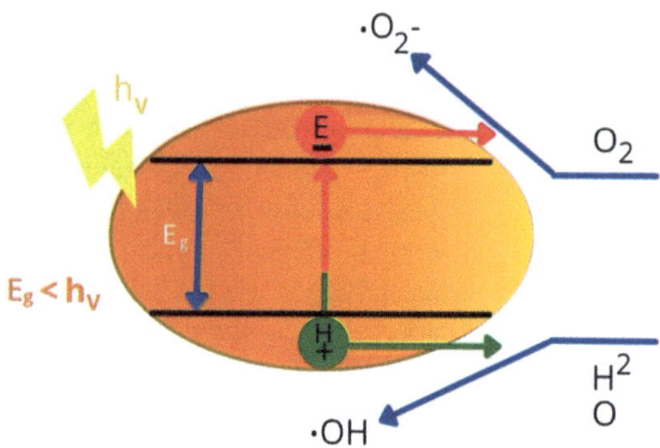

Fig. 9.4 Schematic representation of the energies of ROS formation

oxygen species and the generation of electron–hole pairs. The formation of reactive oxygen species (ROS) destroys the outer membrane of the bacterial cell wall.

Zinc oxide's anti-inflammatory, antibacterial, and re-epithelialization, fibroblast proliferation properties suggest that it could be used as an active stabilizer in wound healing [35]. ZnO is a promising material for a wide range of applications due to its unique optical, chemical sensing, semiconductor, electrical and piezoelectric properties. ZnO NPs can be synthesized in a variety of ways, yielding products with varied sizes and morphologies. Microwave decomposition, simple wet chemical path, deposition process, simple precipitation method, hydrothermal synthesis, solvothermal method, microwave hydrothermal method, and hydrothermal technique are used in the literature for the synthesis of ZnO nanoparticles. ZnO nanoparticles have been demonstrated in the literature to have potent antibacterial properties against a variety of microorganisms. The principal antibacterial mechanism of ZnO NPs has been revealed to be photocatalytic hydrogen peroxide production. In addition, penetration and subsequent disruption of the cell membrane have been shown to alter bacterial growth inhibition by ZnO contact [34].

The antibacterial activity of ZnO-incorporated CS (ZnO-CS) was studied by Visnuvinayagam et al. ZnO-NPs with nanosheet shapes were detected in scanning electron microscope images. MRSA and Pseudomonas aeruginosa cells treated with ZnO showed cell membrane disruption and shrinkage, respectively. The inhibitory zone of CS coupled with ZnO is 5–15 mm larger than that of chitosan alone. Yuvaraja et al. developed CS/PVA/ZnO beads as new antibacterial agents for wound healing using polyvinyl alcohol (PVA). Transmission electron microscopy revealed a hexagonal crystal structure of ZnO. The antibacterial activity and therapeutic effect of CS /PVA/ZnO in mouse skin wounds were much better than those of CS and PVA. The results mainly confirm the usefulness of the nanocomposite system for wound healing [33].

8 Conclusion and Future Outlook

The mechanism of wound-healing and application of chitosan and chitosan-based nanocomposites have been expounded in this chapter. With its noteworthy set of properties such as antimicrobial nature, antioxidant property, biodegradability, facilitation of collagen deposition, biocompatibility, and among others, chitosan is a promising candidate in wound-healing applications. The methods of preparation of various chitosan-based nanocomposites have been elaborated in this chapter in addition to the effects and mechanism of chitosan on skin wounds. A major impediment to realizing the clinical application of chitosan-based nanocomposites and scaffolds for wound healing is their poor mechanical strength. The development of cost-effective, simple wound dressings of chitosan capable of promoting repair of the wound and causing no to least inconvenience to the user, while addressing the current limitations, should be the focus of researchers in this field.

Important Websites

1. https://www.webmd.com/vitamins/ai/ingredientmono-625/chitosan.
2. https://www.gmp-chitosan.com/en/component/tags/tag/wound-dressing.html.
3. https://www.glycoforum.gr.jp/article/23A6.html
4. https://www.wur.nl/en/show/chitin-and-chitosan.htm

References

1. Granick MS, Baetz NW, Labroo P, Milner S, Li WW, Sopko NA (2019) In vivo expansion and regeneration of full-thickness functional skin with an autologous homologous skin construct: clinical proof of concept for chronic wound healing. Int Wound J 16(3):841–846. https://doi.org/10.1111/iwj.13109
2. Abd El-Hack ME et al (2020) Antimicrobial and antioxidant properties of chitosan and its derivatives and their applications: a review. Int J Biol Macromolecules 164:2726–2744, Elsevier B.V. https://doi.org/10.1016/j.ijbiomac.2020.08.153
3. Bhardwaj AK, Sundaram S, Yadav KK, Srivastav AL (2021) An overview of silver nanoparticles as promising materials for water disinfection. Environm Technol Innov 23. Elsevier B.V. https://doi.org/10.1016/j.eti.2021.101721
4. Daristotle JL et al (2020) Sprayable and biodegradable, intrinsically adhesive wound dressing with antimicrobial properties. Bioeng Translational Med 5(1). https://doi.org/10.1002/btm2.10149
5. Rinaudo M (2006) Chitin and chitosan: properties and applications. Progress in Polym Sci (Oxford) 31(7):603–632. https://doi.org/10.1016/j.progpolymsci.2006.06.001
6. Rahman MM, Shahruzzaman M, Islam MS, Khan MN, Haque P (2019) Preparation and properties of biodegradable polymer/nano-hydroxyapatite bioceramic scaffold for spongy bone regeneration. J Polym Eng 39(2):134–142. https://doi.org/10.1515/polyeng-2018-0103
7. Ahmad M, Zhang B, Manzoor K, Ahmad S, Ikram S (2020) Chitin and chitosan-based bionanocomposites. In: Bionanocomposites: green synthesis and applications, Elsevier, pp 145–156. https://doi.org/10.1016/B978-0-12-816751-9.00006-4
8. "Notes to the Editor," (1988)

9. Ramachandran S, Rajinipriya M, Soulestin J, Nagalakshmaiah M (2019) Recent developments in chitosan-based nanocomposites. https://doi.org/10.1007/978-3-030-05825-8_9

10. Hasan MM, Habib ML, Anwaruzzaman M, Kamruzzaman M, Khan MN, Rahman MM (2020) Processing techniques of chitosan-based interpenetrating polymer networks, gels, blends, composites and nanocomposites. Inc. https://doi.org/10.1016/B978-0-12-817968-0.00003-2

11. Ahmad M, Zhang B, Manzoor K, Ahmad S, Ikram S (2020) Chitin and chitosan-based bionanocomposites. In: Bionanocomposites: green synthesis and applications, pp 145–156. https://doi.org/10.1016/B978-0-12-816751-9.00006-4

12. Levengood SKL, Zhang M (2014) Chitosan-based scaffolds for bone tissue engineering. J Mater Chem B 2(21):3161–3184. https://doi.org/10.1039/c4tb00027g

13. Sara M, Rekha T, Koshy R, Mary SK, Thomas S, Pothan LA (2022) Springer briefs in molecular science biobased polymers, pp 123. [Online]. Available: www.epnoe.eu

14. Azmana M, Mahmood S, Hilles AR, Rahman A, Bin Arifin MA, Ahmed S (2021) A review on chitosan and chitosan-based bionanocomposites: promising material for combatting global issues and its applications. Int J Biol Macromolecules 185:832–848. https://doi.org/10.1016/j.ijbiomac.2021.07.023

15. Xue C, Wilson LD (2021) An overview of the design of Chitosan-based fiber composite materials. J Composites Sci 5(6). https://doi.org/10.3390/jcs5060160

16. Devi MG, Kavya R, Dumaran Joefel J (2014) Characterization and stability studies of multilayer nano thin films. Int J Eng Res Technol 3(8):1509–1514

17. Li X, Bai H, Yang Y, Yoon J, Wang S, Zhang X (2019) Supramolecular antibacterial materials for combatting antibiotic resistance. Adv Mater 31(5). Wiley-VCH, Feb 01. https://doi.org/10.1002/adma.201805092

18. Feng et al (2021) Chitosan-based functional materials for skin wound repair: mechanisms and applications. In: Frontiers in bioengineering and biotechnology, vol 9. Frontiers Media S.A., Feb 18. https://doi.org/10.3389/fbioe.2021.650598

19. Bernkop-Schnürch A (2018) Strategies to overcome the polycation dilemma in drug delivery. Adv Drug Delivery Rev 136–137. Elsevier B.V., pp 62–72. https://doi.org/10.1016/j.addr.2018.07.017

20. Xing K, Chen XG, Kong M, Liu CS, Cha DS, Park HJ (2009) Effect of oleoyl-chitosan nanoparticles as a novel antibacterial dispersion system on viability, membrane permeability and cell morphology of *Escherichia coli* and Staphylococcus aureus. Carbohyd Polym 76(1):17–22. https://doi.org/10.1016/j.carbpol.2008.09.016

21. Clifton LA et al (2015) Effect of divalent cation removal on the structure of gram-negative bacterial outer membrane models. Langmuir 31(1):404–412. https://doi.org/10.1021/la504407v

22. Farhadihosseinabadi B, Zarebkohan A, Eftekhary M, Heiat M, Moosazadeh Moghaddam M, Gholipourmalekabadi M (2019) Crosstalk between chitosan and cell signaling pathways. Cellular and Molecular Life Sci 76(14):2697–2718, Birkhauser Verlag AG. https://doi.org/10.1007/s00018-019-03107-3

23. Xing K, Chen XG, Liu CS, Cha DS, Park HJ (2009) Oleoyl-chitosan nanoparticles inhibits *Escherichia coli* and *Staphylococcus aureus* by damaging the cell membrane and putative binding to extracellular or intracellular targets. Int J Food Microbiol 132(2–3):127–133. https://doi.org/10.1016/j.ijfoodmicro.2009.04.013

24. Helander IM, Nurmiaho-Lassila E-L, Ahvenainen R, Rhoades J, Roller S (2001) Chitosan disrupts the barrier properties of the outer membrane of Gram-negative bacteria, 2001. [Online]. Available: www.elsevier.comrlocaterijfoodmicro

25. Butnaru E, Stoleru E, Brebu MA, Darie-Nita RN, Bargan A, Vasile C (2019) Chitosan-based bionanocomposite films prepared by emulsion technique for food preservation. Materials 12(3). https://doi.org/10.3390/ma12030373

26. Qian Z et al (2009) Chitosan-alginate sponge: preparation and application in curcumin delivery for dermal wound healing in rat. J Biomed Biotechnol 2009. https://doi.org/10.1155/2009/595126

27. Ahmed S, Ikram S (2016) Chitosan based scaffolds and their applications in wound healing. Achievem Life Sci 10(1):27–37. https://doi.org/10.1016/j.als.2016.04.001
28. Tanodekaew S, Prasitsilp M, Swasdison S, Thavornyutikarn B, Pothsree T, Pateepasen R (2004) Preparation of acrylic grafted chitin for wound dressing application. Biomaterials 25(7–8):1453–1460. https://doi.org/10.1016/j.biomaterials.2003.08.020
29. Kweon DK, Song SB, Park YY (2003) Preparation of water-soluble chitosan/heparin complex and its application as wound healing accelerator. Biomaterials 24(9):1595–1601. https://doi.org/10.1016/S0142-9612(02)00566-5
30. Wang K et al (2020) Evaluation of new film based on chitosan/gold nanocomposites on antibacterial property and wound-healing efficacy. In: Advances in materials science and engineering, vol 2020. https://doi.org/10.1155/2020/6212540
31. Mohandas A, Deepthi S, Biswas R, Jayakumar R (2018) Chitosan based metallic nanocomposite scaffolds as antimicrobial wound dressings. Bioactive Mater 3(3):267–277, KeAi Communications Co. https://doi.org/10.1016/j.bioactmat.2017.11.003
32. Basumallick S, Rajasekaran P, Tetard L, Santra S (2014) Hydrothermally derived water-dispersible mixed valence copper-chitosan nanocomposite as exceptionally potent antimicrobial agent. J Nanoparticle Res 16(10). https://doi.org/10.1007/s11051-014-2675-9
33. Wang K et al (2020) Recent advances in Chitosan-based metal nanocomposites for wound healing applications. In: Advances in materials science and engineering, vol 2020. Hindawi Limited. https://doi.org/10.1155/2020/3827912
34. Bui VKH, Park D, Lee YC (2017) Chitosan combined with ZnO, TiO_2 and Ag nanoparticles for antimicrobialwound healing applications: a mini review of the research trends. Polymers 9(1). MDPI AG. https://doi.org/10.3390/polym9010021
35. Sakthiguru N, Sithique MA (2020) Preparation and In Vitro biological evaluation of lawsone loaded O-carboxymethyl chitosan/zinc oxide nanocomposite for wound-healing application. ChemistrySelect 5(9):2710–2718. https://doi.org/10.1002/slct.201904159

Chapter 10
Role of Antibacterial Agents Derived from Chitosan-Based Nanocomposites

Neha Dhingra, Anubhuti Mathur, Nishaka, and Kanchan Batra

Abstract Chitosan is an abundant naturally sourced polymer that exhibits coveted properties of biodegradability, bioactivity, non-toxicity, and polycationic character and their significant antimicrobial, antitumor, antioxidative, anticholesterolemic, hemostatic, and analgesic ramifications. Chitosan-based nanoparticles have revolutionized antibacterial dispersions owing to their capacity to amplify the action of the parent chitosan's inherent effect. The degree of deacetylation of parent chitosan, molecular weight as well as concentration and size of nanoparticles is the various factors affecting its activity. The proof of its versatility has been its crucial applications in per-oral administration of drugs, delivery of parenteral drugs, delivery of vaccines, delivery of non-viral genes, delivery of brain targeting drugs, delivery of mucosal drugs, delivery of ocular drugs, instability improvement in controlled drug delivery of drugs, and tissue engineering. Chitosan-based system efficacy in both organic and inorganic formulations has been favored by its polycationic character which interacts with the negatively charged residues of macromolecules at the exterior. There are numerous pieces of evidence enumerating chitosan-based nanoparticles' antibacterial properties involving testing against detrimental microbes like *Escherichia coli*, *Klebsiella pneumoniae*, *Pseudomonas aeruginosa*, and *Staphylococcus aureus*. Currently accelerated development of chitosan-based nanoparticles is being carried out in the medical industry for the creation of improved wound dressings, dental and orthopedic mediums, and drug delivery carriers, in defense against plant pathogens, and in the food packaging industry which has increased its relevance in the mainstream.

Keywords Chitosan · Nanocomposites · Antibacterial · Antimicrobial · Bacteria · Applications

N. Dhingra
Department of Zoology, University of Delhi, Delhi, India

A. Mathur · Nishaka · K. Batra (✉)
Department of Zoology, Kalindi College, University of Delhi, Delhi, India
e-mail: kanchanbatra02@gmail.com

© The Author(s), under exclusive license to Springer Nature Singapore Pte Ltd. 2022 221
S. Gulati (ed.), *Chitosan-Based Nanocomposite Materials*,
https://doi.org/10.1007/978-981-19-5338-5_10

Abbreviations

APDI	antimicrobial photodynamic inactivation
ChNP	chitosan nanoparticles
CM	cell membrane
CS	chitosan
CS	composite films
CSF-Chi-NPs	colony stimulating factor chitosan-based nanoparticles
CS-NPs	chitosan nanoparticles
DD	degree of deacetylation
DS	degree of substitution
EOs	essential oils
GM-CSF	granulocyte-macrophage colony-stimulating factor
HMw	high-molecular-weight
LDPE	low-density polyethylene
LMw	low-molecular-weight
LPS	lipopolysaccharide
MBC	minimum bacterial concentration
MB	methylene blue
MIC	least inhibitory concentration
MWs	molecular weight
NHCS	N-hexanoyl chitosan
NPN	N-phenyl-1-naphthylamine
OM	outer membrane
OmpA	outer membrane protein A
QCh/PVP	quaternized chitosan/polyvinylpyrrolidone
RR	reactive dye
TA	teichoic acid

1 Introduction of Chitosan-Based Nanocomposites

Public health concerns about pathogenic microorganisms have surged recently. This in turn has increased the demand for safe and effective treatments that are less likely to promote resistance development. Due to this urgent need, nanotechnology has emerged to play a significant role, owing to its potential to amplify the efficacy of antibacterial treatment, molecular biology, pharmaceuticals, cell biology, and detection methods that have all seen significant advances. Chitosan-based nanosystems have piqued interest over the years because of their adaptability, biocompatibility, and biodegradability, particularly for the construction of mixed systems with better attributes. Chitosan's antibacterial action has been used in a variety of applications, spanning from agriculture to biomedicine [1].

An ideal antimicrobial material should be

(1) synthesized quickly and cheaply,
(2) robust in repeated applications and preservation at the appropriate application temperature,
(3) be insoluble for the purpose of water treatment,
(4) does not breakdown into harmful compounds or create them,
(5) it should not be poisonous or annoying to individuals who handle it,
(6) regeneratable when activity is lost, and
(7) biocidal to a broad spectrum of bacteria [2, 3].

Chitosan has long been known for its extraordinary characteristics, and it has been utilized in agricultural fields, industry, and medicinal drugs. In the field of agriculture, the characterization of chitosan is an antiviral for plants, a component of liquid multicomponent fertilizers, and an agent for metal-recovery in agriculture and industry [4]. It has been used in cosmetics as an agent for film-formation, a dye binder for textile, a paper strengthening addition, as well as a hypolipidemic diet ingredient. There has been a wide use as a biomaterial owing to its immuno-stimulatory qualities, anti-coagulant capabilities, antibacterial and antifungal activity, and activity serving as a wound-healer in the case of surgery.

Chitosan is nature's second most abundant polysaccharide. It has regained appeal to the fact that bacteria have not developed resistance to it. Chitin and its derivatives are water-insoluble as well as hydrophobic natural bio-polymers that have been proven to be biocompatible and biodegradable. Chitin is formed of -(1,4)-linked 2-acetamido-2 deoxy—D-glucose, whereas chitosan is a 1,4-linked 2-amino-deoxy-D-glucan produced from partial N-deacetylation of chitin [5]. Chitin's deacetylation, a polysaccharide abundantly distributed in nature (like in crustaceans, insects, and fungi), is a simple way to get it. Because of its low solubility in aqueous mediums, chitin is less likely to be appropriate for commercial use. It is a polymer constituted up of acetylamino-D-glucose units, whereas its chemical composition is more difficult to determine. It generally refers to a collection of polymers that have their molecular weight and degree of deacetylation determined by the sugar units' number per polymer molecule (n) [6]. Despite the fact that the degree of deacetylation impacts chitosan's solubility in aqueous mediums, it is soluble in acidic solutions and very feebly soluble in weak alkaline solutions [7]. There are many techniques to investigate the antibacterial activity of the polymer and its derivatives that have resulted in a myriad of physical forms of chitosan, ranging from the original solution employed in agriculture activities to film configuration in the food industry and universal medicinal nanostructured substances. The insertion of a polyanion such as tripolyphosphate (TPP) under constant stirring in a chitosan solution can easily produce chitosan nanoparticles (ChNP). When compared to the parent chitosan, these nanoparticles show more activity. Chitosan has antimicrobial properties that include bacteria, filamentous fungus, yeast, and even viruses. ChNP has been shown to exhibit antibacterial action against *E. coli, S. aureus, S. mutans, Salmonella typhimurium, Salmonella choleraesuis,* and *Pseudomonas aeruginosa* in studies [8–11].

2 Mode of Action

Even though chitosan and its derivatives' exact processes behind the antibacterial characteristics are unknown, its antibacterial action is recognized to also be regulated by several factors that operate sequentially and distinctly. According to common opinion, the existence of amine groups (NH_3^+) in glucosamine is the fundamental property of chitosan's ability to interact with surface components of many bacteria that are negatively charged, producing substantial changes to the cell surface and ultimately cell death [12]. Gram-positive bacteria's teichoic acid and gram-negative bacteria's lipopolysaccharide are both important in chitosan binding in this situation, and disruption of cell wall dynamics resulted in changes and instability in cell membrane function [13]. According to a study by Jeon [14], the antibacterial activity of the chitosan microparticles is executed through microparticle binding to bacteria. In a binding assay examination, the researchers used CM-coated glass slides with the *E. coli* O157:H7 as a bacterial study. Scientists conducted several tests to identify the specific target for chitosan microparticles binding after demonstrating that contact was required for bactericidal effects, hypothesizing that surface-exposed proteins were the primary targets. They discovered that the Omp A protein, which is present on the outer membrane, was involved in the binding process using gene deletion procedures [14].

Antibacterial mechanisms begin at the cell surface and destroy the cell wall or the outer membrane to start with, even though gram-negative and gram-positive bacteria have distinguishable cell walls. Lipoteichoic acid has the potential to form a molecular connection with chitosan on the cell surface, permitting it to damage membrane functions in gram-positive bacteria [15]. Electrostatic exchanges with divalent cations maintain the gram-negative bacteria's outer membrane together, which is required to maintain the outer membrane stable [16]. Because the pH of chitosan and its derivatives is below the pKa, polycations may contest with divalent metals for mixing with polyanions. As the pH climbs over pKa, it shifts to chelation. Mg^{2+} and Ca^{2+} ions exchange in the cell wall is likely to interrupt cell wall stability or alter degradative enzyme function. Several methods have been employed to illustrate the disruption of cell wall integrity. A hydrophobic probe, N-phenyl-1-naphthylamine (NPN), in general, is excluded from the outer membrane. N-phenyl-1- naphthylamine can divide into perturbed outer membrane. When outer membrane is functionally damaged invariably in nanoindentation tests, the slopes of the deflection curves for *E. coli* and *S. aureus* administered with chitosan oligomers were less than for the bacteria which were not treated, indicating that cell indentation or cell compaction had happened as a result of less rigid cells after the treatment [17]. The studies suggest that cell walls are weakened, either as a consequence of cell wall destruction or as a result of cell lysis. When the protective covering of the cell wall is lost, the cell membrane becomes vulnerable to the environment. As a result, the processes of the cell membrane can be modified, and the permeability of the membrane can be substantially compromised [13]. The permeability of the cell membrane, which in itself is a negative charge phospholipid bilayer, can vary significantly when chitosan comes into touch with it. The bacterial surface charge is soon neutralized and even reversed by the binding [18]. Additional interactions might denature protein molecules and allow phospholipid bilayer invasion. Increased membrane permeability causes cell membrane instability and intracellular substance leakage, resulting in cell death. This significant alteration in the nature of membrane proteins is believed to play a role in antibacterial activities as of yet.

3   Factors Affecting chitosan's Antibacterial Activity

Chitosan's antibacterial action is determined by a variety of intrinsic and extrinsic variables. The degree of deacetylation and molecular weight of parent chitosan, along with the concentration and size of the nanoparticles, is all intrinsic variables. pH, temperature, and reaction time are examples of external variables (Fig. 1).

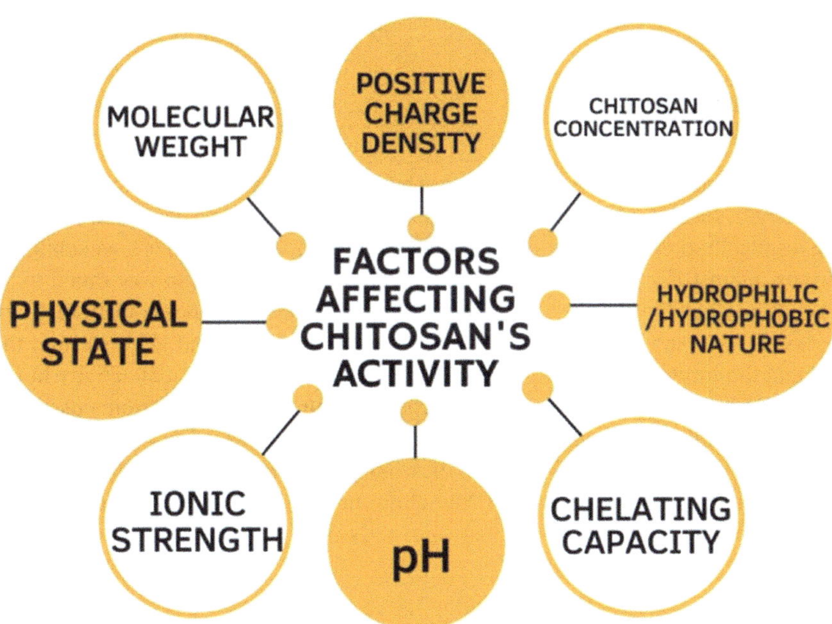

Fig. 1 List of the various factors that affect the activity of chitosan-based nanoparticles

3.1 Concentration of Chitosan

At lower concentrations, chitosan binds to negatively charged cell surfaces and disrupts the cell membrane, enhancing cell death by facilitating intracellular component leakage, whereas, at higher concentrations, protonated chitosan covers the cell surface and blocks intracellular component outflow. Agglutination is also deterred because positively charged bacterial cells resist one other. [19]. By uniformly ingraining 1, 3, and 5% chitosan (w/w) in low-density polyethylene (LDPE) matrix, an antimicrobial packaging material was created [20]. The antibacterial effectiveness of LDPE/chitosan composite (LDPE/CS) films on virgin LDPE films was shown in an *E. coli* assay. Virgin LDPE and 1%, 3%, and 5% LDPE/CS films were tested as chill-stored tilapia packaging films, and samples packaged in LDPE films were rejected by the seventh day, whereas fish sealed in 1%, 3%, and 5% LDPE/CS coatings were fine for up to 15 days. 3% LDPE/CS films had more improved physical and antibacterial qualities and enhanced the maintained quality of tilapia steaks while it was in chilled storage, as compared to other films [21].

3.2 Molecular Weight

When it pertains to the association between bactericidal activity and chitosan's molecular weight, several studies on chitosan's bactericidal effect have generated inconsistent results. In some investigations, increasing chitosan Mw resulted in lower chitosan activity against *E. coli* while in others, high-Mw (HMw) chitosan was found to be more active than low-Mw (LMw) chitosan. Furthermore, regardless of Mw, activity against *E. coli* and *Bacillus subtilis* were shown to be identical [22]. Despite the fact that the limited availability of the results on various bactericidal activity in the case of LMw chitosan was equivalent depending on the bacterial strains used, for the biological testing circumstances, and related chitosan Mw, the results were not consistent. For example, 9.3 kDa chitosan inhibits *E. coli* growth, while 2.2 kDa chitosan promotes it. In addition, LMw chitosan (4.6 kDa) and its derivatives had superior antibacterial, yeast, and fungicidal activities [13].

3.3 Positive Charge Density

The part of the polycationic structure in antibacterial activity has been sufficiently documented. Powerful electrostatic contact is driven by a bigger positive charge density. The positive charge is correlated to the degree of substitution (DS) of chitosan or its derivatives, which influences positive charge density. To some degree, chitosan microspheres with a high DD (97.5%) have a larger positive charge density that causes stronger antibacterial action on *Staphylococcus aureus* at a pH of 5.5 than those with a moderate degree of deacetylation (83.7%) [23]. According to one study, a greater degree of deacetylation with a more positive charge was notably effective in preventing *S. aureus* growth, implying that chitosan's antibacterial efficacy against *S. aureus* was boosted with greater DD [15]. The antimicrobial action of chitosan derivatives is mainly defined through the grafting groups' DS. The antibacterial activity of water-soluble N-alkylated disaccharide chitosan derivatives against *Escherichia coli* and *Staphylococcus aureus* was examined, and it was found that the DS of disaccharides and the kind of disaccharide present in the molecule affect the antibacterial activity of chitosan derivatives [24]. In regard to the exact study, a DS of 30–40% induced the most powerful antibacterial activity against *E. coli* and *S. aureus,* regardless of the type of disaccharide linked to the chitosan molecule, among the myriad chitosan derivatives, the two bacterias are especially vulnerable to cellobiose chitosan derivative DS 30–40% and maltose chitosan derivative DS 30–40%, respectively [13].

3.4   Hydrophilic/Hydrophobic Characteristic

Water is required for antimicrobial agents to act, regardless of their type or amount. For needed to commence contact, completely dry samples are almost impossible to liberate the energy contained within chemical bonds. Hydrophilicity and hydrophobicity are terms that describe how chitosan interacts with bacteria and are also affected by the aqueous environment. Chitosan's hydrophilic properties have a significant impact on its water solubility. Because of its low-water solubility, chitosan's application is restricted [25]. Chemical changes are effective in increasing the solubility of chitosan and its variants in water and expanding its applications [26]. Saccharization, acylation, quaternization, and metallization have all been used to make water-soluble chitosan and its derivatives, which have always been a primary focus of antibacterial studies research. Quaternary ammonium chitosan, for example, could be made by attaching a quaternary ammonium group to a dissociative hydroxyl or amino group. The antibacterial capabilities of chitosan and its derivatives are influenced by the hydrophilic-lipophilic variance. The hydrophobicity of NHCS 0.5 (N-hexanoyl chitosans, with a molar ratio of 0.5 when compared to chitosan residue) is believed to be responsible for its increased inhibitory impact [27, 28]. The inclusion of a long aliphatic chain promoted the absorption and boosted the effects of a substituted LMW chitosan, N/2(3) -(dodec-2enyl) succinoyl/chitosans, onto the cell walls using hydrophobic contact with cell wall proteins in another study [13].

3.5   Chelating Capacity

In acidic circumstances, chitosan has a sufficient chelating ability for numerous metal ions (containing Ni^{+2}, Co^{+2}, Zn^{2+}, Mg^{2+}, Fe^{2+}, and **Cu^{2+}**), and it is been widely used to remove or recover metal ions in a variety of industries [29]. The metal ions that react with the cell wall constituents of the microorganism are important for the stability of the cell wall. The chelation of these metal ions by chitosan has been proposed as a potential method of antibacterial action [30]. Chelation is competent for blending divalent metal ions in neutral conditions and also in acid conditions. Further, the chitosan metal complex is developed by chelating capacity and has substantial antibacterial action [13].

3.6 pH

Chitosan's antibacterial action is pH-dependent. Since chitosan is readily soluble in acidic conditions, thus it becomes polycationic when the pH goes under the molecule's pKa (6.3–6.5), it may be used in a variety of applications [31]. It is been stated that chitosan possesses antibacterial properties exclusively in an acidic environment, although it has not been proven. At lower pHs, chitosan has a stronger inhibitory impact, and as the pH rises, the inhibitory effect weakens. The presence of a large number of positively uncharged amino groups, and also chitosan's poor solubility, might explain why chitosan is unable to kill bacteria at pH 7 [32]. However, according to some workers, chitosan and its derivatives fully lose their antibacterial activity under neutral conditions, which may not be entirely accurate. Under neutral conditions, a novel technique for antibacterial study using chitosan microspheres (CM) in a solid dispersion system revealed that the CM sample with DD of 62.6% exerted the only inhibitory effect among the three DD (97.5, 83.5, 62.6%) [11]. The characteristics of native chitosan were preserved in the CM samples used in this study. Another study found that the antibacterial activity of N-alkylated chitosan derivatives (DS 30–40%) against *E. coli* increased as the pH climbed from 5.0 to 7.0–7.5, with a peak around 7.0–7.5 [33]. These findings also show that a positive charge on amino groups is not the only thing that causes antibacterial activity. However, nothing is known about chitosan's antibacterial action in alkaline environments [13].

3.7 Ionic Strength

Any changes in the ionic strength of a medium can affect the inhibitory effect of chitosan, which is presumably due to two different mechanisms. Firstly, a gain in metal ions, mainly divalent ions, can lower chitosan's chelating capability's efficacy. The inhibitory ratio of chitosan samples reduced dramatically when 0.05 mol/L magnesium ions were added to the medium, resulting in abrogated antibacterial action [23]. In another study, divalent cation doses of 10 and 25 mM reduced shrimp chitosan's antibacterial effectiveness against *E. coli* in comparison with Ba. Furthermore, when compared to Ba^{2+}, Mg^{2+}, Ca^{2+}, and metal ions decreased, the antibacterial action of chitosan was the most successful when the Zn^{2+} and Ca^{2+} ions were added [34]. Secondly, along with polycationic chitosan, cations already present in the medium may compete with the negative components that predominate on the bacterium's cell wall, reducing antibacterial efficacy. The antibacterial effectiveness was also modified by the addition of anion [13].

3.8   Physical State

Chitosan's antibacterial activity is the outcome of many reactions rather than the driver of them. Interaction of chitosan molecules with the cell wall helps in producing processes. The morphology of molecules determines the effectiveness of processes. Similarly, chitosan's physical state, which dictates molecular form, has a massive effect on its antibacterial action [13].

3.8.1   Soluble State Antimicrobial Activity

The dissociating form of soluble chitosan in solution as a dissociating form allows for sufficient reactivity with the counterparts and brings all of the possibilities to life. This explains how soluble chitosan and its variations are often more efficient in inhibiting bacterial growth than insoluble chitosan and its variants [34]. The least inhibitory concentration (MIC) of chitosan derivatives is much lower than that of native chitosan against all tested bacteria. Meanwhile, soluble chitosan and its derivatives are readily affected by both external and intrinsic influences due to their substantial contact with the solution. In one study, Maillard reaction-produced chitosan derivatives (chitosan, fructose, glucose, glucosamine, and maltose) improved the solubility of native chitosan [34]. Incorporating hydrophilic groups into the molecule, such as in quaternary ammonium chitosan, is yet another important way to improve chitosan solubility. After quaternization, the derivative exhibits more solubility in water and better antibacterial activity than chitosan [13].

3.8.2   Solid State Antimicrobial Activity

Solid chitosan, unlike soluble chitosan, only comes into contact with a solution through the surface, such as fibers, hydrogels, membranes, microspheres as well as nanoparticles. The hydrogels are created by covalently cross-linking chitosan with itself. Several initiatives have lately been made to produce chitosan particle systems capable of forming large reactive surface areas as a dispersion in solution. As its physical condition alters, its antibacterial potency will surely vary. Nanoparticles have a weaker inhibitory effect on *S. aureus* ATCC 29737 than polymers in free soluble form because they have less positive charge available to attach to the negative bacterial cell wall [35]. Another study discovered that chitosan nanoparticles have more antibacterial activity than chitosan due to the nanoparticles' unique properties, such as their larger surface area and increased affinity with bacterial cells, resulting in a quantum-size effect [13].

3.9 Temperature and Time

Composing chitosan mixtures in mass and reserving them for future use in commercial applications would have been viable. With preservation, chitosan parameters like viscosity and MW may alter. Due to this, a chitosan solution's viscosity should be regulated since it might further change the solution's other operational properties. Chitosan solution's stability (MW of 2025 and 1110 kDa) and its antibacterial effectiveness on gram-positive (*Listeria monocytogenes* and *S. aureus*) and gram-negative (*E. coli* and *Salmonella enteritidis*) bacteria were analyzed after 15 weeks of storage at the temperatures 4 °C and 25 °C [36]. The antibacterial action of chitosan solutions was greater before storing than after 15 weeks. Chitosan solutions held at 25 °C had an antibacterial effect equivalent to or less than those kept at 4 °C. The sensitivity of *E. coli* to chitosan expanded as the temperature was increased from 4 to 37 °C in one investigation, implying that low-temperature stress might change the cell surface configuration in a manner that diminished the number of surface binding sites (or electronegativity) for chitosan derivatives [13].

3.9.1 Microbial Factors

Microbial Species

Despite its broad antibacterial activity, chitosan has varying inhibitory efficacy against fungi, gram-positive, and gram-negative bacteria. They display antifungal properties by inhibiting the process of spore germination as well as sporulation [37]. By contrast, the antibacterial activity mode is a complicated process that differs between gram-positive and gram-negative bacteria caused by changes in cell surface features. In many investigations, pH values of gram-negative bacteria were found to have stronger antibacterial activity than gram-positive bacteria [38, 39], whereas gram-positive bacteria were found to be more vulnerable in another, probably due to the gram-negative outer membrane barrier. Regardless, multiple studies identified no significant differences in antibacterial activity and microbes. Varied initial reaction materials and settings influence the various outcomes. Based on present data, bacteria appear to be less responsive to chitosan's antibacterial activity than fungus. Chitosan has a higher antifungal effect at low-pH values [13, 40].

- Part of Microorganism

Gram [SG1]-negative bacteria have a lipopolysaccharide (LPS)-containing outer membrane (OM). The bacteria have a hydrophilic outer surface as a result of this, electrostatic interactions between divalent cations alter the overall stability of the LPS layer due to anionic groups (phosphate, carboxyl) in the lipid components and inner core of LPS molecules. When these cations are removed by chelating substances like ethylenediaminetetraacetic acid, the OM is destabilized, resulting in the release of LPS molecules.

Gram-negative bacteria are often resistant to hydrophobic medicines as well as hazardous pharmaceuticals since OM acts as a penetrating barrier against macromolecules and hydrophobic chemicals. As a result, any material that wishes to exhibit bactericidal activity against gram-negative bacteria must first overcome the OM [23] teichoic acid (TA) and peptidoglycan (PG) that make up gram-positive bacteria's cell wall. TA is a polyanionic polymer that traverses the cell wall of gram-positive bacteria to make touch with the PG layer. They can be covalently attached to the N-acetylmuramic acid in the peptidoglycan layer. Teichoic acids and glycolipid remain anchored into the cytoplasmic membrane's outer leaflet forming lipoteichoic acids (LTA).

Because the OM works as a penetrating barrier against macromolecules and hydrophobic substances, gram-negative bacteria are generally immune to hydrophobic medicines and hazardous medications. The poly (glycerol phosphate) anion groups in TA are responsible for the cell wall's structural integrity. Furthermore, it is essential to the functioning of a variety of membrane-bound enzymes. In gram-negative bacteria, TA's counterpart LPS has a comparable effect on the cell wall [15].

Cell Age

The age of a cell can affect antibacterial effectiveness for a certain microbial species. *S. aureus* CCRC 12657, for example, is particularly vulnerable to lactose chitosan derivative mostly in the late exponential phase, with really no viability observed following 10 h of incubation. Cells mostly in mid-exponential and late-stationary stages, on either hand, had less population decline in cell viability. Modifications in cell surface electrical negativity are thought to alter with the growth phase, which may also contribute to variances in cell sensitivity to chitosan. In comparison, mid-exponential *E. coli* O157:H7 cells have been the most vulnerable, while stationary phase cells have been the least sensitive. Because the surface charge of microbial cells depends greatly upon the microbe, the inconsistencies were attributed to the microbial species studied [41].

4   Complexes of Chitosan with Certain Materials

To improve the antibacterial effect, chitosan materials containing certain components might be produced. The use of essential oils (EOs) in chitosan-based coatings has attracted agricultural researchers' attention owing to these volatile molecules' bactericidal and fungicidal properties [42]. EOs like carvacrol, clove, lemongrass, and oregano have recently been properly integrated into the chitosan, demonstrating high-antibacterial activity. Also, researchers [43] showed that chitosan and cinnamon EOs had unique compatibility, with their integration improving the antibacterial characteristics of chitosan. Cinnamon essential oils are effective for covering highly perishable foods like fish and chicken.

The structure of the chitosan/metal complexes is usually determined by the molar ratio of the chitosan to metal ions, the kind of the molecular weight, metal ion and preparation circumstances, as well as the deacetylation of chitosan [21].

4.1 Antimicrobial Activity of Chitosan/Metal Nanocomposites

Chitosan and metals were combined to create novel nanocomposite materials having modified microbicidal characteristics [44]. Silver, gold, and copper-loaded chitosan nanoparticles, in particular, are shown to have a wide range of activity against gram-positive and gram-negative bacteria. Metal ion solutions were added to chitosan nanosuspension to create these nanoparticles, and a soluble metal salt was reduced in the existence of chitosan solutions [45]. The most commonly studied metal/chitosan complexes are silver-based nanocomposites, researchers integrated chitosan films with Ag nanoparticles or Ag^+ ions at varying concentrations and examined those for antibacterial activity against *S. aureus* and *E. coli* to see whether the occurrence of silver as metallic nanoparticles or ions had been responsible for the increased antibacterial effect of the silver-based chitosan composite materials. They demonstrated that chitosan films with 1% w/w silver nanoparticles or 2% w/w silver ions had the highest antibacterial effect, indicating that Ag/chitosan nanoparticles seem to be more effective than Zn ions or Ag^{+2} [46]. Different types of chitosan-based nanocomposites have been mentioned in Table 1.

4.2 Chitosan Nanoparticles' Antimicrobial Activity in Bacterial Biofilms

In biofilms, microbial populations are encased in a sticky extrinsic polymer matrix. Antimicrobial agents are substantially more resilient to microorganisms in biofilms. Antibiofilm activity of natural biological substances is now being investigated to find alternative preventive or treatment strategies. In this regard, anti-biofilm characteristics of chitosan-streptomycin conjugates/gold nanoparticles then were assessed against gram-negative *S. typhimurium* and *P. aeruginosa* as well as gram-positive *Listeria monocytogenes* and *S. aureus* [47]. Gram-negative and gram-positive bacteria biofilms were disrupted by the chitosan-streptomycin gold nanoparticles, which also inhibited the production of gram-negative bacteria biofilms. Streptomycin was conjugated to chitosan and gold nanoparticles, which made it easier

Table 1 List of all the different forms of chitosan nanocomposites with their various applications in different fields

Name of nanomaterial	Form of chitosan nanocomposites created	Distinct application	Crucial properties	References
Ag nanoparticles	Thin film coating on bandage	Antibacterial activity against *E. coli* and *S. aureus*	Inactivation bacterial metabolism	Susilowati and Ashadi [118]
Graphene oxide	Thin film	Antimicrobial against *E. coli* and *R. subtilis*	Improved mechanical and antimicrobial properties	Grande et al. [119]
Bentonite and sepiolite	Thin film	Application in winemaking	Increased immobilization of protease although negatively affected catalyzation	Benucci et al. [120]
Sulfamethoxazole	Nanoparticle drug	Antibacterial activity against *P. aeruginosa*	Synergistic activity with sulfamethoxazole	Tin et al. [121]
Clarithromycin	Nanoparticle drug	Works against *S. aureus*	Shows antibacterial activity	Golmohamadi et al. [122]
Ciprofloxacin Chlortetracycline Hydrochloride Gentamycin sulfate	Nanoparticle drug	Works against *E. coli* and *S. aureus*	Stops the growth of gram-positive and gram-negative bacteria	Sharma et al. [123]
Azithromycin Levofloxacin Tetracycline	Nanoparticle drug	Effective against *E. coli* *S. aureus*	Shows great antibacterial effects	Brasil et al. [124]
Rifampicin Ciprofloxacin Vancomycin Doxycycline Gentamicin	Nanoparticle drug	*S. epidermidis*	Stops bacterial biofilm and demonstrates synergism with antibiotics	Ong et al. [125]

for it to penetrate the biofilm matrix and promote bacterial interaction. Chitosan nanoparticles can also be used for photodynamic activation. The effectiveness of chitosan nanoparticles in methylene blue (MB)-mediated antibiotic photodynamic inactivation (APDI) of *P. aeruginosa* and *S. aureus* biofilms was investigated [48]. Chitosan nanoparticles increased the efficacy of MB-APDI by disturbing biofilm formation and enabling MB to probe deeper and much more efficiently into biofilms of *P. aeruginosa* and *S. aureus* [48]. Biofilm forms on the enamel surface as a result of oral infections, which have been related to human caries, gingivitis, and periodontitis. To combat the biofilm, chitosan nanoparticles were utilized.

4.3 Chitosan Nanoparticles Loaded with Antibiotics or Other Microbicidal Substances Possessing Antimicrobial Activity

Chitosan has been used to transport both synthetic and biological chemicals with the goal of increasing or modulating their antimicrobial properties. Antibiotics were internalized inside cells infected with intracellular bacteria or their efficiency versus multiresistant bacteria was boosted using chitosan nanoparticles. In particular, researchers discovered that ceftriaxone sodium was absorbed better by Caco-2 and J774.2 (macrophages) cells when it was encapsulated in chitosan nanoparticles, as well as that these nanoparticles had a stronger intracellular antibacterial action against *S. typhimurium* than with the drug in the solvent [49]. Tetracycline-loaded O-carboxymethyl chitosan nanoparticles were used in a similar [50] study, and the drug-loaded nanoparticles were reported to boost antibiotic effectiveness versus *S. aureus*. Jamil et al. investigated the efficiency of cefazolin-loaded chitosan nanoparticles against multiresistant gram-negative bacteria such *K. pneumoniae, E. coli,* and *P. aeruginosa.* As per the agar well-shared understanding and microdilution broth test, the drug-loaded chitosan nanoparticles exhibited antibacterial activity against the three pathogens which were stronger than cefazolin in solution. Antimicrobial peptides and proteins have recently been incorporated into chitosan nanoparticles. Among these compounds, lysozyme has attracted much interest since it is used as a preservative in food and medicines [51], while the amphiphilic peptide temporin B has been demonstrated to be beneficial against various bacterial species [51, 52].

5 Applications of Chitosan-Based Nanosystems' Antimicrobial Activity

5.1 Antimicrobial Agent

In the past several years, chitosan-based nano systems have drawn attention for their compatibility, degradability, and diversity, particularly for the formation of mixed systems with better attributes. Chitosan-based systems' antibacterial activity has been used in a variety of applications, spanning from agriculture to biomedicine [53]. The following parts describe the advancements of chitosan-based nanoparticles in the areas of wound healing, textiles, and food packaging. Nanoparticle-based material's antibacterial properties are also useful in a variety of medical applications like burn and wound dressings, filters, medical devices, and materials used in dental plaque reduction [41] (Fig. 2).

When scientists were researching antimicrobials in nature to replace synthetic chemicals, they came across chitosan and chitosan nanoparticles. The antimicrobial

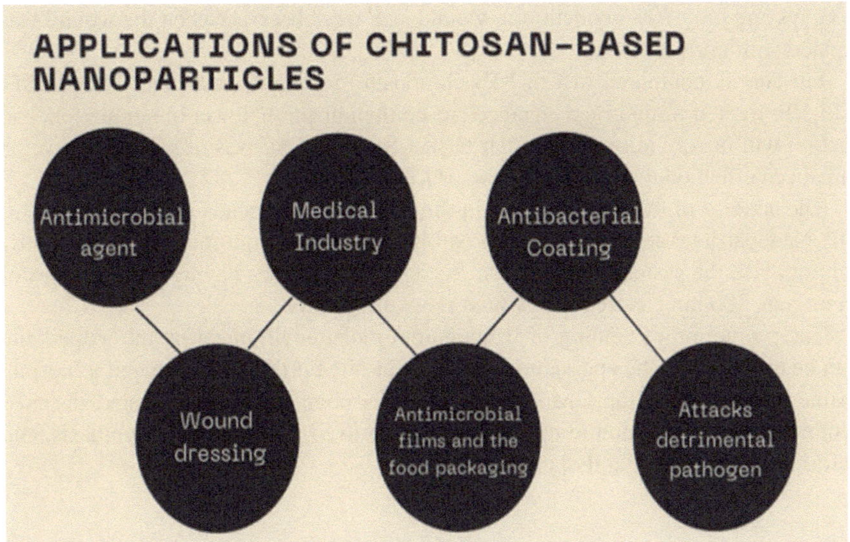

Fig. 2 List of the multifaceted applications of chitosan-based nanoparticles

activity of chitosan is the result of the interaction of negatively charged phospholipid of the plasma membrane with the positively charged chitosan, altering cellular permeability and ultimately resulting in cell death. Chitosan's ability to chelate metal ions, well explains its antibacterial characteristics. Chitosan has the ability to pass through the cell wall and bind with DNA and inhibit transcription. Chitosan nanoparticles can also be formed by decomposing chitosan with hydrogen peroxide. To make antimicrobial paper, after combination with pulp, it was impregnated, dispersed as a coating on hand sheets, and insufflated. *Staphylococcus aureus* and *E. Coli* were discovered to be the most resistant bacterial species to the paper generated by insufflations [54]. Qi et al. (2004) investigated chitosan NP and ChNP loaded with copper against *E. coli, S. aureus, S. choleraesuis, and S. typhimurium,* in vitro [8]. According to the findings, the activity of all the bacteria was suppressed far. Their MBC values exceeded 1 lg/ml, and their MIC values were less than 0.25 lg/ml. *Aspergillus niger* and *Candida albicans* were resistant to low-molecular weight ChNP, and *Fusarium solani* was resistant to high-molecular weight ChNP [55].

5.2 *Wound Dressing*

Studies that were done in vivo revealed that wounds closed completely when coated with the granulocyte–macrophage colony-stimulating factor and sargramostim chitosan-based nanoparticles composites and complete re-epithelialization after

13 days and only 70% reduction in wound size was observed when the wound was treated with normal saline.

Further, as compared to Chi-NPs composite treated wounds, the GM37 CSF-Chi-NPs treated wounds had enhanced re-epithelialization, lower inflammation, and fast growth of formation of granular tissue. So, GM-CSF was observed to have an enhanced effect on the wound healing process [56].

The activity of the nanoparticles in the wound area depends on the size of Chi-NP. Nanoparticles less than 100 nm can be eliminated from the wound region by migrating to the systemic circulation. As a result, Chi-NPs having particle sizes of more than 100 nm can treat the wound more effectively.

The promotion of healing of wounds in a moist environment at the wound site can be achieved by the application of CSF-Chi-NP. Further, as compared to normal saline and Chi-NPs, the GM 37-CSF-Chi-NPs composite demonstrated fastened wound healing, reduction in inflammation, improved granular tissue synthesis, and reformation of epithelial tissue [56].

5.3 Antibacterial Activity Against Plant Pathogens

Chitosan-based nanoparticles may have a long-term and consistent effect on plant development and protection. CS has been used to lower the severity of phytopathogen diseases and to increase plant innate immunity [57, 58]. CS's antibacterial and immune-stimulating qualities make it an effective antimicrobial product for the control of plant diseases. Nanotechnology is a promising field for developing materials to combat pathogens affecting plants. Cu-CS-NPs are believed to be a promising composite for an increase in plant growth and protection. They are a very effective antibacterial agent because of their outstanding ability to endure pathogenic attacks on plant growth. The inhibition of phytopathogenic bacteria by CS and CS-based NPs has been studied [59, 60]. 1129 kDa, DD 85 peCS-NPs may have a long-term and consistent effect on plant growth and development. The bacteria *Xanthomonas, which* causes bacterial leaf spots on *Euphorbia pulcherrima*, was significantly inhibited when treated with chitosan at a dose of 0.10 (mg/mL) [61]. By treating and soaking of seeds of tomato plants at 10.0 (mg/mL), bacterial wilt caused by *Ralstonia solanacearum was* decreased to 48% and 72%, respectively [62]. Similarly, leaf diseases in rice produced by different species of *X. oryzae pv.* were suppressed by treatment of leaves (0.20 mg/mL) with two distinct CS solutions, i.e., solution-A of MW 1129 kDa, DD 85%, and solution-B of 607 kDa, DD 75% [63]. In another study, it was found that spraying solution-A on leaves of *Acidovorax citrulli* which causes fruit blotch of watermelon reduced the disease severity significantly [64].

The synthesis, evaluation, and testing of the antibacterial efficacy of CS/TiO_2 NPs were done in rice pathogen *X. oryzae pv. Oryzae by* Li et al. [60]. Significant inhibition of bacteria *Xanthomonas* species over *E. pulcherrima* was observed [61]. Further, the severity of bacterial speck disease by *P. syringae* pv. in tomato was greatly reduced when tomato seedlings were treated with CS which confirmed the role of CS as a non-toxic biopesticide [58]. CS-NPs have high-antibacterial action against *P. fluorescens* and *Erwinia carotovora*, two bacteria that cause soft rot [65]. CS-NPs effectiveness was observed by Esyanti et al. in chili peppers for lowering the disease severity of *X. campestris* and so can be considered as a bactericidal agent [66]. Antibacterial activity of three strains of pathogenic bacteria like *E. carotovora* subsp. carotovora and one strain of *X. campestris* pv. vesicatoria was studied by Oh et al. [67]. The Ag-NP-CS had remarkable antibacterial activity when tested for *R. solanacearum* which causes tomato wilt [68]. It was also observed in another study that NPs derived from CS including Ag-NPs extracted from leaves were a long-term and good alternative in agriculture. These findings substantiate the use of CS-NPs for better crop yield and for the control of phytopathogens without causing any harm to soil properties or the environment [69]. The potential of ChNP to act as a coating material for prolonging the shelf life of tomato, chili, and brinjal was also studied. The addition of ChNP to soil boosts microbial populations while also improving nutrient availability. ChNP has been approved as a growth promoter, speeding up seed germination, plant growth, and agricultural yield. ChNP is an antimicrobial compound that works against a variety of bacteria, viruses, fungi, and bacteria. As a result, it has the potential to be a viable alternative to many chemical fungicides and bactericides in the control of many plant diseases. ChNP is a great fruit and vegetable coating material due to its non-toxicity and biodegradability. Many fruits, including strawberries, papaya, carrot, cucumber, citrus, apple, kiwifruit, pear, peach, sweet cherry, and strawberry, may benefit from ChNP coating to extend storage life and reduce decay. Functional elements such as antibacterial agents and nutraceuticals could also be included in ChNP coating.

5.4 Antimicrobial Films and the Food Packaging

Chitosan has remarkable antibacterial efficacy due to its polycationic characteristic. As a result, films based on chitosan have a usage in the food packaging industry, as they can protect food from microorganisms and ensure food safety. Although chitosan revealed a broad spectrum of antibacterial action, many factors could influence its antimicrobial activity. Organic acid, molecular weight, concentration, temperature, film diameters, microbial type, etc., influenced the antibacterial activity of films made up of chitosan. In certain cases, the inclusion of other materials, such as Ag-NP and propolis extract, was required to inhibit the tested microorganisms [70, 71]. The antibacterial effectiveness of the films formed on chitosan was proved to be different for gram-negative and gram-positive bacteria and observed to be more effective in gram-negative bacteria. Furthermore, some studies claim that an increase in the molecular weight of chitosan enhanced the antibacterial effect of chitosan against gram-positive bacteria by forming a film that hindered the entry of nutrients into the bacterial cell [72]. However, the antibacterial action of low-molecular weight chitosan was increased against gram-negative bacteria because low-MW chitosan could easily enter the microbial cell and resulted in the breakdown of metabolism [73].

There is evidence that the different gram-positive bacteria showed different antibacterial responses [74]. Due to differences in the outer membrane structure, *Enterococcus faecalis* and *Listeria monocytogenes* were more susceptible than *Staphylococcus aureus* to chitosan-based films [72]. Overall, the produced films demonstrated outstanding antibacterial activity against poisoning microorganisms such as *lactococcus lactis* [74] and *listeria innocua* [75], gram-negative bacteria *Pseudomonas spp.* [76–78], *Salmonella spp.* [79–81], fungal mold [82, 83], and *Candida albicans* [84]. There are some distinctions between fungi and bacteria in terms of antibiotic action. At low temperatures, chitosan inhibits bacteria more effectively than fungus, as per some research. Mold spores were more resistant to cold temperatures than bacteria. Continuation of germination of mold spores even at low temperatures was the reason for less fungistasis. Despite the differences, chitosan films were successful in protecting food from mold.

Therefore, different antimicrobial theories on food state that

1. Chitosan-based films create cellophane-like structures on the food surface, therefore acting as a protective barrier and shielding the food from microbial attacks [85].
2. Chitosan-based coatings can function as an oxygen barrier. Thus, preventing oxygen transport and so inhibit respiratory activity and bacterial development in food [82].
3. Chitosan can diffuse into a microbial surface and barrier made from polymer and create an obstruction. And neither does the membrane chelate nutrients on bacteria's surfaces, but it also prevents nutrients and critical components from entering the microbial cell. This action of chitosan the alters the microbes' physiological activity and thus kill them [86].
4. Chitosan's NH_3^+ groups [87] can disrupt the negative phosphoryl groups on the bacterial cell membrane, causing distortion and then deformation [88, 89].
5. Furthermore, various bacteria were sensitive to the NH_3^+ groups of chitosan in diverse ways. Chitosan can permeate through the cell wall, distort the bacterial membrane, and result in intercellular electrolyte leakage and the death of cells [90, 91].
6. Chitosan can enter the nucleoid, alter the structure of DNA, and prevent DNA replication, RNA transcription, and protein translation [92].
7. Chitosan can easily cause chelation of nutrients and critical metals inside bacteria. The action of these materials can be lost, and the chelation complex will not be available to microorganisms so limiting microbial development. Chitosan can cause the formation of chitinase enzymes in fruits, which destroys the microbial cell walls indirectly. By encouraging the buildup of phenolics and lignin, chitosan-treated wheat seeds were able to promote resistance to *Fusarium graminearum* and increase seed quality [93].

5.5 Application in Medical Industry

Biocidal polymers are disinfectants that may be integrated into fibers, membranes, or hydrogels and employed in wound dressings, orthopedic therapies, hemodialysis, and medication carriers. An authentic dressing material for a wound should be able to soak released liquid from the affected area, allow for controlled water evaporation, and prevent microbiological transmission [94]. The candidate's eligibility and capability are determined by the antimicrobial property assessment, which is a significant measure in wound dressing. The use of polysaccharides with hydrogel-forming properties, such as chitosan, as wound dressing materials, has been deemed beneficial [95]. The products made from chitosan have attracted a lot of attention in this respect.

Chitosan is available in four different forms that provide antibacterial qualities to wound dressing material, i.e., fiber, membrane, hydrogel, and sponge. These methods rely on the chitosan's physicochemical properties. Wound dressings can be made using micro- and nanofiber materials. Electrospinning is one of them, and it is a good way to make continuous polymer threads with nanoscale diameters [96]. Electrospun mats constructed of ultrafine polymer fibers have sparked a lot of attention due to their unusual qualities like nanosize, more porosity, and high-surface-to-volume ratio. In another study, the growth of gram-negative and gram-positive bacteria were found to be inhibited by cross-linked electrospun-like PVC/QCh (polyvinylpyrroliodone/quaternised chitosan) [97]. PVP and PVA both are indeed non-toxic, hydrophilic, and biocompatible, with good complex and film-forming capabilities, making them suitable for wound dressing [14].

5.6 Antibacterial Coating

As discussed earlier, the positive amino group of chitosan can bind with the negatively charged microbial cell membrane. Proteinaceous along with other intracellular substances of the microorganisms pour out as a result. One explanation for chitosan's antibacterial effects could be this [98]. The migration of chitosan on the bacteria's outer surface is another reason. Polycationic chitosan at the concentration of 0.2 mg/ml can bind with negatively charged bacterial surfaces and agglutination can occur. Bacterial surfaces may become positive as a result of the increased number of positive charges, allowing them to remain in suspension at higher concentrations [99, 100]. Chitosan causes *Pythium oaroecandrum* to lose a large quantity of protein content when grown at pH 5.8, as per UV absorption measurements [101]. Chitosan's chelating property helps it to bind with metals and inhibits the formation of toxic chemicals and thus stops the multiplication of microbes. It can serve as a water binder and also block certain enzymes by inducing host tissue defense mechanisms. As a result of its penetration into the microorganism's nuclei interference occurs with transcription and translation [20]. The antibacterial activity of chitosan may be influenced by its own structure, degree of its polymerization, nature of the host, the natural nutrient constituents, the chemical makeup of the substrates, and the variables of the environment. Antimicrobial compounds have been used to cover surfaces in order to prevent the development of pathogenic bacteria.

Chitosan is still used as an antimicrobial film due to its biocompatibility, biodegradability, antibacterial effect, and cytotoxicity [99]. Acidic solutions as compared to neutral and alkaline environments result in the grabbing of dye at higher levels [102]. In acidic solutions, protonation of amino groups of chitosan molecules to produce $-NH_3^+$ groups can occur easily as a large number of protons are available [103, 104].

According to the researchers, the diameter of pores in chitosan beads or their number was decreased by chemical crosslinking. As a result, the transfer of the dye molecule over chitosan was more challenging [100, 102]. Chitosan was used to transform an acrylic resin into an antibacterial covering material. Chitosan was injected into the polymer matrix in two states: solid (powders) and colloid (liquid). The purpose was to establish a homogeneous particle dispersion and to improve particle influence by increasing the number of contact sites.

5.7 Application Against Various Detrimental Bacteria (Animal Pathogens)

The antimicrobial activity of CS-NPs against many pathogenic microorganisms is broad. Ikono et al. 2009 [105] investigated the effects of CS-NP on *Streptococcus mutans,* a dental caries-associated bacterium. The viability of cells reduced considerably with increasing nanoparticle concentrations after CS-NP treatment. Cu-CS-NPs were also tested for antibacterial activities against *S. mutans* [106] Cu-CS-NPs outperformed Cu-NPs in terms of MIC and MBC [107, 108]. Cu-CS-NPs were found to be effective at disrupting *S. mutans* adhesion and biofilm formation [109, 110]. Cu-CS-NPs had a bactericidal effect and were more effective at inhibiting the development of *S. mutans* over the surface and disrupting the biofilm of human teeth [105]. Cu-CS-NPs' bactericidal capabilities should make it a good material for further research into tooth plaque therapies. Hipalaswins et al. reported the antibacterial action of synthesized CS-NPs against clinically harmful bacterial strains such E. coli MTTC 1687, Enterobacter aerogenes MTCC 111, K. pneumoniae MTTC 109, P. fluorescens MTCC 1748, Proteus mirabilis MTCC 1429, and S. aureus MTCC 7443 [111]. The most sensitive bacteria were *E. Coli, E. aerogenes, P. fluorescens, K. pneumoniae,* and *P. mirabilis.* The least harmful to the CS-NPs was *S. aereus.* G- bacteria were more effectively suppressed with CS-NP than G+ bacteria in antibacterial experiments. Antibacterial activity against antibiotic-resistant G+ *S. pneumonia was tested with the use of* CS-NPs and a CS-NP–amoxicillin complex by Nguyen et al. [112].

In comparison with CS-NPs and amoxicillin alone, the CS-NP–amoxicillin injection had higher antibacterial activity. CS–NP Amoxicillin complex was observed to be three times more effective than amoxicillin and prevented the growth of *S. pneumoniae*. Antibacterial action of CS-NP mixed with lime oil was observed against four foodborne pathogens [113], i.e., *S. aureus, E. Coli, L. monocytogenes, and Shigella dysenteriae*. The nanocarriers to treat *S. aureus* were found to be *S. dysenteriae* [114].

According to Tamara et al. [115], CS hybridized with protamine had more antibacterial activity against *E. coli than B. cereus*. CS-NPs loaded with curcumin could also be employed to deliver drugs and as a method to specifically activate antibacterial mechanisms. In mice, curcumin-loaded CS-NPs hindered the development of *S. aureus* and *P. aeruginosa* infections [116].

Four clinically isolated *S. epidermidis* were tested with AMP temporin B (TB)-CS-NPs. CS-NPs showed bactericidal activity within 24 h of incubation and significantly reduced the initial inoculum. Following that, Temporin B was spread across the surface of bacteria, significantly reducing cell viability [117]. It was discovered that CS-NPs inhibited the development of mesophilic bacteria when compared to a conventional coating [63]. The CS-based NPs (110 nm) coating was shown to be significantly more effective in inhibiting microbial growth than conventional coatings (300 nm). There were no pathogenic Salmonella sp. or fecal coliforms detected as a consequence. This analysis shows the use of CS-NPs as edible coatings to limit bacterial growth in fruits and vegetables. As a result, the smaller the NPs are the greater their mobility and surface contact, resulting in increased antibacterial efficacy against animal diseases.

6 Conclusions and Future Outlook

Due to its variety of potential uses and specific features, chitosan has received much interest and attention in recent years. The number of studies on its antibacterial properties is rapidly increasing. Non-toxic and biodegradable compounds from 'natural sources' will become increasingly desirable as a replacement for synthetic compounds due to their wide range of applications and people's environmental consciousness. Chitosan is different from other antibacterial compounds as it is harmless, non-toxic, and environment-friendly. People will be healthier if we use this stuff at various periods in our lives. As a consequence, due to its antibacterial capabilities, it is required to enhance research on this molecule, which will necessitate a mix of disciplines such as Nanotechnology, Physics, Chemistry, Bioinformatics, and Genetics. This will be advantageous for the production of novel biomedicine and the exploration of new antibacterial drugs. Future research could focus on using chitosan in the form of composites to deter lower pH values, as chitosan in solution is employed in an acidic environment. Clarification of the molecular mechanisms and their performance in chitosan's antibacterial action will be advantageous. To date, most research has been conducted in vitro; thus, in situ investigations are essential to developing solutions and alternatives to the issues which both the agricultural and medical areas confront. It is still not known how chitosan-based nanoparticles act against bacteria; thus, it necessitates continued research in the current scenario. Furthermore, it is critical to continue to observe and assess the toxicity associated with the use of chitosan-based nanoparticles when acting against bacteria, as well as to dispense guidance on the regulations and procedures that govern their use and application. To fill the information gap in chitosan-based nanoparticles' antibacterial activity, the following sorts of investigations must be conducted: (1) Determining why chitosan-based nanoparticles are more efficient against gram-negative bacteria as compared to gram-positive bacteria and what is the underlying mechanism; (2) determining why chitosan with medium molecular weight is more effective against bacteria than chitosan with high-molecular weight; (3) designing the technique to reduce the toxicity associated with the synthesis chitosan-based nanoparticles attached to metals and hybrid chitosan-based nanoparticles. These findings will help in the creation of a new class of medicines with efficient and novel antibacterial properties that will be applicable to both animal and plant research.

References

1. Perinelli RD, Fagioli L, Campana R, Lam WKJ, Baffone W, Palmieri FG, Casettari L, Bonacucina G (2018) Chitosan-based nanosystems and their exploited antimicrobial activity. Eur JPharm Sci 117:8–20. https://doi.org/10.1016/j.ejps.2018.01.046, https://www.sciencedirect.com/science/article/pii/S0928098718300617)
2. Mahira S, Jain A, Khan W, Domb A.J (2019) Antimicrobial materials—an overview. In: Antimicrobial materials for biomedical applications. RSC, pp 1–37. https://doi.org/10.1039/9781788012638-00001
3. Kenawy ER, Worley SD, Broughton R (2007) The chemistry and applications of antimicrobial polymers: a state-of-the-art review. Biomacromol 8:1359–1384
4. Hadrami AE, Adam LR, Daayf F (2010) Chitosan in plant protection. Mar Drugs 8(4):968–987. https://doi.org/10.3390/md8040968
5. Badawy MEI, Rabea EI (2011) A biopolymer chitosan and its derivatives as promising antimicrobial agents against plant pathogens and their applications in crop protection. Int J Carbohydr Chem Article ID 460381, 29 pages. https://doi.org/10.1155/2011/460381
6. Komi EA, Daniel Hamblin MR (2016) Chitin and chitosan: production and application of versatile biomedical nanomaterials. Int J Adv Res 4(3):411–427. PMCID: PMC5094803
7. Rinaudo M (2006) Chitin and chitosan: properties and applications. Prog Polym Sci 31:603–632. https://doi.org/10.1016/j.progpolymsci.2006.06.001
8. Qi L, Xu Z, Jiang X, Hu C, Zou X (2004) Preparation and antibacterial activity of chitosan nanoparticles. Carbohydr Res 339(16):2693–700. https://doi.org/10.1016/j.carres.2004.09.007
9. Thakur R, Kaur P, Chaudhury A (2013) Synthesis of Chitosan-silver nanocomposites and their antibacterial activity. Int J Sci Eng Res 4:869–872
10. Huang L, Cheng X, Liu C, Xing K, Zhang J, Sun G, Li X, Chen X (2009) Preparation, characterization, and antibacterial activity of oleic acid-grafted chitosan oligosaccharide nanoparticles. Front Biol China 4:321–327. https://doi.org/10.1007/s11515-009-0027-4
11. Chávez de Paz LE, Resin A, Howard KA, Sutherland DS, Wejse PL (2011) Antimicrobial effect of chitosan nanoparticles on *streptococcus mutans* biofilms. Appl Environ Microbiol 3892–3895. https://doi.org/10.1128/AEM.02941-10
12. Nagy A, Harrison A, Sabbani S, Munson RS Jr, Dutta PK, Waldman WJ (2011) Silver nanoparticles embedded in zeolite membranes: release of silver ions and mechanism of antibacterial action. Int J Nanomed 6:1833–1852. https://doi.org/10.2147/IJN.S24019

13. Kong M, Chen XG, Xing K, Park HJ (2010) Antimicrobial properties of chitosan and mode of action: a state of the art review. Int J Food Microbiol 144(1):51–63. https://doi.org/10.1016/j.ijfoodmicro.2010.09.012 Epub 2010 Oct 15 PMID: 20951455

14. Jeon SJ, Oh M, Yeo W-S, Galvão KN, Jeong KC (2014) Underlying mechanism of antimicrobial activity of chitosan microparticles and implications for the treatment of infectious diseases. PLoS ONE 9(3):e92723. https://doi.org/10.1371/journal.pone.0092723

15. Raafat D, Bargen KV, Haas A, Sahl HG (2008) Insights into the mode of action of chitosan as an antibacterial compound. Appl Environ Microbiol 74(12):3764–3773. https://doi.org/10.1128/AEM.00453-08

16. Sarwar A, Katas H, Samsudin SN, Zin NM (2015) Regioselective sequential modification of Chitosan via Azide-Alkyne click reaction: synthesis, characterization, and antimicrobial activity of chitosan derivatives and nanoparticles. PLoS. https://doi.org/10.1371/journal.pone.0123084

17. Chen HL, Yen CC, Lu CY, Yu CH, Chen CM (2006) Synthetic porcine lactoferricin with a 20-residue peptide exhibits antimicrobial activity against *Escherichia coli, Staphylococcus aureus,* and *Candida albicans.* J Agric Food Chem 3;54(9):3277–3282. https://doi.org/10.1021/jf053031s

18. Chen CZ, Cooper SL (2002) Interactions between dendrimer biocides and bacterial membranes. Biomaterials 23(16):3359–3368. https://doi.org/10.1016/s0142-9612(02)00036-4

19. Lim HS, Hudson MS (2004) Synthesis and antimicrobial activity of water-soluble chitosan derivative with a fiber-reactive group. Carbohydr Res 339:313–319. https://doi.org/10.1016/j.carres.2003.10.024

20. Reesha VK, Panda KS, Bindu J, Varghese OT (2015) Development and characterization of an LDPE/chitosan composite antimicrobial film for chilled fish storage. Int J Biol Macromol 79. https://doi.org/10.1016/j.ijbiomac.2015.06.016

21. Hosseinnejad M, Jafari SM (2016) Evaluation of different factors affecting antimicrobial properties of chitosan. Int J Biol Macromol 85:467–475. https://doi.org/10.1016/j.ijbiomac.2016.01.022

22. Tikhonov EV, Stepnova AE, Babak GV, Yamskov AI, Guerrero PJ, Jansson BH, Lopez-Llorca VL, Salinas J, Gerasimenko VD, Avdienko DI, Varlamov PV (2006) Bactericidal and antifungal activities of a low molecular weight chitosan and its N-/2(3)-(dodec-2-enyl)succinoyl/-derivatives. Carbohydr Polym 64:66–72. https://doi.org/10.1016/j.carbpol.2005.10.021

23. Kong M, Xg C, Xue YP (2008) Preparation and antibacterial activity of chitosan microspheres in a solid dispersing system. Front Mater Sci China 2:214–220. https://doi.org/10.1007/s11706-008-0036-2

24. Wu Y, Yang W, Wang C, Hu J, Fu S (2005) Chitosan nanoparticles as a novel delivery system for ammonium glycyrrhizinate. Int J Pharm 295(1–2):235–245. https://doi.org/10.1016/j.ijpharm.2005.01.042

25. Dutta J, Dutta PK, Tripathi VS (2004) Chitin and chitosan: chemistry, properties and applications. JSIR 63:20–31

26. Xie Y, Liu X, Chen Q (2007) Synthesis and characterization of water-soluble chitosan derivate and its antibacterial activity. Carbohydr Polym 69:0144–8617. https://doi.org/10.1016/j.carbpol.2006.09.010

27. Hu Y (2007) Synthesis, characterization and antibacterial activity of guanidinylated chitosan. Carbohydr Polym 67(1):66–72

28. Hu Y, Du Y, Yang J, Tang Y, Li J, Wang X (2007) Self-aggregation and antibacterial activity of N-acylated chitosan. Polymer 48(11):3098–3106

29. Kurita K (1998) Chemistry and application of chitin and chitosan. Polym Degrad Stab 59:117–120

30. Rabea EI, Badawy MET, Stevens CV, Smagghe G, Steurbaut W (2003) Chitosan as antimicrobial agent: applications and mode of action. Biomacromol 4:1457–1465

31. Lim SH, Hudson SM (2004) Synthesis and antimicrobial activity of a water-soluble chitosan derivative with a fiber-reactive group. Carbohydr Res 339:313–319
32. Sudarshan NR, Hoover DG, Knorr D (1992) Antibacterial action of chitosan. Food Biotechnol 6:257–272
33. Yang TC, Chou CC, Li CF (2005) Antibacterial activity of N-alkylated disaccharide chitosan derivatives. Int J Food Microbiol 97:237–245
34. Chung YC, Kuo CL, Chen CC (2005) Preparation and important functional properties of water soluble chitosan produced through Maillard reaction. Bioresour Technol 96:1473–1482
35. Sadeghi AMM, Dorkoosh FA, Avadi MR, Saadat P, Rafiee-Tehrani M, Junginger HE (2008) Preparation, characterization and antibacterial activities of chitosan, N-trimethyl chitosan (TMC) and N-diethylmethyl chitosan (DEMC) nanoparticles loaded with insulin using both the ionotropic gelation and polyelectrolyte complexation methods. Int J Pharm 355:299–306
36. No HK, Kim SH, Lee SH, Park NY, Prinyawiwatkul W (2006) Stability and antibacterial activity of chitosan solutions affected by storage temperature and time. Carbohydr Polym 65:174–178
37. Hernandez-Lauzardo AN, Bautista-Banos S, Velazquez-del Valle MG, Montealvo MG, SanchezRivera MM, Bello-Perez LA (2008) Antifungal effects of chitosan with different molecular weights on in vitro development of Rhizopus stolonifer Vuill. Carbohydr Polym 73:541–547
38. Chung YC, Su CC, Chen Y, Jia G, Wang JCG, Wu H, Lin JG (2004) Relationship between antibacterial activity of chitosan and surface characteristics of cell wall. Acta Pharmacol Sin 25:932–936
39. No HK, Park NY, Lee SH, Meyers SP (2002) Antibacterial activity of chitosans and chitosan oligomers with different molecular weights. Int J Food Microbiol 74(1–2):65–72. https://doi.org/10.1016/s0168-1605(01)00717-6
40. Roller S, Covill N (1999) The antifungal properties of chitosan in laboratory media and apple juice. Int J Food Microbiol 47(1–2):67–77. https://doi.org/10.1016/s0168-1605(99)00006-9
41. Atay HY (2020) Antibacterial activity of Chitosan-based systems. In: Functional Chitosan: drug delivery and biomedical applications, pp 457–489. https://doi.org/10.1007/978-981-15-0263-7_15
42. Ali A, Noh NM, Mustafa MA (2015) Antimicrobial activity of chitosan enriched with lemongrass oil against anthracnose of bell pepper. Food Packag Shelf Life 3:56–61. https://doi.org/10.1016/j.fpsl.2014.10.003
43. Ojagh MS, Rezaei M, Razavi HS, Hashem Hosseini MS (2010) Development and evaluation of a novel biodegradable film made from chitosan and cinnamon essential oil with low affinity toward water. Food Chem 122:161–166. https://doi.org/10.1016/j.foodchem.2010.02.033
44. Esmaeili A, Ghobadianpour S (2016) Vancomycin loaded superparamagnetic $MnFe_2O_4$ nanoparticles coated with PEGylated chitosan to enhance antibacterial activity. Int J Pharm 501:326–330. https://doi.org/10.1016/j.ijpharm.2016.02.013
45. Rhim JW, Hong SI, Park HM, Ng PKW (2006) Preparation and characterization of chitosan-based nanocomposite films with antimicrobial activity. J Agric Food Chem 54:5814–5822. https://doi.org/10.1021/jf060658h
46. Kumar-Krishnan S, Prokhorov E, Hernández-Iturriaga M, Mota-Morales JD, Vázquez-Lepe M, Kovalenko Y, Sanchez IC, Luna-Bárcenas G (2015) Chitosan/silver nanocomposites: synergistic antibacterial action of silver nanoparticles and silver ions. Eur Polym J 67:242–251. https://doi.org/10.1016/j.eurpolymj.2015.03.066
47. Mu H, Liu Q, Niu H, Sun Y, Duan J (2016) Gold nanoparticles make chitosan streptomycin conjugates effective towards gram-negative bacterial biofilm. RSC Adv 6:8714–8721. https://doi.org/10.1039/C5RA22803D
48. Darabpour E, Kashef N, Mashayekhan S (2016) Chitosan nanoparticles enhance the efficiency of methylene blue-mediated antimicrobial photodynamic inactivation of bacterial biofilms: an 15 Antibacterial Activity of Chitosan-Based Systems in vitro study. Photodiagnosis Photodyn Ther 14:211–217. https://doi.org/10.1016/j.pdpdt.2016.04.009

49. Zaki NM, Hafez MM (2012) Enhanced antibacterial effect of ceftriaxone sodium loaded chitosan nanoparticles against intracellular *Salmonella typhimurium*. AAPS PharmSciTech 13:411–421. https://doi.org/10.1208/s12249-012-9758-7
50. Maya S, Indulekha S, Sukhithasri V, Smitha KT, Nair SV, Jayakumar R, Biswas R (2012) Efficacy of tetracycline encapsulated O-carboxymethyl chitosan nanoparticles against intracellular infections of *Staphylococcus aureus*. Int J Biol Macromol 51:392–399. https://doi.org/10.1016/j.ijbiomac.2012.06.009
51. Wu T, Wu C, Fu S, Wang L, Yuan C, Chen S, Hu Y (2017) Integration of lysozyme into chitosan nanoparticles for improving antibacterial activity. Carbohydr Polym 155:192–200. https://doi.org/10.1016/j.carbpol.2016.08.076
52. Mangoni ME, Nargeot J (2008) Genesis and regulation of the heart automaticity. Physiol Rev 88(3):919–982. https://doi.org/10.1152/physrev.00018.2007
53. Perinelli DR, Fagioli L, Campana R, Lam JKW, Baffone W, Palmieri GF, Casettari L, Bonacucina G (2018) Chitosan-based nanosystems and their exploited antimicrobial activity. Eur J Pharm Sci 117:8–20
54. Gómez-Sequeda N, Ruiz J, Ortiz C, Urquiza M, Torres R (2020) Potent and specific antibacterial activity against *Escherichia coli* O157:H7 and methicillin resistant *Staphylococcus aureus* (MRSA) of G17 and G19 peptides encapsulated into Poly-Lactic-Co-Glycolic Acid (PLGA) nanoparticles. Antibiotics 9(7):384. https://doi.org/10.3390/antibiotics9070384
55. Ing LY, Zin NM, Sarwar A, Katas H (2012) Antifungal activity of chitosan nanoparticles and correlation with their physical properties. Int J Biomater 632–698. https://doi.org/10.1155/2012/632698
56. Dehkordi NK, Minaiyan M, Talebi A, Akbari V, Taheri A (2019) Nanocrystalline cellulose-hyaluronic acid composite enriched with GM-CSF loaded chitosan nanoparticles for enhanced wound healing. Biomed Mater 14(3). https://doi.org/10.1088/1748-605X/ab026c
57. Li B, Liu B, Shan CL, Ibrahim M, Lou Y, Wang Y, Xie G, Li HY, Sun G (2012) Antibacterial activity of two chitosan solutions and their effect on rice bacterial leaf blight and leaf streak. Pest Manag Sci 69:312–320
58. Mansilla AY, Albertengo L, Rodriguez MS, Debbaudt A, Zúñiga A, Casalongué CA (2013) Evidence on antimicrobial properties and mode of action of a chitosan obtained from crustacean exoskeletons on Pseudomonas syringae pv. tomato. DC3000. Appl Microbiol Biotechnol 97:6957–6966
59. Rabea EI, Steurbaut W (2010) Chemically modified chitosans as antimicrobial agents against some plant pathogenic bacteria and fungi. Plant Protect Sci 46:149–158
60. Li B, Zhang Y, Yang Y, Qiu W, Wang X, Liu B, Wang Y, Sun G (2016) Synthesis, characterization, and antibacterial activity of chitosan/TiO$_2$ nanocomposite against *Xanthomonas oryzae* pv. oryzae. Carbohydr Polym 52:825–831. https://doi.org/10.1016/j.carbpol.2016.07.070. Epub . PMID: 27516334
61. Wang Y, Li L, Li B, Wu G, Tang Q, Ibrahim M, Li H, Xie G, Sun G (2012) Action of chitosan against *Xanthomonas* pathogenic bacteria isolated from *Euphorbia pulcherrima*. Molecules 17:7028–7041
62. Algam SAE, Xie G, Li B, Yu S, Su T, Larsen J (2010) Effects of paenibacillus strains and chitosan on plant growth promotion and control of *Ralstonia* wilt in tomato. JPP 92(3):593–600. http://www.jstor.org/stable/41998847
63. Pilon L, Spricigo PC, Miranda M, Moura MRD, Assis OBG, Mattoso LHC, Ferreira MD (2014) Chitosan nanoparticle coatings reduce microbial growth on fresh-cut apples while not affecting quality attributes. Int J Food Sci Technol 50:440–448
64. Li B, Shi Y, Shan C, Zhou Q, Ibrahim M, Wang Y, Wu G, Li H, Xie G, Sun G (2013) Effect of chitosan solution on the inhibition of *Acidovorax citrulli* causing bacterial fruit blotch of watermelon. J Sci Food Agric 93:1010–1015
65. Mohammadi A, Hashemi M, Hosseini S (2016) Effect of chitosan molecular weight as micro and nanoparticles on antibacterial activity against some soft rot pathogenic bacteria. LWT 71. https://doi.org/10.1016/j.lwt.2016.04.010

66. Esyanti RR, Farah N, Bajra BD, Nofitasari D, Martien R, Sunardi S, Safitri R (2020) Comparative study of nano-chitosan and synthetic bactericide application on chili pepper (*Capsicum annuum L.*) infected by *Xanthomonas campestris*. Agrivita 42:13–23. https://doi. org/10.17503/agrivita.v42i1.1283

67. Jw OH, Chun SC, Chandrasekaran M (2019) Preparation and in vitro characterization of chitosan nanoparticles and their broad-spectrum antifungal action compared to antibacterial activities against phytopathogens of tomato. Agronomy 9:21

68. Santiago TR, Bonatto CC, Rossato M, Lopes CAP, Lopes CA, Mizubuti G, ES, Silva LP, (2019) Green synthesis of silver nanoparticles using tomato leaf extract and their entrapment in chitosan nanoparticles to control bacterial wilt. J Sci Food Agric 99:4248–4259

69. Chandrasekaran M, Kim KD, Chun SC (2020) Antibacterial activity of chitosan nanoparticles: a review. Processes 8:1173. https://doi.org/10.3390/pr8091173

70. Siripatrawan U, Vitchayakitti W (2016) Improving functional properties of chitosan films as active food packaging by incorporating with propolis. Food Hydrocoll 61:695–702

71. Nair SB, Alummoottil NJ, Moothandasserry SS (2017) Chitosan-konjac glucomannan-cassava starch-nanosilver composite films with moisture resistant and antimicrobial properties for food-packaging applications. Starch Stärke 69:1600210. https://doi.org/10.1002/STAR.201600210

72. Wang H, Qian J, Ding F (2018) Emerging Chitosan-based films for food packaging applications. J Agric Food Chem 66(2):395–413. https://doi.org/10.1021/acs.jafc.7b04528

73. Cruz-Romero MC, Murphy T, Morris M, Cummins E, Kerry JP (2013) Antimicrobial activity of chitosan, organic acids and nano-sized solubilisates for potential use in smart antimicrobially-active packaging for potential food applications. Food Control 34:393–397

74. Fernández-Saiz P, Sánchez G, Soler C, Lagaron GM, Ocio MJ (2013) Chitosan films for the microbiological preservation of refrigerated sole and hake fillets. Food Control 34:61–68

75. Iturriaga L, Olabarrieta I, Castellan A, Gardrat C, Coma V (2014) Active naringin-chitosan films: impact of UV irradiation. Carbohydr Polym 110:374–381. https://doi.org/10.1016/j.carbpol.2014.03.062

76. Genskowsky E, Puente DL, Pérez Alvarez JA, Fernández LJ, Muñoz LA, Viuda Martos M (2016) Assessment of antibacterial and antioxidant properties of chitosan edible films incorporated with maqui berry (*Aristotelia chilensis*). LWT-Food Sci Technol 64:1057–1062. https://doi.org/10.1016/j.lwt.2015.07.026

77. Song Z, Li F, Guan H, Xu Y, Fu Q, Li D (2017) Combination of nisin and ε-polylysine with chitosan coating inhibits the white blush of fresh-cut carrots. Food Control 74:34–44. https://doi.org/10.1016/j.foodcont.2016.11.026

78. Caro N, Quiñonez ME, Dosque DM, Lopez L, Abugoch EL, Tapia C (2015) Novel active packaging based on films of chitosan and chitosan/quinoa protein printed with chitosan-tripolyphosphate-thymol nanoparticles via thermal ink-jet printing. Food Hydrocoll 52. https://doi.org/10.1016/j.foodhyd.2015.07.028

79. Siripatrawan U, Vitchayakitti W (2016) Improving functional properties of chitosan films as active food packaging by incorporating with propolis. Food Hydrocoll 61:695–702. https://doi.org/10.1016/j.foodhyd.2016.06.001

80. Severino R, Ferrari G, Vu DK, Donsì F, Salmieri S, Lacroix M (2015) Antimicrobial effects of modified chitosan-based coating containing nanoemulsion of essential oils, modified atmosphere packaging and gamma irradiation against *Escherichia coli* O157:H7 and S*almonella Typhimurium* on green beans. Food Control 50: 215–222. https://doi.org/10.1016/j.foodcont.2014.08.029

81. Al-Naamani L, Dobretsov S, Dutta J (2016) Chitosan-zinc oxide nanoparticle composite coating for active food packaging applications. Innov Food Sci Emerg Technol 38:231–237. https://doi.org/10.1016/j.ifset.2016.10.010

82. Dotto GL, Vieira MLG, Pinto LAA (2015) Use of chitosan solutions for the microbiological shelf life extension of papaya fruits during storage at room temperature. Lebensm Wiss Technol 64:126–130. https://doi.org/10.1016/j.lwt.2015.05.042

83. Vimaladevi S, Panda SK, Xavier KA, Bindu J (2015) Packaging performance of organic acid incorporated chitosan films on dried anchovy (*Stolephorus indicus*). Carbohydr Polym 127:189–194. https://doi.org/10.1016/j.carbpol.2015.03.065

84. Xiao G, Zhang X, Zhao Y, Su H, Tan T (2014) The behavior of active bactericidal and antifungal coating under visible light irradiation. Appl Surf Sci 292:756–763

85. Pushpa K, Barman K, Patel VB, Siddiqui MW, Koley B (2015) Reducing postharvest pericarp browning and preserving health promoting compounds of litchi fruit by combination treatment of salicylic acid and chitosan. Sci Hortic 197:555–563. https://doi.org/10.1016/j.scienta.2015.10.017

86. Díez-Pascual AM, Díez-Vicente AL (2015) Antimicrobial and sustainable food packaging based on poly(butylene adipate-co-terephthalate) and electrospun chitosan nanofibers. RSC Adv 5:93095–93107. https://doi.org/10.1039/C5RA14359D

87. De Aquino BA, Blank FA, Santana LCL (2015) Impact of edible chitosan-cassava starch coatings enriched with Lippia gracilis Schauer genotype mixtures on the shelf life of guavas (*Psidium guajava* L.) during storage at room temperature. Food Chem 171:108–116. https://doi.org/10.1016/j.foodchem.2014.08.077

88. Latou E, Mexis SF, Badeka AV, Kontakos S, Kontominas MG (2014) Combined effect of chitosan and modified atmosphere packaging for shelf life extension of chicken breast fillets.LWT-Food. Sci Technol 55:263–268. https://doi.org/10.1016/j.lwt.2013.09.010

89. Hoyo-Gallego DS, Álvarez PL, Galván GF, Lizundia E, Kuritka I, Sedlarik V, Laza MJ, Vila-Vilela LJ (2016) Construction of antibacterial poly(ethylene terephthalate) films via layer by layer assembly of chitosan and hyaluronic acid. Carbohydr Polym 143:35–43. https://doi.org/10.1016/j.carbpol.2016.02.008

90. Shao X,Cao B, X F, Xie S, Yu D, Wang H (2015) Effect of postharvest application of chitosan combined with clove oil against citrus green mold. Postharvest Biol Technol 99:37–43. https://doi.org/10.1016/j.postharvbio.2014.07.014

91. Youssef AM, El-Sayed SM, El-Sayed HS, Salama HH, Dufresne A (2016) Enhancement of Egyptian soft white cheese shelf life using a novel chitosan/carboxymethyl cellulose/zinc oxide bionanocomposite film. Carbohydr Polym 151:9–19. https://doi.org/10.1016/j.carbpol.2016.05.023

92. Youssef AM, Abou-Yousef H, El-Sayed SM, Kamel S (2015) Mechanical and antibacterial properties of novel high performance chitosan/nanocomposite films. Int J Biol Macromol 76:25–32. https://doi.org/10.1016/j.ijbiomac.2015.02.016

93. Bhaskara Reddy MV, Arul J, Angers P, Couture L (1999) Chitosan treatment of wheat seeds induces resistance to *Fusarium graminearum* and improves seed quality. J Agric Food Chem 47(3):1208–1216. https://doi.org/10.1021/jf981225k

94. Yang JM, Lin HT (2004) Properties of chitosan containing PP-g-AA-g-NIPAAm bigraft nonwoven fabric for wound dressing. J Membr Sci 243:1–7

95. Chen S, Wu G, Zeng H (2005) Preparation of high antimicrobial activity thioureachitosan–Ag+ complex. Carbohydr Polym 60:33–38

96. Deitzel JM, Kleinmeyer J, Harris D, Beck TNC (2001) The effect of processing variables on the morphology of electrospun nanofibers and textiles. Polymer 42:261–272

97. Ignatova M, Manolova N, Rashkov I (2007) Novel antibacterial fibers of quaternized chitosan and poly(vinyl pyrrolidone) prepared by electrospinning. Eur Polym J 43:1112–1122

98. Shahidi F, Arachchi JKV, Jeon YJ (1999) Food application of chitin and chitosans. Trends Food Sci Technol 10:37–51

99. Dutta PK, Tripathi S, Mehrotra GK, Dutta J (2009) Perspectives for chitosan based antimicrobial films in food applications. Food Chem 114:1173–1182

100. Atay HY, Çelik E (2016) Investigations of antibacterial activity of chitosan in the polymeric composite coatings. Prog Org Coat. https://doi.org/10.1016/j.porgcoat.2016.10.013

101. Liu H, Du Y, Wang X, Sun L (2004) Chitosan kills bacteria through cell membrane damage. Int J Food Microbiol 95:147–155

102. Hasan M, Ahmad AL, Hameed BH (2008) Adsorption of reactive dye onto cross-linked chitosan/oil palm ash composite beads. Chem Eng J 136:164–172

103. Yoshida H, Fukuda S, Okamato A, Kataoka T (1991) Recovery of direct dye and acid dye by adsorption on chitosan fiber-equilibria. Water Sci Technol 23:1667–1676

104. Chiou MS, Li HY (2003) Adsorption behavior of reactive dye in aqueous solution on chemical cross-linked chitosan beads. Chemosphere 50:1095–1105

105. Wei D, Sun W, Qian W, Ye Y, Ma X (2009) The synthesis of chitosan-based silver nanoparticles and their antibacterial activity. Carbohydr Res 344:2375–2382

106. Covarrubias C, Trepiana D, Corral C (2018) Synthesis of hybrid copper-chitosan nanoparticles with antibacterial activity against cariogenic *Streptococcus mutans*. Dent Mater J 37:379–384

107. Martínez-Robles ÁM, Loyola-Rodríguez JP, Zavala-Alonso NV, Martinez-Martinez RE, Ruiz F, Lara-Castro RH, Donohué-Cornejo A, Reyes-López SY, Espinosa-Cristóbal LF (2009) Antibacterial effect of silver nanoparticles against *Streptococcus mutans*. Mater Lett 63:2603–2606

108. Besinis A, De Peralta T, Handy RD (2014) The antibacterial effects of silver, titanium dioxide and silica dioxide nanoparticles compared to the dental disinfectant chlorhexidine on *Streptococcus mutans* using a suite of bioassays. Nanotoxicology 8:1–16

109. Aliasghari A, Rabbani Khorasgani M, Vaezifar S, Rahimi F, Younesi H, Khoroushi M (2016) Evaluation of antibacterial efficiency of chitosan and chitosan nanoparticles on cariogenic *streptococci*: an in vitro study. Iran J Microbiol 8:93–100

110. Costa EM, Silva S, Tavaria FK, Pintado MM (2013) Study of the effects of chitosan upon *Streptococcus mutans* adherence and biofilm formation. Anaerobe 20:27–31

111. Hipalaswins WM, Balakumaran M, Jagadeeswari S (2016) Synthesis, characterization and antibacterial activity of chitosan nanoparticles and its impact on seed germination. J Acad Ind Res 5:65–71

112. Nguyen TV, Nguyen TTH, Wang SL,Vo TPK, Nguyen AD (2016) Preparation of chitosan nanoparticles by TPP ionic gelation combined with spray drying, and the antibacterial activity of chitosan nanoparticles and a chitosan nanoparticle–amoxicillin complex. Res Chem Intermed 43:3527–3537. https://doi.org/10.1007/s11164-016-2428-8

113. Sotelo-Boyás ME, Correa-Pacheco ZN, Bautista-Baños S, Corona-Rangel ML (2017) Physicochemical characterization of chitosan nanoparticles and nanocapsules incorporated with lime essential oil and their antibacterial activity against food-borne pathogens. LWT 77:15–20

114. Taponen S, Pyörälä S (2009) Coagulase-negative staphylococci as cause of bovine mastitis—not so different from *Staphylococcus aureus*. Veter Microbiol 134:29–36

115. Tamara FR, Lin C, Mi FL, Ho YC (2018) Antibacterial effects of chitosan/cationic peptide nanoparticles. Nanomaterials 8:88

116. Tran HV, Tran LD, Ba CT, Vu HD, Nguyen TN, Pham DG, Nguyen PX (2010) Synthesis, characterization, antibacterial and antiproliferative activities of monodisperse chitosan-based silver nanoparticles. Colloids Surf A Physicochem Eng Asp 360:32–40

117. Piras AM, Maisetta G, Sandreschi S, Gazzarri M, Bartoli C, Grassi L, Esin S, Chiellini F, Batoni G (2015) Chitosan nanoparticles loaded with the antimicrobial peptide temporin B exert a long-term antibacterial activity in vitro against clinical isolates of *Staphylococcus epidermidis*. Front Microbiol 6:372

118. Susilowati E, Ashadi M (2016) Preparation of silver-chitosan nanocomposites and coating on bandage for antibacterial wound dressing application. AIP Conf Proc 1710:030015

119. Grande CD, Mangadlao J, Fan J, De Leon A, Delgado-Ospina J, Rojas JG, Rodrigues DF, Advincula R (2017) Chitosan cross-linked graphene oxide nanocomposite films with antimicrobial activity for application in food industry. Macromol Symp 374:1600114

120. Benucci I, Liburdi K, Cacciotti I, Lombardelli C, Zappino M, Nanni F, Esti M (2018) Chitosan/clay nanocomposite films as supports for enzyme immobilization: an innovative green approach for winemaking applications. Food Hydrocoll 74:124–131

121. Tin S, Sakharkar KR, Lim CS, Sakharkar MK (2009) Activity of chitosans in combination with antibiotics in *Pseudomonas aeruginosa*. Int J Biol Sci 5:153–160. https://doi.org/10.7150/ijbs.5.153

122. Golmohamadi M, Ghorbani HR, Otadi M (2019) Synthesis of chitosan nanoparticles loaded with antibiotics as drug carriers and the study of antibacterial activity. J Nanoanal. https://doi.org/10.22034/JNA.2019.664506

123. Sharma PK, Halder M, Srivastava U, Singh Y (2019) Antibacterial PEG-chitosan hydrogels for controlled antibiotic/protein delivery. ACS Appl Bio Mater 2:5313–5322. https://doi.org/10.1021/acsabm.9b00570

124. Brasil M, Filgueiras A, Campos M, Neves M, Eugênio M, Sena L, Sant'Anna C, da Silva V, Diniz C, Sant'Ana A (2018) Synergism in the antibacterial action of ternary mixtures involving silver nanoparticles, chitosan and antibiotics. J Braz Chem Soc 29:2026–2033. https://doi.org/10.21577/0103-5053.20180077

125. Ong T.H, Chitra E, Ramamurthy S, Ling CCS, Ambu SP, Davamani F (2019) Cationic chitosan-propolis nanoparticles alter the zeta potential of *S. epidermidis*, inhibit biofilm by modulating gene expression and exibit synergism with antibiotics. PLoS ONE 14:e0213079. https://doi.org/10.1371/journal.pone.0213079

Chapter 11
Biomedical Application of Chitosan-Based Nanocomposites as Antifungal Agents

Richa Arora and Upasana Issar

Abstract Chitosan is the second most abundant natural biopolymer that possesses immense applications in the food and agriculture sector, pharmaceuticals, biomedical field, textile industries, etc. The inherent properties of chitosan are enhanced by making its blend with other polymers, metal/metal oxide nanomaterials, essential or non-essential oils, plant extracts, etc. This functionalization of chitosan depends on the concentration, pH, molecular weight, degree of deacetylation of chitosan, and also on the type of functionality. The prepared chitosan-based nanocomposites are shown to possess antifungal activities. The chitosan metal or metal oxide nanoparticle composite is active against a variety of fungal strains. A lot of modulations are being done on the chitosan nanocomposite with the aim to obtain a robust bio-nanocomposite that effectively inhibits the fungal growth.

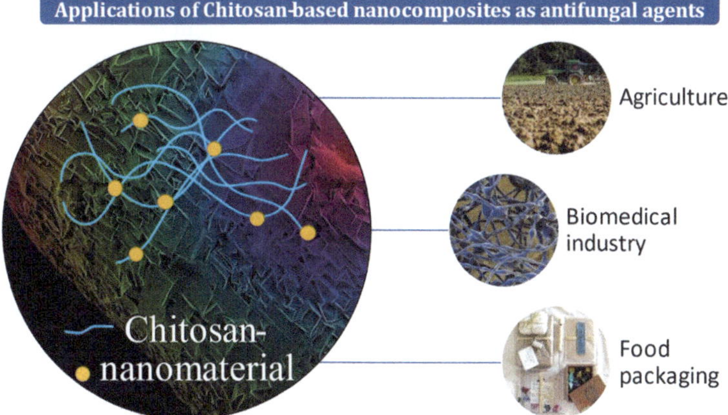

Keywords Chitosan · Nanocomposite · Antifungal · Fungi · Strains · Candida albicans

R. Arora
Department of Chemistry, Shivaji College, University of Delhi, Delhi, India

U. Issar (✉)
Department of Chemistry, Kalindi College, University of Delhi, Delhi, India
e-mail: upasana139@gmail.com

Abbreviations

CMC   Carboxymethyl cellulose
EVA   Poly(ethylene-co-vinyl acetate)
FDA   Food and Drug Administration
GRAS   Generally recognized as safe
ID   Iprodione
MMT   Multifunctional montmorillonite
PEG/PCL   Poly(ethylene glycol)-poly(ε-caprolactone)
ROS   Reactive oxygen species
SEP   Sepiolite

1 Introduction

Chitosan is the second most abundant biopolymer that contains D-glucosamine and N-acetyl-D-glucosamine linked via β-1,4 glycosidic bonds. It is derived from the deacetylation of chitin. Chitin is obtained from the exoskeleton of arthropods and is the main ingredient of crustacean shells, fungi, and insects. Though chitosan in itself possesses valuable characteristics and hence has been used in the field of biotechnology and pharmaceutical [1], still, there are certain drawbacks associated with it that need attention. Chitosan displays poor mechanical strength with low gas/vapor permeability. Being a cationic polymer, chitosan is unstable in the solution having variable pH and ionic strength. The presence of various functional groups on the backbone of chitosan makes it highly vulnerable the attack by other moieties such as polymer and metal ions forming bio-nanocomposites [2]. These chitosan-based nanocomposites are prepared to improve the intrinsic properties of chitosan. Cellulose nanomaterials are used to improve the mechanical properties of chitosan, while nanometals such as copper, silver, zinc, and zinc oxide add to the antimicrobial properties of chitosan [3–8]. Moreover, there are plant extracts and essential oils that are added to chitosan to improve its biocompatibility. Fungal infections are considered the most dangerous infections for agricultural crops [9]. If infected, it is also responsible for the early decay of stored food and harvested crops [10]. There are nearly 300 metabolites of fungi that are reported to be hazardous to animals as well as humans [11]. In order to eradicate fungal infections, a lot of chemical fungicides are used. However, conventional fungicides are harmful and pose serious environmental and health concerns [12]. Natural biopolymers which are biodegradable and biocompatible such as chitin, chitosan, and cellulose have been tested for their fungicidal properties. Chitosan has been shown to possess antifungal potency against fungal strains such as Aspergillus niger, Alternaria alternata, Phomopsis asparagi, Rhizopus oryzae, and Rhizopus stolonifera [13–15]. The antifungal activity of chitosan and its derivatives are affected by the type of substituent, type of fungi, molecular weight, concentration, and pH [15–19]. The main reason behind the antifungal activity of chitosan and its derivatives is their polycationic nature. The cationic backbone of chitosan interacts with the negatively charged phospholipid membrane of fungi. This interaction makes the membrane thin (increases the permeability) thereby causing leakage of cellular fluids from the cell leading to cell death [20, 21]. Chitosan can also bind with the metal/trace elements present in the fungi cell leading to the deficiency of essential nutrients required for the healthy growth of fungi cells [22]. There are reports that chitosan can also penetrate the fungi cell wall to bind with its DNA. This interaction with DNA hinders the translation and transcription process and thereby disrupts the cell growth cycle [23]. Chitosan-based nanocomposites display a wide range of applications. However, this chapter specifically focuses on the antifungal properties shown by chitosan-based nanocomposites. We aim to cover its applications in the biomedical field, in agriculture, in food packaging, as a food preservative, as a paper preservative, in the maritime industry, etc.

2 Antifungal Properties of Chitosan-Based Nanocomposites

Over the years, chitosan–silver nanoparticle composites have shown greater potency toward many fungi. A study comprising of greener route of synthesis of chitosan–silver nanocomposites without using a chemical reducing agent showed enhanced anticandidal properties in comparison to the precursor chitosan. The prepared composite particles are small in size with a high zeta potential value and low quantity of silver nanoparticles making them unique [24]. In another study, an improved synthetic route is proposed for the synthesis of ascorbic acid-chitosan–silver nanocomposites in order to develop a film that can display good antimicrobial activity and is also non-toxic to humans cells. Chitosan is a sought-after biopolymer owing to its intermediate average molecular weight and a high degree of deacetylation. Ascorbic acid acts as a reducing as well as a stabilizing agent. The resultant solid composite shows high antibacterial and antifungal activity against strains such as E. Coli and Candida albicans. This way of production of composite is unique as it ensures controlled release of silver atoms from the silver nanoparticles layer which is finely distributed on the polymer matrix of chitosan allowing complete disruption of biofilm of fungal strains. This methodology could prove to be helpful in biomedical industries in the future [25]. Following the green chemistry principles, Zdenka Marková and co-workers synthesized magnetic nanocomposites of chitosan, silver nanoparticles, and magnetic nanoparticles extracted from magnetotactic bacteria, that display antifungal activity against four Candida species [26].

Candida albicans is an opportunistic fungus that usually occupies and spread over the skin. It also attacks the moist mucosal lining of body organs and cavities such as the mouth, vagina, and stool [27]. In order to inhibit the growth of these fungi, copper–silver–chitosan nanocomposites were synthesized. Initially, silver, copper, and chitosan nanoparticles were synthesized individually, and then, these were mixed to get the desired nanocomposite. The antifungal activity of this nanocomposite was tested on C. albicans. The study revealed that the copper–silver–chitosan nanocomposite displayed comparable antifungal activity with Amphotericin B (a common FDA-approved drug for the treatment of fungal infection) [28]. A greener synthetic route was adopted for the preparation of gold nanoparticles and chitosan nanocomposite. A chitosan–gold nanocomposite is prepared from chitosan and gold chloride trihydrate in the absence of a reducing agent. The antifungal activity of these nanocomposites was tested against C. albicans. This pathogen at tested concentration had the tendency to pass through the muscle tissue of zebrafish. However, results showed that the chitosan–gold nanoparticle nanocomposite displayed substantial damage to the plasma membrane of C. albicans. The treatment with this chitosan–gold nanocomposite acts upon candidiasis and enhances the survival of zebrafish [29].

Silver–chitosan–sepiolite (SEP) nanocomposite synergistically displays excellent antimicrobial properties. According to the proposed mechanism (Fig. 1), the amine group in the nanocomposite reacts with Ag^+ and helps in the formation of silver nanoparticles and silver–chitosan nanocomposite [30, 31]. This silver–chitosan complex reacts favorably with the negatively charged fungi cell wall. This interaction changes the permeability of the cell wall and makes chitosan, silver nanoparticles, and their complex enter the cell wall smoothly without destroying it. The presence of -NH_2 and -OH groups on chitosan help in the controlled dispersion of silver nanoparticles on the chitosan–SEP surface [32]. Within the cell, chitosan–silver nanocomposite induces the production of reactive oxygen species (ROSs). These produced ROS immediately attacks the DNA/RNA/proteins of the cell, thereby disrupting the entire cell membrane [33]. In all, the prepared chitosan–silver–SEP nanocomposite does not directly disrupts the cell wall of Aspergillus niger but rather increases the level of ROS which in turn destroys the cell membrane structure. The hybrid structure thus promises to be an effective material for antimicrobial activity [34]. Another study reports the antifungal activity of chitosan–Foeniculum vulgare Mill. essential oil nanocomposite. It has been found that the coating of the prepared nanocomposite on S. bicolor seeds protects the seeds from pathogens viz. A. flavus and aflatoxin B1. Here, the chitosan nano-matrix acts as a carrier in transporting essential oil having antifungal properties in the food system [35]. In another work, chitosan nanocomposite was prepared by mixing carboxymethyl cellulose (CMC), oleic acid, and zinc oxide nanoparticles. Chitosan is an excellent biopolymer with a high barrier to oxygen and carbon dioxide and hence is used as an active packaging material [36]. However, its low water barrier property imposes a certain restriction on its use [37]. This can be slightly overcome by blending chitosan with water-soluble CMC which improves the appearance of chitosan films and reduces water barrier capacity to a certain extent [38]. Despite this, the chitosan–CMC blend is sensitive to the water, and therefore, some lipophilic moiety has to be introduced to overcome this problem [37]. Oleic acid can be used for this purpose. The prepared nanocomposite of the chitosan–CMC–oleic acid-ZnO nanoparticle is found to be stable, transparent, and inhibits the Aspergillus niger fungal growth. Therefore, this blend can effectively be used in food preservation [39]. In fact, only, chitosan–ZnO nanocomposite

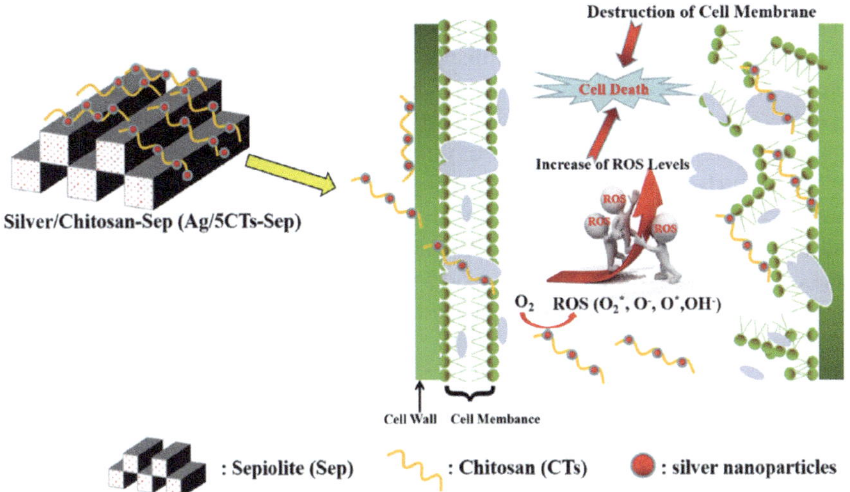

Fig. 1 Proposed antifungal action of chitosan–silver–sepiolite nanocomposite. Reprinted with permission from [34]. Copyright (2022) Elsevier

also displays strong antifungal activity against C. albicans in comparison to ZnO nanoparticles [40].

A biocompatible nanocomposite made up of molybdenum diselenide ($MoSe_2$) nanosheets and cationic chitosan show exceptional antifungal activity tested against many unicellular fungal strains. The minimum fungicidal concentration was found to be between 0.5 to 75 $\mu g\ ml^{-1}$. The activity of this nanocomposite was found to be better than other counter-nanomaterial antifungal agents containing graphene or silver nanoparticles. The mechanism of action revealed that the nanocomposite depolarises and disrupts the cell membrane within three hours of inception. The $MoSe_2$–chitosan nanocomposite also shows high antifungal activity against multi-drug resistant fungi strain Candida auris [41].

Chitosan–silver nanocomposite, prepared using a greener route, shows antifungal activity against chickpea wilt (Fusarium oxysporum f. sp. ciceris) [42]. Over the last decade, special attention has been given to introducing metallic nanoparticles in chitosan creating a nanocomposite that strengthens the biopolymer matrix of chitosan. These metals include silver, gold, copper, and zinc [43, 44]. Mohamed and Madian [45] employed a greener route for the synthesis of silver nanoparticles. They utilized aloe vera plant extract in the synthesis of silver nanoparticles. The mechanical and antifungal activity of chitosan–silver nanocomposite is tested against Staphylococcus aureus and Candida albicans. The results show that the incorporation of silver nanoparticles in chitosan not only inhibits the growth of these strains but also improves the tensile strength and structural property of the chitosan matrix [45].

Hossain et al. [46] studied the antifungal activities of essential oils encapsulated chitosan nanocomposite films against Aspergillus niger, Aspergillus flavus, Aspergillus parasiticus, and Penicillium chrysogenum. The results show that this nanocomposite shows fungal growth reduction up to 51–77%. Chitosan nanoparticles and chitosan–silver nanocomposite display potential activity against F. oxysporum. These fungi have been reported to be responsible for the adverse infection in several aquatic animals [47, 48] as well as in humans [49]. The result of chitosan–silver nanocomposite on the growth of F. oxysporum reveals the effective breakdown of mycelium surface and disintegration of the cell [29]. In another work, a nanocomposite comprising chitosan–silver–zinc oxide nanoparticles was tested on bacterial and fungal strains. It is found that this nanocomposite shows effective control of fungal C. albicans growth [50]. The oxide of cerium (CeO_2) and sulfide of antimony (Sb_2S_3) in conjugation with chitosan–starch nanoparticles is also found to possess antifungal activity [51].

3 Chitosan-Based Nanocomposites Displaying Antifungal Activity in Agriculture

According to United Nations, approximately, half of the globally produced fruits and vegetables are wasted every year (https://www.unep.org/thinkeatsave/get-inf ormed/worldwide-food-waste). Among various factors responsible, the major factor is the lack of storage provisions for food crops. The storage pathogens destroy fruits and vegetables to a significant level and hence must be addressed. Chitosan and its derivatives are the most significant materials in the field of agricultural nanotechnology owing to their biodegradable, non-toxic, and antimicrobial nature. Moreover, the antifungal activity of the chitosan family is due to the favorable interaction of its cationic amino group with the negative cell membrane of the pathogen [52–56]. All these applications are favorable when the amount of chitosan is within limits as the bulk chitosan becomes less soluble in water and therefore does not display antifungal activity to that extent [57]. Chitosan–sodium montmorillonite, chitosan–polyaniline, and chitosan–polyaniline–montmorillonite nanocomposites are used to load two antifungal agents, viz., vanillin and cinnamaldehyde. It has been reported that the release of vanillin and cinnamaldehyde depends on the rate of accumulation of these compounds on various nanocomposites. When these are tested for the antifungal activity against Fusarium oxysporum and Pythium debaryanum, chitosan–polyaniline–cinnamaldehyde and Chitosan–sodium montmorillonite–cinnamaldehyde displayed effective pathogen in tomato seedlings. Therefore, these chitosan-based bio-nanocomposites are sustainable and eco-friendly and can be used as a promising candidate against fungal infections in tomato seedlings [58]. Combating plant diseases and managing them requires proper planning and environment-friendly strategies. Exhaustive literature is available on the use of nanotechnology in addressing plant fungal diseases. Initially, metal nanoparticles were used to treat these infections [59–64]. But lately, due to reports on toxicity in plants and instability of these metal nanoparticles, their use has been reduced [65–70]. Chitosan being biocompatible displays antifungal properties and can be used in agricultural technology. Over the years, chitosan has been exploited for its anti-fungal properties yet there are few limitations associated with its direct use. Chitosan is soluble in an acidic medium and displays antimicrobial properties in acidic solutions. Since the microorganisms grow in a neutral pH medium, it is important for chitosan to work under these conditions. Cationic modifications on chitosan somewhat solve this problem as these chitosan derivatives are soluble in quite a wide range of pH [52–54, 56, 71]. Copper–chitosan nanocomposites are synthesized, and their antifungal properties are investigated in tomatoes (Solanum lycopersicum Mill). Results show that Cu–chitosan nanoparticles display antifungal potency against two diseases of tomato crop, viz., early blight and Fusarium wilt [72]. In recent work, a nanocomposite made from cellulose nanocrystals–chitosan nanoparticles uploaded on polyvinyl alcohol matrix was studied for obtaining a greener packaging material. The resultant blend displays good thermal stability and biodegradability. It also displays exceptional antifungal activity against post-harvest microorganisms making it suitable for use in the food packaging industry [73]. Lately, Xiao and co-workers prepared a nanocomposite consisting of poly(ethylene-co-vinyl acetate) (EVA) on a chitosan surface and then incorporated it on iprodione (ID) encapsulated in poly(ethylene glycol)-poly(ε-caprolactone) (PEG/PCL) (Fig. 2). This assembly is shown to have antifungal activity. When this was tested as a pre-harvest spray on grapes, the quality of fruit was improved a lot. This material can be utilized more in agriculture to reduce the use of harmful pesticides [74].

Chitosan–silica nanocomposites are proposed as an alternative to treat plant diseases. The antifungal activity of this nanocomposite against Botrytis cinerea is investigated. B. cinerea (gray mold) is a destructive fungus that affects many plant species, especially wine grapes. The experimental observations on both artificial and natural infections in table grapes reveal that chitosan–silica nanocomposite 100% reduces the fungal growth. This study opens an alternative path for the use of bio-nanocomposite as a fungicide in controlling gray mold growth on table grapes keeping the quality of fruit intact [75]. In another work, Colletotrichum coccodes, Aspergillus niger, and Pyricularia sp agricultural pathogens are treated using magnetic chitosan-based nanocomposite. These nanocomposites are made up of Fe_3O_4–chitosan–silver nanoparticles. In this study, the chitosan is made from the

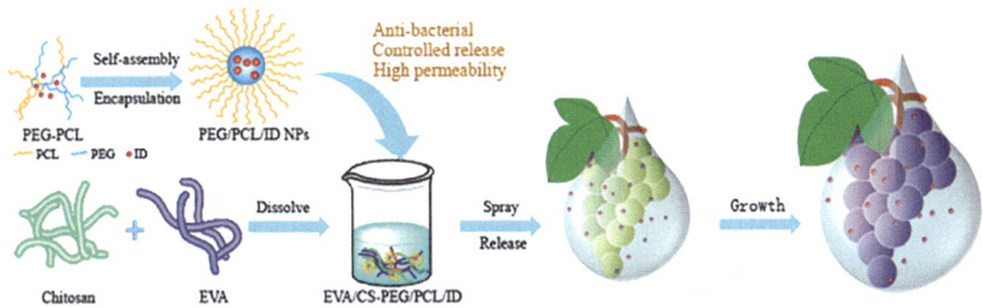

Fig. 2 EVA–chitosan nanocomposite on ID-PEG/PCL nanomiscelle. Reprinted with permission from [74]. Copyright (2020) Elsevier

fish industry waste and then treated with Fe_3O_4. Thereafter, the silver nanoparticles are loaded onto the exterior of this whole assembly. The prepared magnetic nanocomposite displays excellent antifungal properties and hence can be utilized in agricultural applications [76]. Linalool is spicy and floral terpene alcohol with a tertiary and acyclic structure. It is present in more than 200 plants including citrus fruits and lavender. It also possesses antifungal, anti-inflammatory, anticancer, and analgesic properties [77]. In one of the studies linalool incorporated in chitosan, the nanocomposite is used to develop a system that effectively inhibits fungal infection, aflatoxin B1 contamination, and lipid peroxidation in stored rice. It is found that this nanocomposite acts as an eco-friendly and smart preservative in agricultural industries [78]. A nanoemulsion is made from chitosan, and essential oils are also used in inhibiting the fungal growth in stored rice. It also reduces the aflatoxin B1 secretion and lipid peroxidation in it. The essential oils chosen are Pimpinella anisum and Coriandrum sativum [79].

Blending chitosan with metal nanoparticles not only synergistically improves its antimicrobial properties but also helps in the reduction of the toxic effects of metal nanoparticles [80]. Therefore, making a nanocomposite of chitosan with metal nanoparticles improves its fungicidal activity [72]. Chitosan–copper oxide and chitosan–zinc oxide nanocomposite are used in treating fungal infections in chickpea. Fusarium oxysporum f. sp. Cicero is a common fungus that attacks the growth of the chickpea plant [81, 82]. It is a seed and soil-born disease called Fusarium wilt disease. The results report that the introduction of Chitosan–copper oxide and chitosan–zinc oxide nanocomposite not only reduces the wilt disease but also increases the healthy growth of the chickpea plant [83]. Moreover, antifungal properties of chitosan-silver nanoparticle composites have also been used on chickpea seeds [84]. There is a synergistic relation between copper and chitosan that controls the growth of F. gramimearum [85]. The redox properties of copper are responsible for producing reactive hydroxyl radicals in fungi cells which ultimately disrupt the growth/functioning of fungi [86].

4 Antifungal Activity of Chitosan-Based Nanocomposites in Food Packaging Industries

Food packaging is a significant aspect of storing food and increasing its shelf life. It is important to store the food in such material that keeps the food fresh and inhibits the bacterial and fungal attacks on it for a substantial period of time. Worldwide, the food packaging industry is dominated by the use of petroleum-based products. These plastics generate serious environmental hazards, and hence, there is an urgent need to use materials made up of biodegradable polymers. Chitosan and its derivatives have been studied in this regard as they display antifungal and antibacterial properties. These are also non-toxic and biodegradable in nature. It has been found that chitosan derivatives inhibit sporangia germination and mycelium growth in fungi [16, 54,

87–90]. However, some of the drawbacks such as the small retention time of coating on fruits and vegetables and low mechanical properties (tensile strength, elasticity, etc.) [91, 92] many other chitosan composites are explored. These chitosan-based nanocomposites are believed to improve the properties of chitosan synergistically thus increasing the application scope of this biopolymer [93].

The antimicrobial properties of chitosan have been found to increase with the addition of fruit extracts and essential oils. The major challenge encountered by the food packaging industry is to prepare the material used in a coating that shows resistance to microbial growth. Nowadays, microorganisms are becoming more and more resistant to antibiotics and antifungal agents. A worldwide serious problem is a foodborne disease arising from contamination in food thereby producing pathogens. Researchers have constantly worked on the materials/coating that display effective antimicrobial activity without causing any major side effects of pollution in the environment. In this regard, a greener synthetic route was employed to develop a nanocomposite comprising silver nanoparticles and chitosan. One-pot synthesis of nanocomposite comprised of silver nitrate mixed with N-quaternized chitosan and N,N, N-trimethyl chitosan chloride. These water-soluble chitosan derivates act as reducing as well as a stabilizing agents during the synthesis of silver nanoparticles. The results show the successful formation of a layer of silver nanoparticles on the N,N,N-trimethyl chitosan chloride surface as well as in the polymeric matrix. Antimicrobial activity of prepared nanocomposite against Salmonella Typhimurium bacteria, Bacillus subtilis bacteria, and Aspergillus fumigatus fungus revealed effective inhibition in the growth of these pathogens. In fact, the study shows that the silver nanoparticle–chitosan derivate is a better antimicrobial agent in comparison to N,N, N-trimethyl chitosan chloride. Therefore, this eco-friendly method can be adopted for food preservation [94]. An edible antifungal coating film is made from chitosan-zinc oxide nanoparticles and Indonesian sandalwood oil to investigate the growth of Penicillium italicum. Zinc oxide nanoparticles have attained "generally recognized as safe" (GRAS) status due to their low toxicity, antimicrobial property, and biocompatibility [95]. Hence, it displays great utility in food processing and agriculture. In sandalwood oil, the presence of α- and β-santalol have proven to be responsible for inhibiting the growth of fungi such as T. mentagrophytes [96], M. canis [97], and T. rubrum [98]. Therefore, the amalgamation of ZnO nanoparticles with chitosan-sandalwood oil presents large scope in the food packaging industry. The results show inhibition of mycelium growth as well as spore germination of Penicillium italicum by the prepared nanocomposite. The coating of the film is applied to tangerine fruits, and it is observed that the nanocomposite protected the fruit from blue mold fungal decay [99]. In another study, a nanocomposite biofilm is made from corn starch, chitosan, nanoclay, sorbitol, and grapefruit seed extract. The aim of taking so many important constituents is to extract peculiar properties from each constituent. The prepared bio-nanocomposite film displays best in class crystallinity, tensile strength, and thermal stability. This film effectively inhibited the fungal growth on the food sample for up to 20 days in comparison to the synthetic plastic film which displayed the inhibition for up to only 6 days. Therefore, corn starch–chitosan–nanoclay–sorbitol–grapefruit seed extract bio-nano composite coating increases the shelf life and quality of food and hence can be used as a potential material in the food packaging industry [100]. Coating by bio-nanocomposite-based essential oils (extracted from plants) shows the reduction in contamination and spread of fungal infections in processed food [101]. The nanocomposite films made up of cellulose nanocrystal, chitosan, and essential oils of oregano, thyme, tea tree, and peppermint were investigated for their antifungal activities. The results revealed that these bio-nanocomposite films displayed an insignificant change in color, odor, taste, and quality of food. This indicates that these films exhibit controlled fungal growth in stored food items and hence can be utilized in food storage industries [46]. Chitosan-derived carbon-quantum dots have been synthesized using a greener methodology. The hydrothermal approach is utilized to prepare a stable 7.8-nm-sized quantum dot. This prepared carbon-quantum dot loaded over CMC is tested against fungal strains A. niger and P. chrysogenum. The prepared material shows great antifungal activity tested over lemon fruit. Coating from this film protected the lemon fruit from mold growth for about 21 days thereby increasing its shelf life [102].

5 Biomedical Applications of Chitosan-Based Nanocomposites as Antifungal Agents

Wound healing is a critical process, and its ignorance could lead to severe bacterial and fungal infections leading to huge damage to the injured area. It has been shown that wounds of low exuding can be healed effectively by thin-layer polymer films having low swelling properties, while wounds of high exuding can be treated with polymer films having gel-forming properties [103]. There is another kind of water-non-soluble polymeric film that can be utilized as a wound dressing. These are crosslinked films that are made up of materials in between solid and liquid and can take up more water from the wound. This is done by forming a three-dimensional polymeric network around the wound thereby preventing the leakage of water from it [104]. Biopolymers like gelatin, cellulose, chitin, chitosan, and pectin have gained enormous popularity in the manufacturing of wound dressing material as these are all hydrophilic carbohydrate polymers. Chitosan is one of the most widely used carbohydrate polymers owing to the presence of polar amino functional groups in it [105]. It is biocompatible and has antimicrobial properties making it useful in biomedical applications [106].

Among several benefits of chitosan, there is one limitation associated with its use. It is due to the minute amount of acetic acid that remains in the network of chitosan which causes irritation/itching in the wound [107]. To overcome this issue, scientists tried to modify chitosan by adding quaternary groups to it, making the polysaccharide backbone more water-soluble even in the high pH range [108–110]. Furthermore, the wound dressing abilities of quaternary chitosan derivatives can be enhanced by introducing metal or metal oxide nanoparticles components to it. Traditionally, silver or zinc oxide nanoparticles have been used in wound dressing applications due to their high antimicrobial potency [7, 111–116]. A nanocomposite made from quaternized chitosan, multifunctional montmorillonite (MMT) nanosheet, 5-fluorocytosine, and metallic copper ions has shown to powerfully inhibit the activities of S. aureus, E. coli, and C. albicans. Copper ions coordinate with 5-fluoro cytosine and deposit on quaternised chitosan. This whole assembly then attaches with MMT forming the desired nanocomposite. The formed nanocomposite shows the enhanced killing of various microorganisms and also displays extended continuous release inhibition. Therefore, these can be used in diverse infections in wounds [117].

In another study quaternised imidazolium–chitosan and silver nanoparticle composites are used in wound dressing to effectively treat the microbial infections on wounds. The nanocomposites were tested on E. Coli, S. aureus, P. aeruginosa, C. albicans, and a multi-drug-resistance P. aeruginosa. The results showed that both quaternized chitosan film, as well as its silver nanocomposite, showed good antimicrobial activities [118].

Chitosan helps in the natural coagulation process by stopping the nerve endings. This hemostatic property is essential for it to be a potential material in wound management [119, 120]. The positively charged amine groups in chitosan can attract negatively charged species as well as various antibodies present in blood and then exudate. Proof of this can be seen from the fact that there are Food and Drug Administration (FDA) approved drugs such as ChitoGauze and ChitoFlex, which are currently being employed in clinical use for wound dressing, are all chitosan based.

The addition of nanomaterial such as silver or zinc oxide to chitosan further enhances the applications of chitosan. The chitosan–zinc oxide nanocomposite has been found to increase antimicrobial activity. ZnO nanoparticles improve the flexibility and tensile strength of dressing making the nanocomposite more robust [121]. Another metal nanoparticle that has been studied for wound healing applications is the copper nanoparticles. These nanoparticles have been shown to control cytokines and growth factors responsible for wound infection in a much better way than simple copper ions. The amalgamation of chitosan and copper nanoparticles can certainly enhance the wound healing potential, and therefore, [122] prepared and studied chitosan-based copper nanocomposite on wounds in rats. The results showed that the combination of copper nanoparticles and chitosan promotes hemostasis, inflammation, proliferation, and remodeling phases of the wound healing process and helps in modulating the cytokines and growth factors systematically [122].

There is a lot of significance of chitosan-based nanocomposite in dentistry. False teeth or artificial dentures are made up of various dental materials such as nylon, plastic, or acrylic resins and are fixed very carefully in the mouth. On facing any difficulty in removing these dentures from the mouth, the tissue conditioners are applied to the mucus surface of the dentures. These conditioners act as a buffer and help in the smooth removal and placement of dentures in the mouth when needed [123]. In healthy humans, it is normal to have a small amount of fungus Candida albicans in the oral cavity [124]. However, in denture wearers and also in individuals with poor oral hygiene, the magnitude of this fungal infection increases rapidly causing denture stomatitis. Especially, the large accumulation of this fungus takes place on the tissue-fitting lining of the denture base in denture wearers causing severe

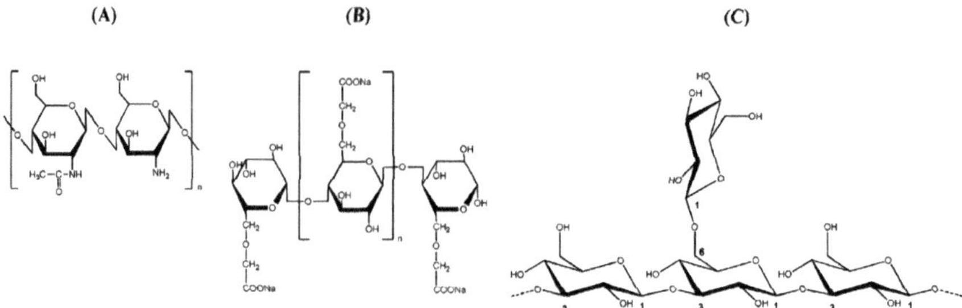

Fig. 3 Molecular structure of chitosan (**a**), CMC (**b**) and scleroglucan (**c**). Reprinted with permission from Bozoğlan et al. [130]. Copyright (2021) Elsevier

fungal infection [125]. In this regard, the addition of chitosan nanoparticles in tissue conditioners has been tested for their antibacterial and antifungal properties. The results showed that a concentration of 5 to 10% of chitosan nanoparticles in a tissue conditioner is sufficient in inhibiting the complete growth of all microorganisms [126]. Another application of chitosan-based nanocomposite is its use in the treatment of onychomycosis. Onychomycosis is a fungal disease of nails where the fungi from the families of dermatophytes, yeasts, and non-dermophytes attack nails resulting in their discoloration, splitting, and roughening. Moreover, the subungual thickening under the nail plate results in the dispositioning or lifting of the nail bed. This is known as onycholysis [127–129]. In order to treat this fungal infection, hydrogels are being utilized. Hydrogels are made up of polymers that are hydrophilic in nature. It has a three-dimensional structure with large porosity, swelling characteristics, and good biocompatibility.

Recently, Bozoğlan and co-workers synthesized a hydrogel containing chitosan, carboxymethylcellulose, and scleroglucan (Fig. 3).

The hydrogel is then taken on a montmorillonite nanosheet making a chitosan–carboxymethylcellulose–scleroglucan–montmorillonite nanocomposite. The formed nanocomposite is found to be thermosensitive. Moreover, the drug-releasing ability of nanocomposites was tested by taking oxiconazole nitrate, an antifungal drug, loaded on nanocomposites differing in montmorillonite concentration. It is found that controlled release of the drug could be made by altering the montmorillonite concentration in the hydrogel system. The drug-loaded nanocomposite displays brilliant antifungal properties against dermatophytes. One of the major findings of this work is that the smart hydrogels display the ability to alter their physical form when applied to the nail. It is found that these hydrogel nanocomposites get converted from liquid to gel revealing their thermosensitive character and making them potential candidates for the treatment of onychomycosis [130].

6 Chitosan-Based Nanocomposites Displaying Antifungal Activity in the Maritime Sector

Biofouling in marine is a process of accumulation of marine plants, algae, and animal waste on unwanted surfaces. This is a global concern as it affects the maritime industry on a larger scale. To remove biofouling, coating with antifouling agents is done. Conventional antifouling agents are toxic paints that harm marine life. In this regard, eco-friendly chitosan–zinc oxide nanocomposites are developed and examined for the antifouling properties against Navicula sp. and antibacterial potency against marine bacterium Pseudoalteromonas nigrifaciens. The results showed that the growth of biofilm is efficiently inhibited by the ZnO–chitosan nanocomposite [131]. There is one more aspect by which marine pollution can be addressed. It is by utilizing the waste generated from oceanic biomass. Salaberria and co-workers have made value-added bio-nanocomposite films from the biopolymers obtained from the wastes of shellfish. In particular, the authors isolated chitin nanofillers (nanocrystals and nanofibers) and chitosan from lobster wastes. The solvent evaporation casting method was then employed on these raw materials to obtain two types of bio-nanocomposite films. One is chitin nanocrystal–chitosan composite, and other one is chitin nanofiber–chitosan composite. The inhibitory action of these bio-nanocomposite films against the growth of A. niger reveals that the action depends upon the concentration and nature of chitin nanofiller in the chitosan matrix. In comparing the antifungal activity of two bio-nanocomposite films, chitin nanofiber–chitosan nanocomposite is found to be better than chitin nanocrystal–chitosan nanocomposite [132].

7 Miscellaneous Applications of Antifungal Properties of Chitosan-Based Nanocomposites

The fungal contamination in water is a serious health hazard and a major concern for the environment. Nanomaterials can be successfully employed to treat the infected water. In this regard, chitosan-based nanocomposites are also found to be useful in treating fungal contamination in water. In one of the studies, three metal oxides, viz., ZnO, CuO, and Ag_2O are used to form three nanocomposites with chitosan, namely ZnO–chitosan, CuO–chitosan, and Ag_2O–chitosan. These nanocomposites are then coated over cellulose filter paper and are tested against Aspergillus niger, Aspergillus flavus, and Rhizopus oryzae fungal strains. A stable coating of nanocomposite is obtained on the filter paper. The result of antifungal activity of these films suggests that ZnO-chitosan displayed better antifungal activity in comparison to CuO-chitosan and Ag_2O-chitosan nanocomposites [133]. Munnawar and co-workers investigated the antifouling properties of chitosan–ZnO nanocomposite. The nanocomposite is added to the casting solution of polyethersulfone membranes. The prepared membrane is stable and shows substantial water permeability. This filter is also observed to have an antifouling property and shows effective inhibition toward S. typhi, A. fumigatus, and F. solani fungal strains [134].

Another interesting application of chitosan-based nanocomposite is in preserving the quality of the paper. Ariafar and co-workers prepared a TiO_2-chitosan nanocomposite that displays antifungal activity and hence can be utilized to optimize the CMC adhesive applied on paper. This optimized adhesive applied on cellulose paper was dried as a thin film and then tested for the antifungal properties, yellowness, chemical oxidation, and tensile strength. The presence of TiO_2-chitosan nanocomposite on paper induces the fungal inhibition of Aspergillus flavus and Aspergillus niger strains. Moreover, it even shows antifungal activity in darkness. The treated paper also displays a reduction in the yellowness of paper. This study opens a new path in utilizing chitosan-based nanomaterials as paper preservatives and, therefore, preserving the precious artworks [135].

8 Conclusions

Chitosan-based nanocomposites possess diverse functional properties that help in improving the intrinsic properties and applications of chitosan. The blend with other polymers, metal oxide, and essential oils improves the physical and mechanical properties of chitosan. The preparation of chitosan-based nanocomposite depends on various factors physical and chemical factors. The most important ones are the degree of deacetylation of chitosan, pH, and concentration. In this chapter, emphasis was given to the antifungal potency of chitosan-based nanocomposites and their use in various fields such as biomedical, agricultural, pharmaceutics, food packaging, and maritime industries. Chitosan nanocomposite possesses a high ability to destroy the cell membrane of a variety of fungal strains. In the future, more focused and dedicated research is required to prepare the chitosan nanocomposite in a greener manner and then exploit that bio-nanocomposite as an antifungal agent over a wider range of fungi metabolites.

References

1. Yang Y, He Q, Duan L, Cui Y, Li J (2007) Assembled alginate/chitosan nanotubes for biological application. Biomaterials 28:3083–3090
2. Denkbas EB, Kilicay E, Birlikseven C, Öztürk E (2002) Magnetic chitosan microspheres: preparation and characterization. React Funct Polym 50:225–232
3. Biao L, Tan S, Wang Y, Guo X, Fu Y, Xu F, Zu Y, Liu Z (2017) Synthesis, characterization and antibacterial study on the chitosan-functionalized Ag nanoparticles. Mater Sci Eng C 76:73–80
4. Chen C, Lee SH, Cho M, Kim J, Lee Y (2016) Cross-linked chitosan as an efficient binder for Si anode of Li-ion batteries. ACS Appl Mater Interfaces 8:2658–2665
5. Elwakeel KZ, Guibal E (2015) Arsenic (V) sorption using chitosan/Cu(OH)2 and chitosan/CuO composite sorbents. Carbohydr Polym 134:190–204

6. Ou C, Chen S, Liu Y, Shao J, Li S, Fu T, Fan W, Zheng H, Lu Q, Bi X (2016) Study on the thermal degradation kinetics and pyrolysis characteristics of chitosan-Zn complex. J Anal Appl Pyrolysis 122:268–276

7. Wang M, Sun R, Li C, Wang Q, Zhang B (2017) MicroRNA expression profiles during cotton (Gossypium hirsutum L) fiber early development. Sci Rep 7:44454

8. Wen Y, Ma J, Chen J, Shen C, Li H, Liu W (2015) Carbonaceous sulfur-containing chitosan–Fe (III): a novel adsorbent for efficient removal of copper (II) from water. Chem Eng J 259:372–380

9. Rossman AY (2008) The impact of invasive fungi on agricultural ecosystems in the United States. In: Langor DW, Sweeney J (eds) ecological impacts of non-native invertebrates and fungi on terrestrial ecosystems. Springer, Dordrecht

10. da Cruz CL, Fernández Pinto V, Patriarca A (2013) Application of plant derived compounds to control fungal spoilage and mycotoxin production in foods. Int J Food Microbiol 166:1–14

11. Nesic K, Ivanovic S, Nesic V (2014) Fusarial toxins: secondary metabolites of Fusarium fungi. Rev Environ Contam Toxicol 228:101–120

12. Fishbein L (1976) Environmental health aspects of fungicides. I. Dithiocarbamtes. J Toxicol Environ Health 1:713–735

13. Guerra-Sánchez MG, Vega-Pérez J, Velázquez-del Valle MG, Hernández-Lauzardo AN (2009) Antifungal activity and release of compounds on Rhizopus stolonifer (Ehrenb.:Fr.) Vuill. by effect of chitosan with different molecular weights. Pestic Biochem Phys 93:18–22

14. Zhong Z, Chen R, Xing R, Chen X, Liu S, Guo Z, Ji X, Wang L, Li P (2007) Synthesis and antifungal properties of sulfanilamide derivatives of chitosan. Carbohydr Res 342:2390–2395

15. Ziani K, Fernández-Pan I, Royo M, Maté JI (2009) Antifungal activity of films and solutions based on chitosan against typical seed fungi. Food Hydrocoll 23:2309–2314

16. Guo Z, Chen R, Xing R, Liu S, Yu H, Wang P, Li C, Li P (2006) Novel derivatives of chitosan and their antifungal activities in vitro. Carbohydr Res 341:351–354

17. Seyfarth F, Schliemann S, Elsner P, Hipler UC (2008) Antifungal effect of high- and low-molecular-weight chitosan hydrochloride, carboxymethyl chitosan, chitosan oligosaccharide and N-acetyl-D-glucosamine against Candida albicans, Candida krusei and Candida glabrata. Int J Pharm 353:139–148

18. Tayel AA, Moussa S, El-Tras WF, Knittel D, Opwis K, Schollmeyer E (2010) Anticandidal action of fungal chitosan against Candida albicans. Int J Biol Macromol 47:454–457

19. Tikhonov VE, Stepnova EA, Babak VG, Yamskov IA, Palma-Guerrero J, Jansson HB, Lv L-L, Salinas J, Dv G, Avdienko ID, Varlamov VP (2006) Bactericidal and antifungal activities of a low molecular weight chitosan and its N-/2(3)-(dodec-2-enyl)succinoyl/-derivatives. Carbohydr Polym 64:66–72

20. García-Rincón J, Vega-Pérez J, Guerra-Sánchez MG, Hernández-Lauzardo AN, Peña-Díaz A, Velázquez-Del Valle MG (2010) Effect of chitosan on growth and plasma membrane properties of Rhizopus stolonifer (Ehrenb.:Fr.) Vuill. Pestic Biochem Phys 97:275–278

21. Liu H, Du Y, Wang X, Sun L (2004) Chitosan kills bacteria through cell membrane damage. Int J Food Microbiol 95:147–155

22. Roller S, Covill N (1999) The antifungal properties of chitosan in laboratory media and apple juice. Int J Food Microbiol 47:67–77

23. Kong M, Chen XG, Xing K, Park HJ (2010) Antimicrobial properties of chitosan and mode of action: a state-of-the-art review. Int J Food Microbiol 144:51–63

24. Dananjaya SHS, Kulatunga DCM, Godahewa GI, Lee J, de Zoysa M (2016) Comparative study of preparation, characterization and anticandidal activities of a chitosan silver nanocomposite (CAgNC) compared with low molecular weight chitosan (LMW-chitosan). RSC Adv 6:33455–33461

25. Regiel-Futyra A, Kus-Liśkiewicz M, Sebastian V, Irusta S, Arruebo M, Kyzioł A, Stochel G (2017) Development of noncytotoxic silver-chitosan nanocomposites for efficient control of biofilm forming microbes. RSC Adv 7:52398–52413

26. Marková Z, Šišková K, Filip J, Šafářová K, Prucek R, Panáček A, Kolář M, Zbořil R (2012) Chitosan-based synthesis of magnetically-driven nanocomposites with biogenic magnetite core, controlled silver size, and high antimicrobial activity. Green Chem 14:2550–2558

27. Vázquez-González D, Perusquía-Ortiz AM, Hundeiker M, Bonifaz A (2013) Opportunistic yeast infections: candidiasis, cryptococcosis, trichosporonosis and geotrichosis. J Dtsch Dermatol Ges 11:381–393
28. Ashrafi M, Bayat M, Mortazavi P, Hashemi SJ, Meimandipour A (2020) Antimicrobial effect of chitosan–silver–copper nanocomposite on Candida albicans. J Nanostruct Chem 10:87–95
29. Dananjaya SHS, Udayangani RMC, Oh C, Nikapitiya C, Lee J, de Zoysa M (2017) Green synthesis, physio-chemical characterization and anti-candidal function of a biocompatible chitosan gold nanocomposite as a promising antifungal therapeutic agent. RSC Adv 7:9182–9193
30. Li D, Gao X, Huang X, Liu P, Xiong W, Wu S, Hao F, Luo H (2020) Preparation of organic-inorganic chitosan@silver/sepiolite composites with high synergistic antibacterial activity and stability. Carbohydr Polym 249:116858
31. Kumar-Krishnan S, Prokhorov E, Hernández-Iturriaga M, Mota-Morales JD, Vázquez-Lepe M, Kovalenko Y, Sanchez IC, Luna-Bárcenas G (2015) Chitosan/silver nanocomposites: synergistic antibacterial action of silver nanoparticles and silver ions. Eur Polym J 67:242–251
32. Liu Y, Liu Y, Liao N, Cui F, Park M, Kim HY (2015) Fabrication and durable antibacterial properties of electrospun chitosan nanofibers with silver nanoparticles. Int J Biol Macromol 79:638–643
33. Sharma S (2017) Enhanced antibacterial efficacy of silver nanoparticles immobilized in a chitosan nanocarrier. Int J Biol Macromol 104(Pt B):1740–1745
34. Li D, Liu P, Hao F, Lv Y, Xiong W, Yan C, Wu Y, Luo H (2022) Preparation and application of silver/chitosan-sepiolite materials with antimicrobial activities and low cytotoxicity. Int J Biol Macromol 210:337–349
35. Kumar A, Pratap Singh P, Prakash B (2020) Unravelling the antifungal and anti-aflatoxin B_1 mechanism of chitosan nanocomposite incorporated with Foeniculum vulgare essential oil. Carbohydr Polym 236:116050
36. Wang Z, Zhou J, Xuan WX, Zhang N, Xiu SX, Su MZ (2014) The effects of ultrasonic/microwave assisted treatment on the water vapor barrier properties of soybean protein isolate-based oleic acid/stearic acid blend edible films. Food Hydrocoll 35:51–58
37. Perdones Á, Vargas M, Atarés L, Chiralt A (2014) Physical, antioxidant and antimicrobial properties of chitosan-cinnamon leaf oil films as affected by oleic acid. Food Hydrocoll 36:256–264
38. Vargas M, Albors A, Chiralt A, González-Martínez C (2011) Water interactions and microstructure of chitosan-methylcellulose composite films as affected by ionic concentration. LWT—Food Sci Technol 44:2290–2295
39. Noshirvani N, Ghanbarzadeh B, Mokarram RR, Hashemi M, Coma V (2017) Preparation and characterization of active emulsified films based on chitosan-carboxymethyl cellulose containing zinc oxide nano particles. Int J Biol Macromol 99:530–538
40. Dananjaya SHS, Kumar RS, Yang M, Nikapitiya C, Lee J, De Zoysa M (2018) Synthesis, characterization of ZnO-chitosan nanocomposites and evaluation of its antifungal activity against pathogenic Candida albicans. Int J Biol Macromol 108:1281–1288
41. Saha S, Gilliam MS, Wang QH, Green AA (2022) Eradication of fungi using MoSe2/chitosan nanosheets. ACS Appl Nano Mater 5:133–148
42. Mirajkar S, Rathod P, Pawar B, Penna S, Dalvi S (2021) γ-irradiated chitosan mediates enhanced synthesis and antimicrobial properties of chitosan-silver (Ag) nanocomposites. ACS Omega 6:34812–34822
43. Bui VKH, Park D, Lee YC (2017) Chitosan combined with ZnO, TiO_2 and Ag nanoparticles for antimicrobial wound healing applications: a mini review of the research trends. Polymers (Basel) 9:21
44. Qin Y, Liu Y, Yuan L, Yong H, Liu J (2019) Preparation and characterization of antioxidant, antimicrobial and pH-sensitive films based on chitosan, silver nanoparticles and purple corn extract. Food Hydrocoll 96:102–111
45. Mohamed N, Madian NG (2020) Evaluation of the mechanical, physical and antimicrobial properties of chitosan thin films doped with greenly synthesized silver nanoparticles. Mater Today Commun 25:101372

46. Hossain F, Follett P, Salmieri S, Vu KD, Fraschini C, Lacroix M (2019) Antifungal activities of combined treatments of irradiation and essential oils (EOs) encapsulated chitosan nanocomposite films in vitro and in situ conditions. Int J Food Microbiol 295:33–40

47. Cutuli MT, Gibello A, Rodriguez-Bertos A, Blanco MM, Villarroel M, Giraldo A, Guarro J (2015) Skin and subcutaneous mycoses in tilapia (Oreochromis niloticus) caused by Fusarium oxysporum in coinfection with Aeromonas hydrophila. Med Mycol Case Rep 9:7–11

48. Shahbazian N, Ebrahimzadeh Mousavi HA, Soltani M, Khosravi AR, Mirzargar S, Sharifpour I (2010) Fungal contamination in rainbow trout eggs in Kermanshah province propagations with emphasis on Saprolegniaceae. Iran J Fish Sci 9:151–160

49. Peterson A, Pham MH, Lee B, Commins D, Cadden J, Giannotta SL, Zada G (2014) Intracranial fusarium fungal abscess in an immunocompetent patient: case report and review of the literature. J Neurol Surg Rep 75:e241-245

50. Thaya R, Malaikozhundan B, Vijayakumar S, Sivakamavalli J, Jeyasekar R, Shanthi S, Vaseeharan B, Ramasamy P, Sonawane A (2016) Chitosan coated Ag/ZnO nanocomposite and their antibiofilm, antifungal and cytotoxic effects on murine macrophages. Microb Pathog 100:124–132

51. Hosseini M, Pourabadeh A, Fakhri A, Hallajzadeh J, Tahami S (2018) Synthesis and characterization of Sb2S3-CeO2/chitosan-starch as a heterojunction catalyst for photo-degradation of toxic herbicide compound: Optical, photo-reusable, antibacterial and antifungal performances. Int J Biol Macromol 118:2108–2112

52. Badawy MEI, Rabea EI (2011) A Biopolymer chitosan and its derivatives as promising antimicrobial agents against plant pathogens and their applications in crop protection. Int J Carbohydr Chem 2011:1–29

53. Bautista-Baños S, Hernández-Lauzardo AN, Velázquez-Del Valle MG, Hernández-López M, Ait Barka E, Bosquez-Molina E, Wilson CL (2006) Chitosan as a potential natural compound to control pre and postharvest diseases of horticultural commodities. Crop Prot 25:108–118

54. El Hadrami A, Adam LR, El Hadrami I, Daayf F (2010) Chitosan in plant protection. Mar Drugs 8:968–987

55. Meng X, Yang L, Kennedy JF, Tian S (2010) Effects of chitosan and oligochitosan on growth of two fungal pathogens and physiological properties in pear fruit. Carbohyd Polym 81:70–75

56. Qin Y, Liu S, Xing R, Yu H, Li K, Meng X, Li R, Li P (2012) Synthesis and characterization of dithiocarbamate chitosan derivatives with enhanced antifungal activity. Carbohydr Polym 89:388–393

57. Saharan V, Mehrotra A, Khatik R, Rawal P, Sharma SS, Pal A (2013) Synthesis of chitosan-based nanoparticles and their in vitro evaluation against phytopathogenic fungi. Int J Biol Macromol 62:677–683

58. Elsherbiny AS, Galal A, Ghoneem KM, Salahuddin NA (2022) Novel chitosan-based nanocomposites as ecofriendly pesticide carriers: synthesis, root rot inhibition and growth management of tomato plants. Carbohydr Polym 282:119111

59. He L, Liu Y, Mustapha A, Lin M (2011) Antifungal activity of zinc oxide nanoparticles against Botrytis cinerea and Penicillium expansum. Microbiol Res 166(3):207–215

60. Jayaseelan C, Ramkumar R, Rahuman AA, Perumal P (2013) Green synthesis of gold nanoparticles using seed aqueous extract of Abelmoschus esculentus and its antifungal activity. Ind Crop Prod 45:423–429

61. Kim SW, Jung JH, Lamsal K, Kim YS, Min JS, Lee YS (2012) Antifungal effects of silver nanoparticles (AgNPs) against various plant pathogenic fungi. Mycobiol 40:53–58

62. Lamsal K, Kim SW, Jung JH, Kim YS, Kim KS, Lee YS (2011) Application of silver nanoparticles for the control of colletotrichum species in vitro and pepper anthracnose disease in field. Mycobiol 39:194–199

63. Park HJ, Kim SH, Kim HJ, Choi SH (2006) A new composition of nanosized silica-silver for control of various plant diseases. Plant Pathol J 22:295–302

64. Wani IA, Ahmad T (2013) Size and shape dependant antifungal activity of gold nanoparticles: a case study of Candida. Colloids Surf B: Biointerfaces 101:162–170

65. Dimkpa CO, McLean JE, Latta DE, Manangón E, Britt DW, Johnson WP, Boyanov MI, Anderson AJ (2012) CuO and ZnO nanoparticles: phytotoxicity, metal speciation, and induction of oxidative stress in sand-grown wheat. J Nanopart Res 14:1125–1140

66. Kumari M, Mukherjee A, Chandrasekaran N (2009) Genotoxicity of silver nanoparticles in Allium cepa. Sci Total Environ 407:5243–5246

67. Lee WM, Jil K, An YJ (2012) Effect of silver nanoparticles in crop plants Phaseolus radiatus and Sorghum bicolor: media effect on phytotoxicity. Chemosphere 86:491–499

68. Lin D, Xing B (2008) Root uptake and phytotoxicity of ZnO nanoparticles. Environ Sci Technol 42:5580–5585

69. Thuesombat P, Hannongbua S, Akasit S, Chadchawan S (2014) Effect of silver nanoparticles on rice (Oryza sativa L. cv. KDML 105) seed germination and seedling growth. Ecotoxicol Environ Saf 104:302–309

70. Thul ST, Sarangi BK, Pandey RA (2013) Nanotechnology in agroecosystem: implications on plant productivity and its soil environment. Expert Opin Environ Biol 02(01)

71. Meng X, Tian F, Yang J, He CN, Xing N, Li F (2010) Chitosan and alginate polyelectrolyte complex membranes and their properties for wound dressing application. J Mater Sci Mater Med 21:1751–1759

72. Saharan V, Sharma G, Yadav M, Choudhary MK, Sharma SS, Pal A, Raliya R, Biswas P (2015) Synthesis and in vitro antifungal efficacy of Cu-chitosan nanoparticles against pathogenic fungi of tomato. Int J Biol Macromol 75:346–353

73. Dey D, Dharini V, Selvam SP, Sadiku ER, Kumar MM, Jayaramudu J, Gupta UN (2021) Physical, antifungal, and biodegradable properties of cellulose nanocrystals and chitosan nanoparticles for food packaging application. Mater Today: Proc 38

74. Xiao NY, Zhang XQ, Ma XY, Luo WH, Li HQ, Zeng QY, Zhong L, Zhao WH (2020) Construction of EVA/chitosan-based PEG-PCL micelles nanocomposite films with controlled release of iprodione and its application in pre-harvest treatment of grapes. Food Chem 331:127277

75. Youssef K, de Oliveira AG, Tischer CA, Hussain I, Roberto SR (2019) Synergistic effect of a novel chitosan/silica nanocomposites-based formulation against gray mold of table grapes and its possible mode of action. Int J Biol Macromol 141:247–258

76. Tomke PD, Rathod VK (2020) Facile fabrication of silver on magnetic nanocomposite (Fe_3O_4@Chitosan -AgNP nanocomposite) for catalytic reduction of anthropogenic pollutant and agricultural pathogens. Int J Biol Macromol 149:989–999

77. Aprotosoaie AC, Hǎncianu M, Costache II, Miron A (2014) Linalool: a review on a key odorant molecule with valuable biological properties. Flavour Fragr J 29:193–219

78. Das S, Singh VK, Chaudhari AK, Dwivedy AK, Dubey NK (2021) Fabrication, physico-chemical characterization, and bioactivity evaluation of chitosan-linalool composite nano-matrix as innovative controlled release delivery system for food preservation. Int J Biol Macromol 188:751–763

79. Das S, Singh VK, Chaudhari AK, Deepika DAK, Dubey NK (2022) Co-encapsulation of Pimpinella anisum and Coriandrum sativum essential oils based synergistic formulation through binary mixture: Physico-chemical characterization, appraisal of antifungal mechanism of action, and application as natural food preservative. Pestic Biochem Physiol 184:105066

80. Ahmed TA, Aljaeid BM (2016) Preparation, characterization, and potential application of chitosan, chitosan derivatives, and chitosan metal nanoparticles in pharmaceutical drug delivery. Drug Des Devel Ther 10:483–507

81. Haware M, Nene Y, Natarajan M (1996) The survival of Fusarium oxysporum f. sp. ciceri in the soil in the absence of chickpea. Phytopathol Mediterr 35:9–12

82. Jalali BL, Chand H (1992) In: Singh US, Chaube HS, Kumar J, Mukhopadhyay AN (eds) Diseases of international importance, diseases of cereals and pulses, vol 1. Prentice Hall, Englewood Cliff, New Jersey, pp 429–444

83. Kaur P, Duhan JS, Thakur R (2018) Comparative pot studies of chitosan and chitosan-metal nanocomposites as nano-agrochemicals against fusarium wilt of chickpea (Cicer arietinum L.). Biocatal Agric Biotechnol 14:466–471

84. Anusuya S, Banu KN (2016) Silver-chitosan nanoparticles induced biochemical variations of chickpea (Cicer arietinum L.). Biocatal Agric Biotechnol 8:39–44
85. Brunel F, El Gueddari NE, Moerschbacher BM (2013) Complexation of copper(II) with chitosan nanogels: toward control of microbial growth. Carbohydr Polym 92:1348–1356
86. Borkow G, Gabbay J (2005) Copper as a biocidal tool. Curr Med Chem 12:2163–2175
87. Badawy MEI (2010) Structure and antimicrobial activity relationship of quaternary N-alkyl chitosan derivatives against some plant pathogens. J Appl Polym Sci 117:960–969
88. Tan H, Ma R, Lin C, Liu Z, Tang T (2013) Quaternized chitosan as an antimicrobial agent: antimicrobial activity, mechanism of action and biomedical applications in orthopedics. Int J Mol Sci 14:1854–1869
89. Tan W, Zhang J, Mi Y, Dong F, Li Q, Guo Z (2018) Synthesis, characterization, and evaluation of antifungal and antioxidant properties of cationic chitosan derivative via azide-alkyne click reaction. Int J Biol Macromol 120(Pt A):318–324
90. Xing K, Zhu X, Peng X, Qin S (2015) Chitosan antimicrobial and eliciting properties for pest control in agriculture: a review. Agron Sustain Dev 35:569–588
91. Cazón P, Vázquez M (2020). Mechanical and barrier properties of chitosan combined with other components as food packaging film. In: Environmental chemistry letters, vol 18, no 2, pp 257–267. Springer
92. Oh DX, Hwang DS (2013) A biomimetic chitosan composite with improved mechanical properties in wet conditions. Biotechnol Prog 29:505–512
93. Saraji M, Tarami M, Mehrafza N (2019) Preparation of a nano-biocomposite film based on halloysite-chitosan as the sorbent for thin film microextraction. Microchem J 150:104171
94. Abu Elella MH, Shalan AE, Sabaa MW, Mohamed RR (2022) One-pot green synthesis of antimicrobial chitosan derivative nanocomposites to control foodborne pathogens. RSC Adv 12:1095–1104
95. Reddy KM, Feris K, Bell J, Wingett DG, Hanley C, Punnoose A (2007) Selective toxicity of zinc oxide nanoparticles to prokaryotic and eukaryotic systems. Appl Phys Lett 90:2139021–2139023
96. Inouye S, Uchida K, Abe S (2006) Vapor activity of 72 essential oils against a Trichophyton mentagrophytes. J Infect Chemother 12:210–216
97. Nardoni S, Giovanelli S, Pistelli L, Mugnaini L, Profili G, Pisseri F, Mancianti F (2015) In vitro activity of twenty commercially available, plant-derived essential oils against selected dermatophyte species. Nat Prod Commun 10:1473–1478
98. Kim TH, Hatano T, Okamoto K, Yoshida T, Kanzaki H, Arita M, Ito H (2017) Antifungal and Ichthyotoxic sesquiterpenoids from santalum album heartwood. Molecules 22:1139
99. Wardana AA, Kingwascharapong P, Wigati LP, Tanaka F, Tanaka F (2022) The antifungal effect against Penicillium italicum and characterization of fruit coating from chitosan/ZnO nanoparticle/Indonesian sandalwood essential oil composites. Food Packag Shelf Life 32:100849
100. Jha P (2020) Effect of plasticizer and antimicrobial agents on functional properties of bionanocomposite films based on corn starch-chitosan for food packaging applications. Int J Biol Macromol 160:571–582
101. Hossain F, Follett P, Dang VK, Harich M, Salmieri S, Lacroix M (2016) Evidence for synergistic activity of plant-derived essential oils against fungal pathogens of food. Food Microbiol 53(B):24–30
102. Riahi Z, Rhim JW, Bagheri R, Pircheraghi G, Lotfali E (2022) Carboxymethyl cellulose-based functional film integrated with chitosan-based carbon quantum dots for active food packaging applications. Prog Org Coat 166:106794
103. Kavianinia I, Plieger PG, Kandile NG, Harding DR (2015) Preparation and characterization of an amphoteric chitosan derivative employing trimellitic anhydride chloride and its potential for colon targeted drug delivery system. Mater Today Commun 3:78–86
104. Hoffman AS (2002) Hydrogels for biomedical applications. Adv Drug Deliv Rev 54:3–12
105. Rahimi M, Safa KD, Salehi R (2017) Co-delivery of doxorubicin and methotrexate by dendritic chitosan-g-mPEG as a magnetic nanocarrier for multi-drug delivery in combination chemotherapy. Polym Chem 8:7333–7350

106. Bano I, Arshad M, Yasin T, Ghauri MA, Younus M (2017) Chitosan: a potential biopolymer for wound management. Int J Biol Macromol 102:380–383
107. Khan TA, Peh KK, Ch'ng HS (2000) Mechanical, bioadhesive strength and biological evaluations of chitosan films for wound dressing. J Pharm Pharm Sci 3:303–311
108. Hu D, Wang H, Wang L (2016) Physical properties and antibacterial activity of quaternized chitosan/carboxymethyl cellulose blend films. LWT—Food Sci Technol 65
109. Luo J, Wang X, Xia B, Wu J (2010) Preparation and characterization of quaternized chitosan under microwave irradiation. J Macromol Sci—Pure Appl Chem 47:952–956
110. Zhu D, Cheng H, Li J, Zhang W, Shen Y, Chen S, Ge Z, Chen S (2016) Enhanced water-solubility and antibacterial activity of novel chitosan derivatives modified with quaternary phosphonium salt. Mater Sci Eng C 61:79–84
111. Dong Y, Hassan WU, Kennedy R, Greiser U, Pandit A, Garcia Y, Wang W (2014) Performance of an in situ formed bioactive hydrogel dressing from a PEG-based hyperbranched multifunctional copolymer. Acta Biomater 10:2076–2085
112. Kohsari I, Shariatinia Z, Pourmortazavi SM (2016) Antibacterial electrospun chitosan-polyethylene oxide nanocomposite mats containing bioactive silver nanoparticles. Carbohydr Polym 140:287–298
113. Pawar MD, Rathna GVN, Agrawal S, Kuchekar BS (2015) Bioactive thermoresponsive polyblend nanofiber formulations for wound healing. Mater Sci Eng C 48:126–137
114. Rakhshaei R, Namazi H (2017) A potential bioactive wound dressing based on carboxymethyl cellulose/ZnO impregnated MCM-41 nanocomposite hydrogel. Mater Sci Eng C Mater Biol Appl 73:456–464
115. Vichery C, Nedelec JM (2016) Bioactive glass nanoparticles: from synthesis to materials design for biomedical applications. Materials (Basel) 9:288
116. Wang C, Zhu F, Cui Y, Ren H, Xie Y, Li A, Ji L, Qu X, Qiu D, Yang Z (2017) An easy-to-use wound dressing gelatin-bioactive nanoparticle gel and its preliminary in vivo study. J Mater Sci Mater Med 28:10
117. Sun B, Xi Z, Wu F, Song S, Huang X, Chu X, Wang Z, Wang Y, Zhang Q, Meng N, Zhou N, Shen J (2019) Quaternized chitosan-coated montmorillonite interior antimicrobial metal-antibiotic in situ coordination complexation for mixed infections of wounds. Langmuir 35:15275–15286
118. Rahimi M, Ahmadi R, Samadi Kafil H, Shafiei-Irannejad V (2019) A novel bioactive quaternized chitosan and its silver-containing nanocomposites as a potent antimicrobial wound dressing: structural and biological properties. Mater Sci Eng C Mater Biol Appl 101:360–369
119. Dai T, Tanaka M, Huang YY, Hamblin MR (2011) Chitosan preparations for wounds and burns: antimicrobial and wound-healing effects. Expert Rev Anti Infect Ther 9(7):857–879
120. Khan MA, Mujahid M (2019) A review on recent advances in chitosan based composite for hemostatic dressings. Int J Biol Macromol 124:138–147
121. Kumar PT, Lakshmanan VK, Anilkumar TV, Ramya C, Reshmi P, Unnikrishnan AG, Nair SV, Jayakumar R (2012) Flexible and microporous chitosan hydrogel/nano ZnO composite bandages for wound dressing: in vitro and in vivo evaluation. ACS Appl Mater Interfaces 4:2618–2629
122. Gopal A, Kant V, Gopalakrishnan A, Tandan SK, Kumar D (2014) Chitosan-based copper nanocomposite accelerates healing in excision wound model in rats. Eur J Pharmacol 731:8–19
123. McCarthy JA (1984) Tissue conditioning and functional impression materials and technique. Dent Clin North Am 28:239–251
124. Nater JP, Groenman NH, Wakkers-Garritsen BG, Timmer LH (1978) Etiologic factors in denture sore mouth syndrome. J Prosthet Dent 40:367–373
125. Myatt GJ, Hunt SA, Barlow AP, Winston JL, Bordas A, El Maaytah M (2002) A clinical study to assess the breath protection efficacy of denture adhesive. J Contemp Dent Pract 3:1–9
126. Mousavi SA, Ghotaslou R, Kordi S, Khoramdel A, Aeenfar A, Kahjough ST, Akbarzadeh A (2018) Antibacterial and antifungal effects of chitosan nanoparticles on tissue conditioners of complete dentures. Int J Biol Macromol 118(Pt A):881–885

127. Paterson DL, Bonomo RA (2005) Extended-spectrum β-lactamases: a clinical update. Clin Microbiol Rev 18:657–686
128. WHO. Antibiotic resistance [Internet]. Available from: https://www.who.int/news-room/fact-sheets/detail/antibiotic-resistance
129. Wright GD (2005) Bacterial resistance to antibiotics: enzymatic degradation and modification. Adv Drug Deliv Rev 57:1451–1470
130. ancı Bozoğlan B, Duman O, Tunç S (2021) Smart antifungal thermosensitive chitosan/carboxymethylcellulose/scleroglucan/montmorillonite nanocomposite hydrogels for onychomycosis treatment. Colloids Surf A: Physicochem Eng Asp 610
131. Al-Naamani L, Dobretsov S, Dutta J, Burgess JG (2017) Chitosan-zinc oxide nanocomposite coatings for the prevention of marine biofouling. Chemosphere. 168:408–417
132. Salaberria AM, Diaz RH, Labidi J, Fernandes SCM (2015) Preparing valuable renewable nanocomposite films based exclusively on oceanic biomass—Chitin nanofillers and chitosan. React Funct Polym 89:31–39
133. Jain S, Nehra M, Dilbaghi N, Singhal NK, Marrazza G, Kim KH, Kumar S (2022) Insight into the antifungal effect of chitosan-conjugated metal oxide nanoparticles decorated on cellulosic foam filter for water filtration. Int J Food Microbiol 372:109677
134. Munnawar I, Iqbal SS, Anwar MN, Batool M, Tariq S, Faitma N, Khan AL, Khan AU, Nazar U, Jamil T, Ahmad NM (2017) Synergistic effect of chitosan-zinc oxide hybrid nanoparticles on antibiofouling and water disinfection of mixed matrix polyethersulfone nanocomposite membranes. Carbohydr Polym 175:661–670
135. Ariafar AA, Afsharpour M, Samanian K (2018) Use of TiO2/chitosan nanoparticles for enhancing the preservative effects of carboxymethyl cellulose in paper-art-works against biodeterioration. Int Biodeterior Biodegradation 131:67–77
136. Dananjaya SHS, Edirisinghe SL, Thao NTT, Kumar RS, Wijerathna HMSM, Mudiyanse-lage AY, De Zoysa M, Choi D (2020) Succinyl chitosan gold nanocomposite: preparation, characterization, in vitro and in vivo anticandidal activity. Int J Biol Macromol 165(Pt A):63–70
137. Gao K, Qin Y, Liu S, Xing R, Yu H, Chen X, Li K, Li P (2018) Synthesis, characterization, and anti-phytopathogen evaluation of 6-oxychitosan derivatives containing n-quaternized moieties in its backbone. Int J Polym 2018
138. Peh K, Khan T, Ch'ng H (2000) Mechanical, bioadhesive strength and biological evaluations of chitosan films for wound dressing. J Pharm Pharm Sci 3:303–311

Chapter 12
Antiviral Potency of Chitosan, Its Derivatives, and Nanocomposites

Upasana Issar and Richa Arora

Abstract Chitosan is a linear polysaccharide consisting of D-glucosamine and *N*-acetyl-D-glucosamine units. It is manufactured from chitin (a naturally abundant polysaccharide), extracted from the exoskeleton of crustaceans, squids, or fungi walls. Chitosan has a large number of amino groups making it positively charged moiety and, hence, is soluble in neutral as well as acidic solutions. It possesses a great deal of physical, chemical, and biological properties including biodegradability, biocompatibility, non-toxicity, display of antibacterial, antifungal and antiviral effects, and adsorption activity for heavy metal ions leading to its variety of applications in fields like water-waste management and treatment, cosmetic industry, food industry, drug and/or gene delivery, and wound healing and dressing. However, it has some shortcomings like high density because of extensive hydrogen bonding and high viscosity, hampering its activity. For this, many researchers over the years have functionalized chitosan and made use of nanoparticles resulting in chitosan nanocomposites, enhancing its properties. In this chapter, the focus would be on the noteworthy application of chitosan, modified chitosan, and its nanocomposites as viable antiviral agents against various animal and plant viruses. In fact, they have also been studied as a potential antiviral agent against severe acute respiratory syndrome coronavirus-2 (SARS-CoV-2) which caused the Coronavirus Disease of 2019 (COVID-19) pandemic, which will also be discussed in detail.

Keywords Chitosan · Chitin · Coronavirus · COVID-19 · HIV · HPV · Plant virus

Abbreviations

ACE-2   Angiotensin converting enzyme-2

U. Issar
Department of Chemistry, Kalindi College, University of Delhi, Delhi, India

R. Arora (✉)
Department of Chemistry, Shivaji College, University of Delhi, Delhi, India
e-mail: aroraricha007@gmail.com

S. Gulati (ed.), *Chitosan-Based Nanocomposite Materials*,
https://doi.org/10.1007/978-981-19-5338-5_12

AIDS	Acquired immune deficiency syndrome
AMV	Alfalfa mosaic virus
BYMV	Bean yellow mosaic virus
CMV	Cucumber mosaic virus
COVID-19	Coronavirus Disease of 2019
HA	Hemagglutinin
HCMV	Human cytomegalovirus
HCoV-NL63	Human coronavirus NL63
HCV-4a	Hepatitis C virus genotype 4a
HIV-1	Human immunodeficiency virus type-1
HPV	Human papillomavirus
HSV	Herpes simplex virus
HTCC	N-(2-hydroxypropyl)-3-trimethylammonium chitosan chloride
IL-10	Interleukin-10
MD	Molecular dynamics
NBCs	Nanobiocomposites
NO	Nitric oxide
PPE	Personal protective equipment
PSTV	Potato spindle tuber viroid
PSV	Peanut stunt virus
PVX	Potato virus X
RBD	Receptor binding domain
ROS	Reactive oxygen species
RSV	Respiratory syncytial virus
RVFV	Rift valley fever virus
SARS-Cov-2	Severe acute respiratory syndrome coronavirus-2
TCID$_{50}$	Tissue culture infectious dose
THI	T helper 1
TMV	Tobacco mosaic virus
TNV	Tobacco necrosis virus

1 Introduction

Chitosan (Fig. 1) is a polysaccharide and a co-polymer of β-(1 → 4)-linked D-glucosamines and N-acetyl–D-glucosamines. It is usually manufactured from the deacetylation of chitin, which is a homopolymer of β-(1 → 4) linkages of N-acetyl-glucosamine. Chitin is a natural polymer present in the exoskeletons of shrimp, shellfish and crabs, cuticles of some insects, and fungi, whereas chitosan is a man-made polymer. Chitosan has many free amino groups making it soluble in neutral as well as acidic solutions. Its solubility depends on the source of chitin, the degree of deacetylation of chitin, and its molecular weight [60]. The ease of production of chitosan and the raw material used for its production makes it a low cost, environmentally benign, biodegradable, biocompatible, non-toxic biopolymer. Chitosan is a positively charged polysaccharide having a variety of applications in the field of pharmacy, medicine [60], tissue engineering [3], 3D bioprinting [32], dental specialties [82], drug delivery [56], and diagnosis of ailments, especially if coated by nanoparticles [77]. Apart from this, chitosan is an excellent antiviral [2] and antimicrobial agent [11, 62]. Chitosan and its derivatives have also been tested successfully against Coronavirus Disease of 2019 (COVID-19) [22]. Few other applications of chitosan are in wastewater treatment, air filtration, cosmetic industry, food industry, and drug delivery. But, chitosan has some drastic limitations related to its reactivity and processability. This could be owed to its high density because of the hydrogen bonding interactions between chains of chitosan and high viscosity because of the intramolecular electrostatic interactions. Such problems of chitosan could be overcome, without changing its original properties, by chemically modifying it and introducing functional groups via phosphorylation [70], quaternization [28], carboxyalkylation [75], and hydroxyalkylation [15], to name a few [20].

Chitosan exhibits antimicrobial activity against a broad range of plant and animal microbes [44, 63], but its insolubility in the aqueous medium hampers its antimicrobial activity. Hence, for enhancing its solubility, dispersity, and antimicrobial activity in agriculture, its solution is prepared in an acidic medium which induces toxicity [68]. Chitosan has been modified chemically using halogenated acetate, like *N,N,N*-trimethyl chitosan chloroacetate and *N,N,N*-trimethyl chitosan trifluoroacetate, which has been shown to exhibit higher solubility and, hence, higher antimicrobial activity as compared to the unmodified chitosan [84]. Another problem with chitosan is that it is easily degraded physically and biologically. The instability of chitosan in acidic and basic pH and solutions with different ionic strengths is another major cause of concern, hampering its activity. It gets dissolved in acidic pH owing to the presence of amino groups but precipitates out in basic pH due to the de-protonation of amino groups [50]. Hence, the only way out is to modify chitosan, enhancing its physicochemical properties without affecting its basic structure or bioactivity. One

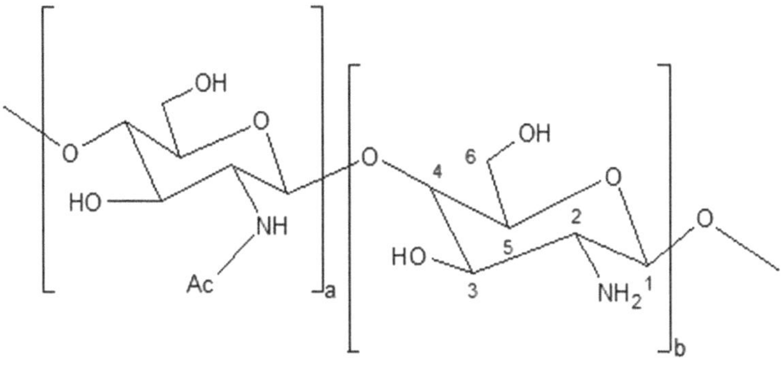

Fig. 1 Structure of chitosan

way of modifying it is by forming chelates with organic and inorganic compounds enhancing its stability, solubility, and biocidal activity [46, 67].

There is an array of factors that govern the antimicrobial activity of chitosan. Some of them include the type of synthesis process, conditions like temperature and pH, the extent of deacetylation from chitin, nature of substituents used to prepare modified chitosan, the molecular weight of chitosan, and/or its nanocomposites and derivatives as well as its charge, and also the source of chitin [29]. Chitosan, in itself, displays limited antiviral activity over a wide range of viruses, but modifications of the amino group and the hydroxyl group of chitosan have proved quite effective in improving its antiviral efficacy. For example, sulfated chitosan possesses better antiviral properties against human immunodeficiency virus type-1 (HIV-1) [71]. Since the antimicrobial activity of chitosan has vast applications against a lot of microbes, in this chapter, we have focused on and discussed the antiviral property of chitosan and its derivatives.

2 Use of Chitosan and Its Nanocomposites as Antiviral Agents

Chitosan and its composites are known to inhibit viral infection directly by damaging the viral cell membrane due to the electrostatic interaction of the positive charge of chitosan with negatively charged protein on the viral surface or/and by binding viral tail fibers. Chitosan can also inhibit viral infection indirectly by stimulating the immune response against viruses affecting plants, animals, and microorganisms. Chitosan and its derivatives also act as carriers of antiviral drugs (Fig. 2). A lot of factors influence the antiviral activity of chitosan, namely the extent of its polymerization and deacetylation, its charge, and the nature of the chemical changes. In fact, oligomers of chitosan were found to be more effective than their polymeric forms [13, 66].

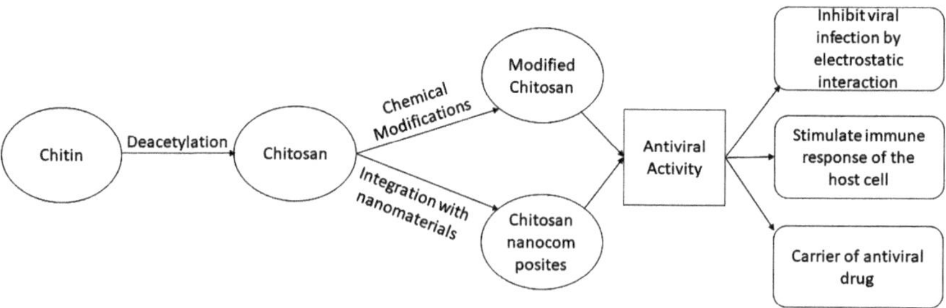

Fig. 2 Antiviral action of chitosan and its derivatives

2.1 Action of Chitosan, Its Derivatives, and Nanocomposites Against Animal Viruses

Acquired immune deficiency syndrome (AIDS) has been a major cause of concern over the years and is caused by HIV-1 retrovirus [18]. Many anti-HIV-1 agents are available, but their use is limited owing to many reasons like the emergence of mutated viruses resistant to such anti-HIV agents and their toxicity toward healthy cells [69]. Researchers have been constantly working toward developing new, alternative, and natural materials and their derivatives to improve the efficacy of drugs against HIV and reduce toxicity. Chitosan and its derivatives have emerged as a good candidate against HIV since they are biodegradable and non-toxic. *N*-carboxymethylchitosan *N,O*-sulfate was prepared and found to inhibit HIV-1 by inhibiting reverse transcriptase enzyme and hampering the binding of viral cells to human CD4$^+$ target cells (host cell receptor) [75]. Sulfated chitooligosaccharides with molecular weight in the range of 3–5 kDa, being non-toxic and water-soluble, were found to block entry of HIV and virus-cell fusion by hampering the binding of HIV-1 to CD4 cell surface receptor [7, 20]. Heparan sulfate binding receptor is the cell-binding receptor to which some viruses bind leading to their entry into the host cell. Such viruses are HIV-1, herpes simplex virus (HSV), human cytomegalovirus, and papillomaviruses [71]. Sulfate substituted succinyl chitosan was found to be effective against Moloney murine leukemia virus (having heparan sulfate binding receptor), as studied on infected rodent SC-1 and NIH-3T3 cell lines with the concentration of 0.01–100 μg mL^{-1}. This study revealed that high molecular weight and a high degree of substitution increase the antiviral efficacy of chitosan [76].

Chitosan oligomers having low molecular weight can be modified with peptides like tryptophan and glutamine and are found to have good anti-HIV-1 activity. This is achieved by the reduction in the amount of p24 protein (having an active role in maintaining the virus structure), envelope, and viral infection factor proteins on the viral surface, inhibiting the replication of the virus [41]. Another safe and non-toxic complex of chitosan, effective against HIV-1, is hyaluronan, stabilized with Zn(II) ions [83]. Chitosan complex with Ni(II) ions leads to an effective binding of Ni(II) with enterovirus-71 (a cause of infections in children and responsible for neurological diseases) [49, 71].

Human cytomegalovirus (HCMV) belongs to the betaherpesvirinae family of double-stranded DNA viruses with 55–100% infection in humans [65]. HCMV is asymptomatic in most healthy humans but could prove lethal for humans having weak immune responses, children, organ transplant recipients, and HIV-positive patients [24]. The available drugs like ganciclovir, cidofovir, and foscarnet [14] are toxic, have low bioavailability, and face resistance from the virus [79]. For developing a probable drug against HCMV, certain factors have to be accounted for, which have been integrated into a nanocomposite of ZnO nanoparticles with phenyloxy functionalized chitosan. Such nanocomposites exhibited excellent antiviral activity against HCMV in vitro with low cytotoxicity, which could be used as an independent drug or as a combination with other available anti-HCMV drugs. The prepared nanocomposites of 4-hydroxybenzaldehyde, 4-(benzyloxy)benzaldehyde-modified chitosan with ZnO nanoparticles decreased the viral load interfering with the active replication of the virus [39].

A lot of research has been done by different research groups to study the antiviral property of chitosan and its derivatives. Chitosan displayed antiviral activity against adenovirus [59] and human noroviruses [16]. Silver nanoparticle–chitosan nanocomposite is quite effective against the H1N1 influenza A virus. Such a composite not only reduced the tissue culture infectious dose ($TCID_{50}$) ratio of virus suspension but reduced the toxicity as well by preventing the release of silver in the environment as it is fixed on chitosan [55]. The nanocomposite of chitosan with curcumin works against hepatitis C virus genotype 4a (HCV-4a) and prevents their entry and replication in the human hepatoma Huh7 cells [51].

Chitosan is known to enhance antiviral immune responses. They stimulate the innate immune cells by increasing the number of leukocytes and macrophages, reactive oxygen species (ROS), nitric oxide (NO), and enhancing the cell-mediated responses, to fight the pathogen. Macrophages can lead to phagocytosis of chitosan and, hence, leads to the production of ROS which induces the synthesis of gamma interferon in spleen cells, which in turn shows its antiviral activity by inhibiting the translation of RNAs of the viral cells [72].

A lot of studies were conducted on different murine models, which showed that the size of chitin particles is of immense importance in inducing the innate immune response of the host. Big-sized chitosan when enters the host, the immune response is stimulated which tends to break the chitosan into small parts with the help of chitinases. The intermediate-sized chitin particles send an alarm signal leading to the inflammation by activating the recognition receptors. This would continue until the virus infection is done with, and chitin is further oxidized into much smaller fragments. The small-sized chitin leads to the production of interleukin-10 (IL-10), which works toward controlling the inflammation. Similar response is by chitosan and mixture of chitin and chitosan [47]. Therefore, chitosan can induce innate immune responses against viral infections.

If vaccines are administered with chitin and chitosan-based nanoparticles, a drastic enhancement in the innate and adaptive immune response occurs. DNA vaccine in complexation with chitin microparticles was administered against HIV infection in BALB/c mice. Chitin was found to enhance the adaptive immune response as DNA/chitin microparticles were found to increase antibodies against HIV and their T cells [25]. Some other examples to illustrate this involve the administration of chitin along with chitosan, in BALB/c mice, for the inactivation of hemagglutinin (HA) protein (which mainly helps in binding host cells) of H1N1 influenza virus [26] and recombinant HA protein of H5N1 influenza virus [8]. Mucosal, as well as systemic humoral immune responses, was stimulated. Chitin along with chitosan was also found to increase the production of secretory IgA antibodies in the nasal wash and IgG antibodies in the serum causing T helper 1 (TH1) immune response against viruses [8, 26]. Chitin microparticles having a size of 10 μm, when administered in H5N1 and H1N1-infected mice through the intranasal route resulted in an increase in the number of natural killer cells in the cervical lymph tissues. This led to a decrease in the viral load and enhanced survival rate after infection [31]. In another study, microparticles of chitin having a size of 3.72 μm enhanced the immune response of the host cell against the influenza A virus and also enhanced the production of T cells which helps in fighting against the virus [9].

Chitosan polymers could be used as a solution, in the form of powder, as microparticles, and nanoparticles, depending on their physical and chemical properties. They can also be used as carriers for antigen and/or adjuvants in mucosal vaccines as they can induce cellular and humoral immune responses. Certain examples of chitosan to illustrate this include its activity against the influenza virus. When chitosan is used as a solution (0.5%) along with inactivated influenza vaccines, the immunogenicity of the vaccine increases by six to ten times [23]. Again, chitosan solution when administered intranasally with influenza matrix protein induces and increases the amount of IgG and IgA antibodies [73, 74]. Chitosan solution has also been found effective with live attenuated influenza vaccine [81].

A range of chitosan derivatives like sialyl lactose substituted chitosan [48] and chitosan-sialyl oligosaccharides (obtained from bovine colostrum) complex have been found to be effective and selective against influenza virus A [12]. Such modified chitosans block the interaction of the host cell receptors with the viral surface by acting as virus adsorbents. This is because sialyl lactose and sialyl oligosaccharides are the receptors present in the host cells to which the virus binds and enter the host cells. By modifying the chitosan with such receptors present on the host cell, the virus gets adsorbed on the modified chitosan and, hence, is prevented from binding to the host cell receptors. Sulfated chitosan has proved to be quite effective against many widespread viral infections. For example, 3,6-*O*-sulfated chitosan with an average molecular weight of 58.3 kDa and having 45.8% sulfate content inhibits human papillomavirus (HPV), responsible for cervical cancer, by directly binding the viral proteins [21]. Apart from this, in vitro study in dunni cells of friend murine leukemia helper virus and HSV-1 revealed antiviral activity of sulfated carboxyl methyl chitin backbone (with 7.66% degree of sulfation) against these viruses [36]. Both these inhibitory activities of sulphated chitosan are dose dependent.

Chitosan nanoparticles made using *Lucilia cuprina* maggots showed effective antiviral activity in terms of the reduction in viral infection by 24.9% against Rift Valley fever virus (RVFV), 26.1% against Coxsackie viruses and 18.8% in HSV-1 on a Vero cell line of adult African green monkey kidney [27]. Noroviruses, usually present in contaminated food and water, could lead to gastroenteritis and diarrheal diseases if they enter the human body. Chitosan microparticles, yet again, could reduce the viral load in murine norovirus and bacteriophage MS2 of *Escherichia coli* at 0.3% concentration [71, 85].

Respiratory syncytial virus (RSV) affects the lower respiratory system and could be fatal for newborn children, elderly people, and people with low immune responses. Here again, the use of chitosan with vaccines made from RSV-coded DNA can increase the number of IgG antibodies in serum, IgA antibodies in mucous, and T cells which secrete interferon-γ and induces T lymphocyte response which could be lethal for viral cells. This leads to a decrease in the number of viral cells, an increase in the antiviral response, and a reduction in the inflammation of the lungs [45, 66].

Among bacteriophage viruses, chitosan and its derivatives were studied against coliphages T2, T4, and T7 in *E. coli*. The degree of polymerization and functionalization, charge, and molecular weight of 6-*O*-sulfate and *N*-succinate-6-*O*-sulfate-modified chitosan has a profound effect on their activity against phage viruses. Studies revealed that the high degree of polymerization and high concentration of chitosan (100 and 10 μg mL^{-1}) have higher antiviral activity against phage viruses [42, 43, 71].

2.2 Action of Chitosan, Its Derivatives, and Nanocomposites Against Coronaviruses

COVID-19 pandemic has been widespread and has affected many countries around the world. It is a highly contagious disease and spreads quite fast through the air. The aerosol particles containing severe acute respiratory syndrome coronavirus-2 (SARS-CoV-2) viruses, present in the air via sneezing, breathing, coughing, and talking by a person carrying SARS-CoV-2 infection, could be inhaled by a healthy person in the vicinity leading to the spread of infection. Also, the virus particles may spread through the surfaces on which they settle [78]. The main problem with SARS-CoV-2 is that it has size and properties similar to a nanoparticle which is responsible for its high transmission rate. It possesses spike protein having polybasic arginine-rich motifs, making it more infectious than its previous SARS sequences [38, 66].

During this pandemic, frontline workers, including paramedics, doctors, and healthcare workers, who constantly come in touch with infected people on a daily basis, need to be protected. In this regard too, nanotechnology has come to the rescue. Chitosan nanoparticles being positively charged could be integrated into the clothes of frontline workers, which imparts electrostatic repulsion to the positively charged SARS-CoV-2 particles resulting in less viral load on and around them and, hence, the lower transmission of the virus. Chitosan nanofibers having a positive charge including single and double *N*-quaternized chitosan derivatives, like *N*,*N*,*N*-trimethyl chitosan, have been proposed to be useful in the production of personal protective equipment (PPE) for frontline workers, owing to their repulsive interactions with coronavirus [28].

Since the emergence of the COVID-19 pandemic, scientists around the globe have been in search of potent and non-toxic drugs against it since till now, no licensed drug is available against coronavirus. For coronavirus too, chitosan nanoparticles, its derivative, polymer, and nanocomposites could be effective, being biocompatible.

A large number of studies have been conducted and are still underway to prepare antiviral food coatings and fabric to prevent the spread and effect of coronavirus. In this regard too, chitosan has been proved quite useful. This inhibition of coronavirus by chitosan and their derivatives could be due to the electrostatic repulsive interaction between the positively charged chitosan nanocomposite and the spike protein of coronaviruses [80]. A cross-linked derivative of chitosan with genipin (obtained from the plant), i.e., N-(2-hydroxypropyl)-3-trimethyl chitosan, was found to adsorb and inhibit the activity of human coronavirus NL63 (HCoV-NL63), reducing the infection potential of the virus by $(7.2 \pm 0.8) \times 106$ copies mL^{-1}. This was attributed to the electrostatic repulsive interaction between the positively charged chitosan derivative and the viral protein [15].

The addition of chitosan along with DNA, for example, plasmid DNA-loaded biotinylated chitosan nanoparticles, could enhance the immune response to a vaccine against SARS-CoV and could also reduce the side effects like inflammation of the lungs. A study was done by a group of researchers in 2012, [64], well before COVID-19 pandemic, wherein they studied the immune response of a DNA vaccine having chitosan nanoparticles against SARS-CoV nucleocapsid protein and found an increased number of IgG and IgA antibodies (an indicator of strong immune response of the host cell) against nucleocapsid protein, responsible for its enhanced immune response.

Chitosan has also been reported as an effective carrier for drugs. Apart from this, it has an ability to stimulate the immune system, is non-toxic, biocompatible, and degrades in the body releasing non-toxic by-products; it could easily open up tight intersections, and it could be modified into different shapes and sizes. All these useful properties make chitosan-based nanomaterials an effective tool to fight the COVID-19 pandemic [37]. Aerosol formulations of Novochizol™ nanoparticles, made of chitosan, were found to selectively deliver the potential anti-coronavirus drug to the lungs. When administered to a patient, Novochizol™ nanoparticles remain attached to the mucosal membranes of the respiratory tract. The chitosan molecule present is degraded slowly, releasing the active drug molecule in effective amounts [66].

Spike protein of SARS-CoV, SARS-CoV-2, and HCoV-NL63 (common cold virus) has receptor-binding domain (RBD) which binds the angiotensin-converting enzyme-2 (ACE-2) receptors of the human respiratory tract and enter the host cells causing infection. A chitosan-based polymer, N-(2-hydroxypropyl)-3-trimethylammonium chitosan chloride (HTCC), and its hydrophobic derivative were studied and found to exhibit inhibitory activity against not only HCoV-NL63 but against human murine hepatitis virus too. The main action of the polymer includes its inhibitory action for the interaction between the receptors of the host cell and the RBD of the viral spike protein. This could be achieved by aggregating spike proteins so that their interaction with the host cell reduces. The degree of substitution of chitosan in HTCC also had a varied effect on different human coronaviruses like HCoV-OC43, HCoV-229E, HCoV-NL63, and HCoV-HKU1 as revealed by the study conducted on LLC-MK2 (*Macaca mulatta* kidney epithelial) cell line [53, 54]. HTCC has also been an effective inhibitor of SARS-CoV-2 and Middle East respiratory syndrome coronavirus (MERS-CoV) [52], as demonstrated by in vitro studies conducted using Vero and Vero E6 cell lines and ex vivo studies using human airway epithelium mode, through the electrostatic interactions between HTCC and viruses. A molecular dynamics (MD) study also revealed the excellent potential of chitosan as a drug and in vaccines against coronavirus. It demonstrated a new target on SARS-CoV-2 for chitosan apart from RBD, which is a homotrimer pocket of spike protein, which too has a strong affinity for ACE-2 receptors of the host cell [40]. β-chitosan was found to inhibit the binding of the spike protein of SARS-CoV-2 with ACE-2 receptors by itself binding with ACE-2. Experimental studies like in vitro and in vivo analysis and immunofluorescence demonstrated the downregulation of ACE-2 in Vero E6 cells of mice. A decrease in the immunofluorescence intensity from 33.6 to 6.91% was observed in the presence of β-chitosan. This has led to a drastic decrease in the colocalization of RBD and ACE-2 receptors, and thereby, their binding is inhibited [4, 71].

2.3 Action of Chitosan, Its Derivatives, and Nanocomposites Against Plant Viruses

Evidence is available for the inhibitory response of chitosan and its derivatives against plant viruses as well, but it needs to be explored more for their effective use in agriculture [17, 35]. Few studies showed the reduction of necrotic lesions caused by alfalfa mosaic virus (AMV), tobacco necrosis virus (TNV), tobacco mosaic virus (TMV), peanut stunt virus (PSV), cucumber mosaic virus (CMV), potato virus X (PVX), and potato spindle tuber viroid (PSTV) in *Phaseolus vulgaris, Pisum sativum, Nicotiana tabacum, Nicotiana glutinosa, Nicotiana paniculata, Lycopersicon esculentum, Chenopodium quinoa*, on spraying 0.1% chitosan [46, 61].

Although chitosan has remarkable antimicrobial properties owing to the electrostatic stacking of chitosan over the virus surface, the treatment of plants with chitosan increases the permeability of the membranes of plant cells disrupting their stability and, hence, cell death. However, in some viral infections, changes in the plant membrane could actually support their fight against the virus [6, 58].

Let us try to understand how chitosan derivatives fight against bean yellow mosaic virus (BYMV), responsible for mosaic and malformation of numerous plants. Safe and economical, carboxymethyl chitosan–titania nanobiocomposites (NBCs) were prepared and used to treat faba bean plants against BYMV. Here, *N,O*-substituted carboxymethyl (NBC1), and *O*-substituted carboxymethyl (NBC2) derivatives of chitosan, having low molecular weight, were synthesized. The severity of the disease was found to reduce drastically in the NBCs-treated faba beans by 10.66% and 19.33%, when administered with NBC1 and NBC2, respectively. Further, NBCs improved a lot of growth factors in plants, increased the number of photosynthetic pigments and water content, enhanced the stability of plant cell membrane, and increased the quantity of enzymatic and non-enzymatic antioxidants and soluble protein. Not only this, it reduced various phenomena which could adversely impact plant health like the leakage of electrolytes, amount of hydrogen peroxide, and the peroxidation process of lipids. Apart from impacting a lot of internal processes to maintain the stability and physiology of plant cells, NBCs were able to strengthen the immune response and systematic resistance of faba bean plants against BYMV. NBC1 was found to be more effective against BYMV owing to its higher hydrophilicity, biocompatibility, and chelation capacity as compared to NBC2. The antiviral action of NBCs is due to its constituents; TiO_2 nanoparticles and carboxyalkyl substituted chitosan, which provides protection against viral infection [74].

If one talks about the antiviral activity of chitosan in plants, they not only prevent the spread of viral infection but stimulate the resistance genes and a range of defense responses, as well, in plants. Here again, a range of factors including concentration, time of exposure, and poly-cationic nature of chitosan play an important role against the viral infection. The inhibitory activity of chitosan against plant viruses was also found to depend on its structure as well as its molecular weight. Chitosan with high molecular weight in the range of 100–120 kDa exhibited better and enhanced antiviral activity than those with lower molecular weight in the range of 3–36 kDa, against PVX infection in potatoes [13, 35]. This correlation between the molecular weight and the antiviral activity of chitosan does not always hold good. Sometimes, chitosan, having low molecular weight, displays better antiviral activity than its high molecular weight counterparts. One example to illustrate this is the inhibition of TMV by the chitosan having low molecular weight. The method of formation of the chitosan, having low molecular weight, from its high molecular weight precursor was found to influence its antiviral activity. Also, the source of chitosan and even the process used to purify chitosan affect their activity. The chemical hydrolysis (using H_2O_2) of chitosan producing low molecular weight chitosan was found to exhibit much better antiviral properties as compared to those obtained from the enzymatic hydrolysis (using lysozyme) [17, 66].

Not only the chitosan but the type of plant being affected and many times, the type of virus affecting the plant plays a noteworthy role in the extent of the response for chitosan. It is a well-known fact that the activity of lectin proteins, present in plants, is enhanced under biotic stress like low temperature, wound stress, or osmotic stress or when a virus or bacteria attacks a signaling plant, to resist the foreign organism. Chitosan can help augment the lectin activity and also develop resistance, as has been seen in the chitosan-treated tobacco plant and potato tuber trying to fight against TMV [10]. Chitosan is supposedly present in the cell membrane of the pathogen. When a pathogen interacts with the plant, due to structural changes, the chitosan molecule is degraded, and the products are released to stimulate a defense response against a pathogen, as revealed by a study conducted on tobacco against TNV [19]. The treatment of plants by chitosan leads to the deposition of callose, oxidative bursts, and hypersensitive responses, triggering the defense system of the plant against viruses [34]. Abscisic acid also leads to callose accumulation, and here also, chitosan plays a significant role in the accumulation of abscisic acid. This increase in the concentration of callose reduces viral-induced lesion sites and, hence, inhibits the spread of viral infection [33]. Chitosan is also responsible for an increase in the number of proteolytic enzymes like RNAses resulting in thinning of the virions and, hence, reducing their binding capacity, as depicted by the study conducted on leaves of *Nicotiana tabacum* L. cv. Samsun against TMV [57].

Chitosan alone and in combination with glycine betaine has also been evaluated for imparting resistance against CMV in cucumber plant. Gene expression analysis showed the enhancement of defense genes. Apart from this, there is an enhancement of leaf chlorophyll content, hormones like salicylic acid and jasmonic acid, osmoprotectants like soluble sugars and proline, both enzymatic antioxidants like superoxide dismutase and peroxidase, and non-enzymatic antioxidants like ascorbic acid and glutathione in plants, in the presence of chitosan. On the other hand, the amount of malondialdehyde (an indicator of oxidative stress) and abscisic acid (stress hormone) reduces significantly in the presence of chitosan. The treatment of plants with the combination of glycine betaine and chitosan led to far better results as compared to the one using only chitosan [73]. Chitosan-N induces resistance against papaya mosaic virus (which causes the distortion of papaya leaves) in papaya leaf (*Carica papaya L*) by altering the metabolic pathways involving the metabolism of starch and sucrose, synthesis of phenylpropane, and transduction of plant hormone signal [5]. Chitosan actually induces systemic resistance against plant viruses by altering a lot of signaling pathways.

The duration of exposure to chitosan derivatives also impacts their antiviral activity. For example, exposure to 2 mg L^{-1} of guanidinylated chitosan hydrochloride for a short span of time leads to direct damage to the virus [30]. Another example includes effective inhibition by a high dosage of chitosan with different strains of rhizobacteria, which actually promotes the plant growth, in a time-dependent manner, against papaya ringspot virus-W and tomato chlorotic spot virus which affects cucurbits and tomatoes [1, 71].

3 Conclusion

In this chapter, the antiviral activity of micro- and nano-structures of chitosan and its derivatives along with its nanocomposites have been examined. Detailed study of chitosan against a plethora of animal viruses like HIV, HSV, HPV, influenza virus, and plant viruses like TNV and TMV showed that chitosan takes a triple approach in terms of its inhibitory activity. The main inhibition by chitosan is associated with its ability to inhibit the binding of the viral proteins with the receptors of the host cells and also reduce the viral load by strong electrostatic repulsive interactions with the virus, chitosan being cationic in nature. Chitosan also has a tendency to stimulate the immune response of the healthy cells against foreign organisms like a virus. Lastly, chitosan and its derivatives also act as good, selective, and efficient carriers for antiviral drugs.

The activity of chitosan depends on a lot of factors like the degree of polymerization, degree of substitution, the process of deacetylation, source of chitin, molecular weight, concentration, pH, and also on the type of virus in question.

Owing to a number of physicochemical properties of chitosan like cost-effectiveness, environment friendliness, biodegradability, biocompatibility, and non-toxicity, studies have been done to employ it as an anti-SARS-CoV molecule not only as a drug but as a component of nanofiber for the PPE kits. But, as far as its activity against COVID-19 is concerned, there is still a long way to go in terms of the applicability of chitosan-based materials and their mechanism of action.

References

1. Abdalla OA, Bibi S, Zhang S (2017) Integration of chitosan and plant growth-promoting rhizobacteria to control Papaya ringspot virus and Tomato chlorotic spot virus. Arch Phytopathol Pflanzenschutz 50:997–1007
2. Alarcón B, Lacal JC, Fernández-Sousa JM, Carrasco L (1984) Screening for new compounds with antiherpes activity. Antivir Res 4:231–243
3. Alinejad Y, Adoungotchodo A, Hui E, Zehtabi F, Lerouge S (2018) An injectable chitosan/chondroitin sulfate hydrogel with tunable mechanical properties for cell therapy/tissue engineering. Int J Biol Macromol 113:132–141
4. Alitongbieke G, Li X-M, Wu Q-C, Lin Z-C, Huang J-F, Xue Y, Liu J-N, Lin J-M, Pan T, Chen Y-X, Su Y, Zhang G-G, Leng B, Liu S-W, Pan Y-T (2020) Effect of β-chitosan on the binding interaction between SARS-CoV-2 S-RBD and ACE2 (Pre-Print)
5. An N, Lv J, Zhang A, Xiao C, Zhang R, Chen P (2020) Gene expression profiling of papaya (Carica papaya L.) immune response induced by CTS-N after inoculating PLDMV. Gene 144845
6. Angelim AL, Costa SP, Farias BCS, Aquino LF, Melo VMM (2013) An innovative bioremediation strategy using a bacterial consortium entrapped in chitosan beads. J Environ Manag 127:10–17
7. Artan M, Karadeniz F, Karagozlu MZ, Kim M-M, Kim S-K (2010) Anti-HIV-1 activity of low molecular weight sulfated chitooligosaccharides. Carbohydr Res 345:656–662
8. Asahi-Ozaki Y, Itamura S, Ichinohe T, Strong P, Tamura S-i, Takahashi H, Sawa H, Moriyama M, Tashiro M, Sata T, Kurata T, Hasegawa H (2006) Intranasal administration of adjuvant-combined recombinant influenza virus HA vaccine protects mice from the lethal H5N1 virus infection. Microbes Infect 8:2706–2714
9. Baaten BJG, Clarke B, Strong P, Hou S (2010) Nasal mucosal administration of chitin microparticles boosts innate immunity against influenza A virus in the local pulmonary tissue. Vaccine 28:4130–4137
10. Babosha AV (2004) Changes in lectin activity in plants treated with resistance inducers. Biol Bull 31(1):51–55
11. Carrion CC, Nasrollahzadeh M, Sajjadi M, Jaleh B, Soufi GJ, Iravani S (2021) Lignin, lipid, protein, hyaluronic acid, starch, cellulose, gum, pectin, alginate and chitosan-based nanomaterials for cancer nanotherapy: challenges and opportunities. Int J Biol Macromol 178:193–228
12. Cheng S, Zhao H, Xu Y, Yang Y, Lv X, Wu P, Li X (2014) Inhibition of influenza virus infection with chitosan-sialyloligosaccharides ionic complex. Carbohydr Polym 107:132–137
13. Chirkov SN (2002) The antiviral activity of chitosan (review). Prikl Biokhim Mikrobiol 38(1):12–13
14. Chou S, Lurain NS, Thompson KD, Miner RC, Drew WL (2003) Viral DNA polymerase mutations associated with drug resistance in human cytomegalovirus. J Infect Dis 188:32–39
15. Ciejka J, Wolski K, Nowakowska M, Pyrc K, Szczubiałka K (2017) Biopolymeric nano/microspheres for selective and reversible adsorption of coronaviruses. Mater Sci Eng C 76:735–742
16. Davis R, Zivanovic S, D'Souza DH, Davidson PM (2012) Effectiveness of chitosan on the inactivation of enteric viral surrogates. Food Microbiol 32:57–62
17. Davydova VN, Nagorskaya VP, Gorbach VI, Kalitnik AA, Reunov AV, Solov'eva TF, Ermak IM (2011) Chitosan antiviral activity: dependence on structure and depolymerization method. Appl Biochem Microbiol 47:103–108

18. Deeks SG, Overbaugh J, Phillips A, Buchbinder S (2015) HIV infection. Nat Rev Dis Primers 1:1–22
19. Dewen Q, Yijie D, Yi Z, Shupeng L, Fachao S (2017) Plant immunity inducer development and application. Mol Plant Microbe Interact 30:355–360
20. Dimassi S, Tabary N, Chai F, Blanchemain N, Martel B (2018) Sulfonated and sulfated chitosan derivatives for biomedical applications: a review. Carbohydrate Polym 202:382–396
21. Gao Y, Liu W, Wang W, Zhang X, Zhao X (2018) The inhibitory effects and mechanisms of 3,6-O-sulfated chitosan against human papillomavirus infection. Carbohydrate Polym 198:329–338
22. Ghaemi F, Amiri A, Bajuri MY, Yuhana NY, Ferrara M (2021) Role of different types of nanomaterials against diagnosis, prevention and therapy of COVID-19. Sustain Cities Soc 72:103046
23. Ghendon Y, Markushin S, Krivtsov G, Akopova I (2008) Chitosan as an adjuvant for parenterally administered inactivated influenza vaccines. Arch Virol 153:831–837
24. Griffiths P, Reeves M (2021) Pathogenesis of human cytomegalovirus in the immunocompromised host. Nat Rev Microbiol 19:759–773
25. Hamajima K, Kojima Y, Matsui K, Toda Y, Jounai N, Ozaki T, Xin K-Q, Strong P, Okuda K (2003) Chitin micro-particles (CMP): a useful adjuvant for inducing viral specific immunity when delivered intranasally with an HIV-DNA vaccine. Viral Immunol 16:541–547
26. Hasegawa H, Ichinohe T, Strong P, Watanabe I, Ito S, Tamura S-i, Takahashi H, Sawa H, Chiba J, Kurata T, Sata T (2005) Protection against influenza virus infection by intranasal administration of hemagglutinin vaccine with chitin microparticles as an adjuvant. J Med Virol 75:130–136
27. Hassan MI, Mohamed AF, Taher FA, Kamel MR (2016) Antimicrobial activities of Chitosan nanoparticles prepared from Lucilia Cuprina Maggots (Diptera: Calliphoridae) J Egypt Soc Parasitol 46:563–570
28. Hathout RM, Kassem DH (2020) Positively charged electroceutical spun chitosan nanofibers can protect health care providers from COVID-19 infection: an opinion. Front Bioeng Biotech 8:885
29. Hosseinnejad M, Jafari SM (2016) Evaluation of different factors affecting antimicrobial properties of chitosan. Int J Biol Macromol 85:467–475
30. Hu V, Cai J, Yumin D, Lin J, Wang C, Xiong K (2009) Preparation and anti-TMV activity of guanidinylated chitosan hydrochloride. J Appl Polym Sci 112:3522–3528
31. Ichinohe T, Nagata N, Strong P, Tamura S-i, Takahashi H, Ninomiya A, Imai M, Odagiri T, Tashiro M, Sawa H, Chiba J, Kurata T, Sata T, Hasegawa H (2007) Prophylactic effects of chitin microparticles on highly pathogenic H5N1 influenza virus. J Med Virol 79:811–819
32. Intini C, Elviri L, Cabral J, Mros S, Bergonzi C, Bianchera A, Flammini L, Govoni P, Barocelli E, Bettini R, McConnell M (2018) 3D-printed chitosan-based scaffolds: an in vitro study of human skin cell growth and an in-vivo wound healing evaluation in experimental diabetes in rats. Carbohydrate Polym 199:593–602
33. Iriti M, Faoro F (2008) Abscisic acid is involved in chitosan-induced resistance to tobacco necrosis virus (TNV). Plant Physiol Biochem 46:1106–1111
34. Iriti M, Sironi M, Gomarasca S, Casazza AP, Soave C, Faoro F (2006) Cell death-mediated antiviral effect of chitosan in tobacco. Plant Physiol Biochem 44:893–900
35. Iriti M, Varoni EM (2015) Chitosan-induced antiviral activity and innate immunity in plants. Environ Sci Pollut Res 22:2935–2944
36. Ishihara C, Yoshimatsu K, Tsuji M, Arikawa J (1993) Anti-viral activity of sulfated chitin derivatives against Friend murine leukaemia and herpes simplex type-1 viruses. Vaccine 11(6):670–674
37. Itani R, Tobaiqy M, Faraj AA (2020) Optimizing use of theranostic nanoparticles as a life-saving strategy for treating COVID-19 patients. Theranostics 10:5932–5942
38. Jaimes JA, Millet JK, Whittaker GR (2020) Proteolytic cleavage of the SARS-CoV-2 spike protein and the role of the novel S1/S2 site. iScience 23:101212

39. Jana B, Chatterjee A, Roy D, Ghorai S, Pan D, Pramanik SK, Chakraborty N, Ganguly J (2022) Chitosan/benzyloxy-benzaldehyde modified ZnO nano template having optimized and distinct antiviral potency to human cytomegalovirus. Carbohydrate Polym 278:118965
40. Kalathiya U, Padariya M, Mayordomo M, Lisowska M, Nicholson J, Singh A, Baginski M, Fahraeus R, Carragher N, Ball K, Haas J, Daniels A, Hupp TR, Alfaro JA (2020) Highly conserved homotrimer cavity formed by the sars-cov-2 spike glycoprotein: a novel binding site. J Clin Med 9:1473
41. Karagozlu MZ, Karadeniz F, Kim SK (2014) Anti-HIV activities of novel synthetic peptide conjugated chitosan oligomers. Int J Biol Macromol 66:260–266
42. Kochkina ZM, Chirkov SN (2000) Effect of chitosan derivatives on the reproduction of coliphages T2 and T7. Microbiol 69(2):257–260
43. Kochkina ZM, Surgucheva NA, Chirkov SN (2000) Inactivation of coliphages by chitosan derivatives. Microbiol 69(2):261–265
44. Kong M, Chen XG, Xing K, Park HJ (2010) Antimicrobial properties of chitosan and mode of action: a state of the art review. Int J Food Microbiol 144:51–63
45. Kumar M, Behera AK, Lockey RF, Zhang J, Bhullar G, De La Cruz CP, Chen L-C, Leong KW, Huang S-K, Mohapatra SS (2002) Intranasal gene transfer by chitosan-DNA nanospheres protects BALB/c mice against acute respiratory syncytial virus infection. Hum Gene Ther 13:1415–1425
46. Kumaraswamy RV, Kumari S, Choudhary RC, Pal A, Raliya R, Biswas P, Saharan V (2018) Engineered chitosan based nanomaterials: bioactivities, mechanisms and perspectives in plant protection and growth. Int J Biol Macromol 113:494–506
47. Lee CG, da Silva CA, Lee J-Y, Hartl D, Elias JA (2008) Chitin regulation of immune responses: an old molecule with new roles. Curr Opin Immunol 20:684–689
48. Li X, Wu P, Gao GF, Cheng S (2011) Carbohydrate-functionalized chitosan fiber for influenza virus capture. Biomacromol 12:3962–3969
49. Lin YC, Lin ST, Chen CY, Wu SC (2012) Enterovirus 71 adsorption on metal ion-composite chitosan beads. Biotechnol Prog 28:206–214
50. López-León T, Carvalho ELS, Seijo B, Ortega-Vinuesa JL, Bastos-González D (2005) Physicochemical characterization of chitosan nanoparticles: electrokinetic and stability behavior. J Colloid Interface Sci 283:344–351
51. Loutfy SA, Elberry MH, Farroh KY, Mohamed T, Mohamed AA, Mohamed EB, Ibrahim AH, Faraag MSA (2020) Antiviral activity of chitosan nanoparticles encapsulating curcumin against hepatitis C virus genotype 4a in human hepatoma cell lines. Int J Nanomed 15:2699–2715
52. Milewska A, Chi Y, Szczepanski A, Barreto-Duran E, Dabrowska A, Botwina P, Obloza M, Liu K, Liu D, Guo X, Ge Y, Li J, Cui L, Ochman M, Urlik M, Rodziewicz-Motowidlo S, Zhu F, Szczubialka K, Nowakowska M, Pyrc K (2021) HTCC as a polymeric inhibitor of SARS-CoV-2 and MERS-CoV. J Virol 95
53. Milewska A, Ciejka J, Kaminski K, Karewicz A, Bielska D, Zeglen S, Karolak W, Nowakowska M, Potempa J, Bosch BJ, Pyrc K, Szczubialka K (2013) Novel polymeric inhibitors of HCoV-NL63. Antivir Res 97:112–121
54. Milewska A, Kaminski K, Ciejka J, Kosowicz K, Zeglen S, Wojarski J, Nowakowska M, Szczubiałka K, Pyrc K (2016) HTCC: broad range inhibitor of coronavirus entry. PLoS ONE 11(6):e0156552
55. Mori Y, Ono T, Miyahira Y, Nguyen VQ, Matsui T, Ishihara M (2013) Antiviral activity of silver nanoparticle/chitosan composites against H1N1 influenza A virus. Nanoscale Res Lett 8:93
56. Nadimi AE, Ebrahimipour SY, Afshar EG, Falahati-pour SK, Ahmadi Z, Mohammadinejad R, Mohamadi M (2018) Nano-scale drug delivery systems for antiarrhythmic agents. Eur J Med Chem 157:1153–1163
57. Nagorskaya V, Reunov A, Lapshina L, Davydova V, Yermak I (2014) Effect of chitosan on tobacco mosaic virus (TMV) accumulation, hydrolase activity, and morphological abnormalities of the viral particles in leaves of N. tabacum L. cv. Samsun. Virol Sin 29(4):250–256

58. Pal P, Pal A, Nakashima K, Yadav BK (2021) Applications of chitosan in environmental remediation: a review. Chemosphere 266:128934
59. Pauls T (2016) Chitosan as an antiviral (Thesis)
60. Peniche H, Peniche C (2011) Chitosan nanoparticles: a contribution to nanomedicine. Polym Int 60:883–889
61. Pospieszny H, Chirkov S, Atabekov J (1991) Induction of antiviral resistance in plants by chitosan. Plant Sci 79:63–68
62. Qin C, Li H, Xiao Q, Liu V, Zhu J, Du Y (2006) Water-solubility of chitosan and its antimicrobial activity. Carbohydr Polym 63:367–374
63. Qing W, Zuo J-H, Qian W, Yang N, Gao L-P (2015) Inhibitory effect of chitosan on growth of the fungal phytopathogen, sclerotinia sclerotiorum, and sclerotinia rot of carrot. J Integr Agric 14:691–697
64. Raghuwanshi D, Mishra V, Das D, Kaur K, Suresh MR (2012) Dendritic cell targeted chitosan nanoparticles for nasal DNA immunization against SARS CoV nucleocapsid protein. Mol Pharm 9:946–956
65. Rozhnova G, Kretzschmar ME, van der Klis F, van Baarle D, Korndewal MJ, Vossen AC, van Boven M (2020) Short- and long-term impact of vaccination against cytomegalovirus: a modeling study. BMC Med 18:174
66. Safarzadeh M, Sadeghi S, Azizi M, Rastegari-Pouyani M, Pouriran R, Hoseini MHM (2021) Chitin and chitosan as tools to combat COVID-19: a triple approach. Int J Biol Macromol 183:235–244
67. Saharan V, Mehrotra A, Khatik R, Rawal P, Sharma SS, Pal A (2013) Synthesis of chitosan based nanoparticles and their in vitro evaluation against phytopathogenic fungi. Int J Biol Macromol 62:677–683
68. Saharan V, Sharma G, Yadav M, Choudhary MK, Sharma SS, Pal A, Raliya R, Biswas P (2015) Synthesis and in vitro antifungal efficacy of Cu-chitosan nanoparticles against pathogenic fungi of tomato. Int J Biol Macromol 75:346–353
69. Salehi B, Anil Kumar NV, Şener B, Sharifi-Rad M, Kılıç M, Mahady GB, Vlaisavljevic S, Iriti M, Kobarfard F, Setzer WN, Ayatollahi SA, Ata A, Sharifi-Rad J (2018) Medicinal plants used in the treatment of human immunodeficiency virus. Int J Mol Sci 19:1459
70. Shanmugam A, Kathiresan K, Nayak L (2016) Preparation, characterization and antibacterial activity of chitosan and phosphorylated chitosan from cuttlebone of Sepia kobiensis (Hoyle 1885). Biotechnol Rep 9:25–30
71. Sharma N, Modak C, Singh PK, Kumar R, Khatri D, Singh SB (2021) Underscoring the immense potential of chitosan in fighting a wide spectrum of viruses: a plausible molecule against SARS-CoV-2? Int J Biol Macromol 179:33–44
72. Shibata Y, Foster LA, Metzger WJ, Myrvik QN (1997) Alveolar macrophage priming by intravenous administration of chitin particles polymers of N-acetyl-D-glucosamine, in Mice. Infect Immun 65(5):1734–1741
73. Sofy AR, Dawoud RA, Sofy MR, Mohamed I, Hmed AA, El-Dougdoug NK (2020) Improving regulation of enzymatic and non-enzymatic antioxidants and stress-related gene stimulation in cucumber mosaic cucumovirus-infected cucumber plants treated with glycine betaine chitosan and combination. Molecules 25:2341
74. Sofy AR, Hmed AA, Alnaggar AE-AM, Dawoud RA, Elshaarawy RFM, Sofy MR (2020) Mitigating effects of Bean yellow mosaic virus infection in faba bean using new carboxymethyl chitosan-titania nanobiocomposites. Int J Biol Macromol 163:1261–1275
75. Sosa MAG, Fazely F, Koch JA, Vercellotti SV, Ruprecht RM (1991) NCarboxymethylchitosan-N, O-sulfate as an anti-HIV-1 agent. Biochem Biophys Res Commun 174(2):489–496
76. Stepanov OA, Prokof'eva MM, Stocking K, Varlamov VP, Levov AN, Vikhoreva GA, Spirin PV, Mikhailov SN, Prassolov VS (2012) Replication-competent gamma-retrovirus Mo-MuLV expressing green fluorescent protein as efficient tool for screening of inhibitors of retroviruses that use heparan sulfate as primary cell receptor. Mol Biol 46:457–466
77. Swierczewska M, Han HS, Kim K, Park JH, Lee S (2016) Polysaccharide-based nanoparticles for theranostic nanomedicine. Adv Drug Deliv Rev 99:70–84

78. van Doremalen N, Bushmaker T, Morris DH, Holbrook MG, Gamble A, Williamson BN, Tamin A, Harcourt JL, Thornburg NJ, Gerber SI, Lloyd-Smith JO, de Wit E, Munster VJ (2020) Aerosol and surface stability of SARS-CoV-2 as compared with SARS-CoV-1. N Engl J Med 382(16):1564–1567
79. Venton G, Crocchiolo R, Fürst S, Granata A, Oudin C, Faucher C, Coso D, Bouabdallah R, Berger P, Vey N, Ladaique P, Chabannon C, le Merlin M, Blaise D, El-Cheikh J (2014) Risk factors of Ganciclovir-related neutropenia after allogeneic stem cell transplantation: a retrospective monocentre study on 547 patients. Clin Microbiol Infect 20:160–166
80. Vijayan PP, Chithra PG, Abraham P, George JS, Maria HJ, Thomas S (2022) Nanocoatings: universal antiviral surface solution against COVID-19. Prog Org Coat 163:106670
81. Wang X, Zhang W, Liu F, Zheng M, Zheng D, Zhang T, Yi Y, Ding Y, Luo J, Dai C, Wang H, Sun B, Chen Z (2012) Intranasal immunization with live attenuated influenza vaccine plus chitosan as an adjuvant protects mice against homologous and heterologous virus challenge. Arch Virol 157:1451–1461
82. Wieckiewicz M, Boening KW, Grychowska N, Paradowska-Stolarz A (2017) Clinical application of chitosan in dental specialities. Mini Rev Med Chem 17:401–409
83. Wu D, Ensinas A, Verrier B, Primard C, Cuvillier A, Champier G, Paul S, Delair T (2016) Zinc-stabilized colloidal polyelectrolyte complexes of chitosan/hyaluronan: a tool for the inhibition of HIV-1 infection. J Mater Chem B 4:5455–5463
84. Zhang J, Tan W, Luan F, Yin X, Dong F, Li Q, Guo Z (2018) Synthesis of quaternary ammonium salts of chitosan bearing halogenated acetate for antifungal and antibacterial activities. Polymers 10:530
85. Zhu S, Barnes C, Bhar S, Hoyeck P, Galbraith AN, Devabhaktuni D, Karst SM, Montazeri N, Jones MK (2020) Survival of human norovirus surrogates in water upon exposure to thermal and non-thermal antiviral treatments. Viruses 12:461

Chapter 13
Modifications and Applications of Chitosan-Based Nanocomposites in Orthopaedics and Dentistry

Taruna Singh, Parul Pant, and Sarthak Kaushik

Abstract Chitosan has been studied extensively and widely used in the applications of biopharmaceutical and biomedical fields. Further, chitosan can be modified into nanocomposites. The possible applications of chitosan are quite vast, and fascinating properties of chitosan-based nanocomposites like biocompatibility, antimicrobial, antioxidant, anti-inflammatory, good mechanical strength, etc., make chitosan very promising in the sector of orthopaedics and dentistry in tissue engineering and regenerative medicine. This chapter focusses on the different strategies adopted for the modification of chitosan as chitosan nanocomposites in the field of dentistry and orthopaedics. Biomedical applications of chitosan-based nanocomposites in orthopaedics and dentistry have also been covered in the chapter. Lastly, the advantages of chitosan nanocomposites in the fields are also highlighted.

Keywords Chitosan-based nanocomposites · Orthopaedics · Dentistry

Abbreviations

3D	Three dimensional
ABC	Acrylic bone cements
CNT	Carbon nanotubes
CTS	Chitosan
GO	Graphene oxide
GTR	Guided tissue regeneration
HAp	Hydroxyapatite
MMT	Montmorillonite
ZnO	Zinc oxide

T. Singh · P. Pant (✉)
Department of Chemistry, Hansraj College, University of Delhi, Delhi 110007, India
e-mail: parulpant@hrc.du.ac.in

S. Kaushik
Department of Chemistry, University of Delhi, Delhi 110007, India

1 Introduction

Chitosan is a naturally occurring biomaterial that is extracted mainly from chitin, which is derived from the exoskeleton of crabs and shrimps. Chitosan is obtained by deacetylation of chitin, the second most abundant natural biopolymer (Fig. 1) [1].

The need for biocompatible, antimicrobial, antibacterial, and non-toxic derivatives is being fulfilled by nanocomposites obtained from different biopolymers in biomedical and pharmaceutical fields. Since, chitosan has proved itself to be helpful

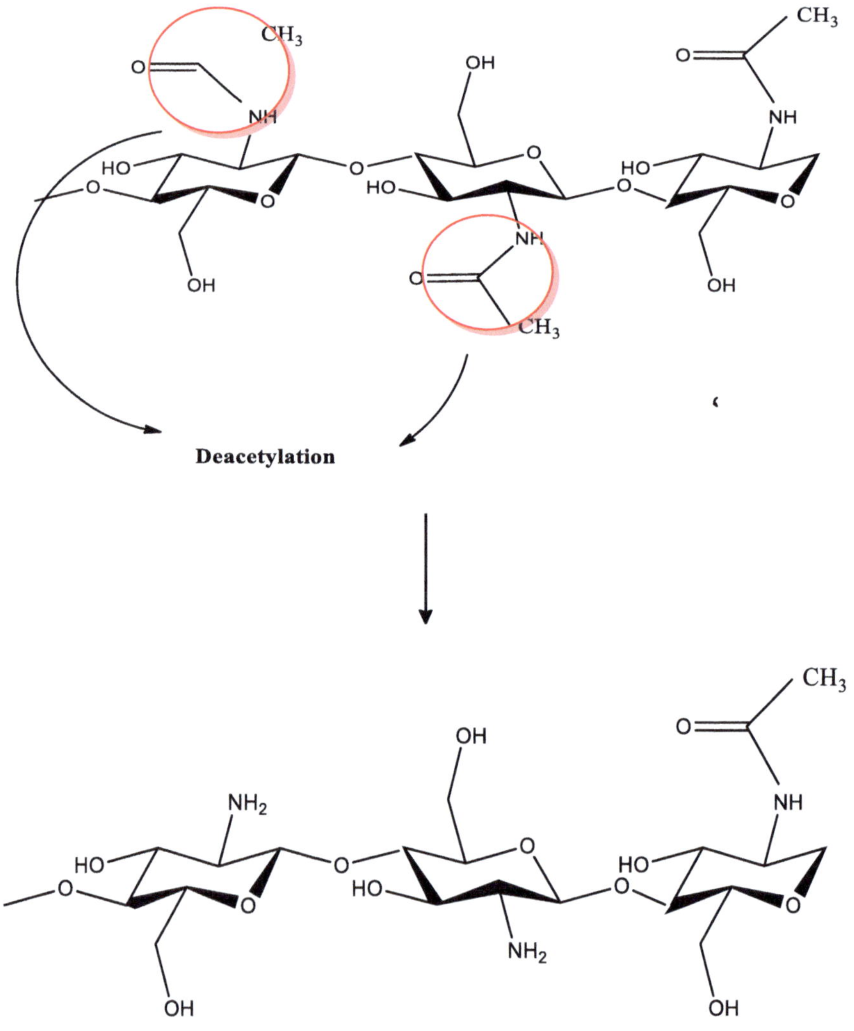

Fig. 1 Preparation of chitosan from deacetylation of chitin

in the fields of biopharmaceuticals and biomedicine having a large number of applications due to its properties, (i.e. antiviral, antibacterial, and antifungal) its nanocomposites have also been showing evident results in the field of biopharmaceuticals and biomedical applications. The synthesis of nanocomposites of chitosan can be achieved using different chemical methods depending upon the field of application [1]. High bioactivity being a prominent property of chitosan and its nanocomposites make it appropriate to be used in the fields of orthopaedics and dentistry [2]. These nanocomposites can be moulded or modified into different formulations for their potential applications in the field of tissue engineering or in the studies of bone regeneration in the field of orthopaedics [3]. It also serves well in enamel repair, oral drug delivery, implants for dental complications, etc., in the field of dentistry.

Presently, the demand for chitosan and its nanocomposites in these fields is high, as ongoing research in the field of nano-scale biomedical applications is growing at an enormous rate. The study of modifications and bio-application of chitosan-based nanocomposites have shown tremendous outcomes.

2 Modifications of Chitosan-Based Nanocomposites

Chitosan-based nanocomposites play a vital role in the fields of biomedicine and biopharmaceuticals. However, the biocompatibility and drug delivery of nanocomposites based on chitosan only are up to a lower extent in the fields of orthopaedics and dentistry. To uplift the bioactivity and compatibility of chitosan-based nanocomposites for increasing the properties like mechanical strength, antibacterial characteristics, thermal stability, etc., these nanocomposites have been modified using different nano-scaled or layered (in the form of thin films) chemical reagents to enhance these properties in orthopaedics and dentistry.

3 Modification of Chitosan-Based Nanocomposites in the Field of Orthopaedics

Various chemical modifications have been done to enhance the ability of chitosan nanocomposites in the orthopaedics sector. These nanocomposites have been modified with hydroxyapatite(HAp), carbon nanotubes (CNT), and acrylic bone cements (ABC) with graphene oxide (GO), etc.

3.1 Hydroxyapatite (HAp)

Chitosan (CTS)-based nanocomposites alone cannot mimic all the needed properties of a natural bone in a human body, but along with modifications in their chemical parts, it shows many similar properties of a natural bone.

Many studies have shown that calcium and phosphate-based materials show osteoconduction (i.e. growing of a bone on a surface), imitating the best properties of a natural bone, becoming materials of great interest in the studies of bone tissue engineering. Therefore, hydroxyapatite (HAp) with the molecular formula $[Ca_{10}(PO_4)_6(OH)_2]$ is used in CTS-based nanocomposites, with HAp being one of the most stable calcium phosphate forms with the abundance of its occurrence (approx. 60–67%) in the natural bone making it a major component. With the variety of applications of Hap in the field of orthopaedics, it also serves its purpose as an essential compound in the sector of preparation of artificial bones in the human body. With CTS-based nanocomposites alone, the cell growth of osteoblast cells (cells responsible for the synthesis of bone) has been found to be slightly increased in orthopaedic treatment, whilst CTS/HAp (chitosan and hydroxyapatite)-based nanocomposites with a quite stable chemical interaction (Fig. 2) significantly increased the osteoblastic cell growth in a natural bone giving its skeletal structure good stability and greater molecular density along with antibacterial properties. Also, CTS/HAp-based nanocomposites fulfil the role of the inorganic as well as the organic portion of a natural bone [4].

3.2 Carbon Nanotubes (CNT)

Studies have shown that the modification of chitosan (CTS) combined with CNT (carbon nanotubes) results in a significant increase in the mechanical strength and structural properties of the nanocomposite serving its role in the natural bone. CNT is also associated with the study of biomaterials, mainly those materials which are used to position the bone in the human body, like—materials used in prosthetics, plates, and screws in case of fractures, drug delivery systems, preparation of scaffolds for regeneration of bone, etc. According to reports the interaction of CNT with CTS, where CNT is uniformly distributed along with the CTS matrix (Fig. 3).

The observable change in properties like tensile strength and mechanical structural integrity is significantly enhanced with the addition of even small amounts of CNT in the CTS matrix. Numerically, with the introduction of about 0.8% of CNT in the CTS matrix improves the tensile properties of nanocomposites by about 93%.

Also, the use of CNT with CTS nanocomposites in bone treatment as an alternative material has been reported, with a possibility of natural bone regrowth. Hence, cells with similar properties to osteoblastic cells were grown on CNT (electrically neutral in nature) to produce a natural bone [4].

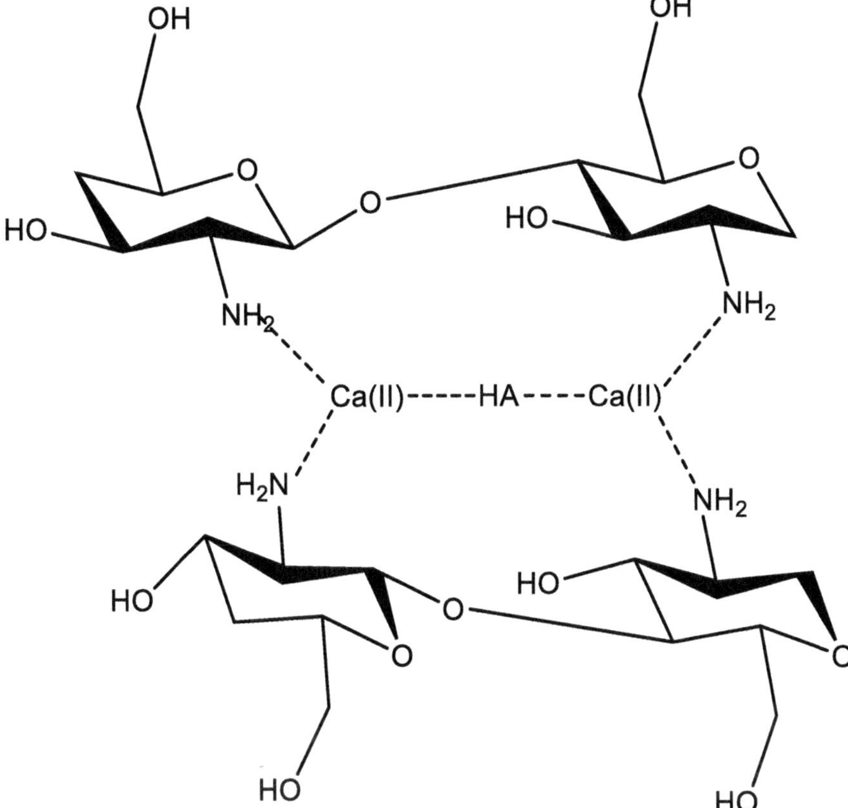

Fig. 2   Chemical interaction of CTS with HAp

3.3   Acrylic Bone Cements (ABC) with Graphene Oxide (GO)

The role of acrylic bone cements (ABC) in orthopaedic surgeries is quite significant, but an increase in their use in other applications has been recorded, such as cancer-causing bone remodelling, cranioplasty (the surgical repair of a bone defect in the skull), and vertebroplasty (for stabilizing compression fractures in the spine), etc. The introduction of ABC along with graphene oxide (GO) in chitosan (CTS)-based nanocomposites leads to changes in mechanical, biological, thermal, and physicochemical properties of the nanocomposites acting upon the bones.

It is observed that using the modification with the formulation of CTS/ABC/GO, i.e. CTS containing ABCs and GO simultaneously, results in the enhancement of thermal stability, osteogenic activities (i.e. the ability of the material to induce bone formation in non-bone-forming sites) and antibacterial properties, making it an essential modification in orthopaedic applications. Moreover, this formulation resulted in enhanced biocompatibility showing potential use in biomedical applications, and GO

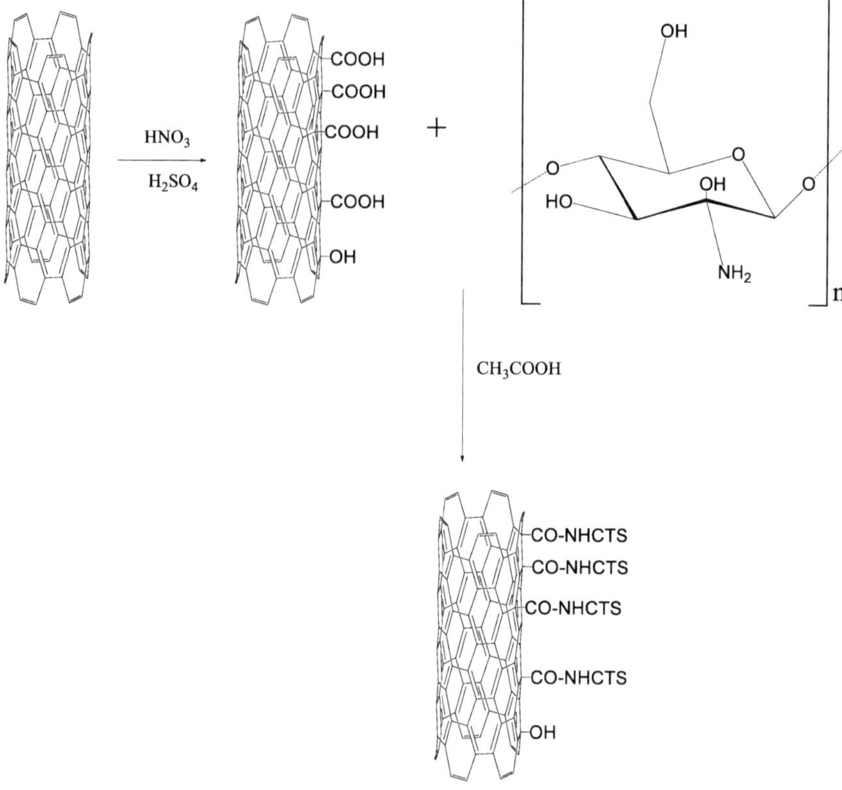

Fig. 3 Chemical interaction between CTS and CNT

present in the formulation was mainly responsible for the mechanical augmentation and antibacterial characteristics [5].

4 Modification of Chitosan-Based Nanocomposites in the Field of Dentistry

In the dentistry sector, these nanocomposites have been modified with montmorillonite, zinc oxide (Table 1).

Table 1 Modifications of chitosan-based nanocomposites, their inferences, and applications

S. No.	Modification of chitosan-based nanocomposites	Inferences and applications	Reference
1	Hap-CTS (modification of chitosan-based nanocomposites with hydroxyapatite in orthopaedics)	Stable chemical interaction between CTS and Hap results in increased osteoblastic cell growth, stable skeletal structure, greater molecular density, and antibacterial properties Also, fulfilling the role of organic and inorganic part of a natural bone	[4]
2	CNT-CTS (modification of chitosan-based nanocomposites with carbon nanotubes in orthopaedics)	Interaction of CNT and CTS results in enhancing tensile strength and structural integrity, possibility of natural bone regrowth	[4]
3	ABC with GO-CTS (modification of chitosan-based nanocomposites with acrylic bone cements and graphene oxide in orthopaedics)	The formulation of CTS/ABC/GO enhances thermal stability, osteogenic activities, and antibacterial properties Also, enhanced biocompatibility for the use of biomedical applications	[5]
4	MMT-CTS (modification of chitosan-based nanocomposites with montmorillonite clay in dentistry)	Increase in tensile properties, material strength, and elasticity (depends upon the concentration of the clay particles) were observed Enhanced antimicrobial properties with the addition of $AGNO_3$ Also, electrostatic interaction between CTS and silicate layers increases the thermal stability	[6]
5	ZnO-CTS (modification of chitosan-based nanocomposites with size-controlled ZnO nanoparticles in dentistry)	This formulation results in antibacterial, antidiabetic, antioxidant, and aytotoxic agents Also, ZnO-CTS interaction results in increased bond strength and mechanical properties	[7]

4.1 MMT-CLAY (Montmorillonite)

The preparation of chitosan-based nanocomposites loaded with clay (montmorillonite) (CS/MMT) is done by mixing chitosan biopolymer and sodium montmorillonite particles physically. The process of physical mixing is done to get thin transparent films that serve as a dental material in dental applications and dental surgeries. Different concentration of clay particles is used to prepare these thin films, and the measurement of antibacterial and mechanical properties takes place. The studies have shown a significant amount of increase in tensile properties, material strength, and elasticity as per the increase in the concentration of clay particles. Also, it was observed that the addition of $AgNO_3$ (silver nitrate) nanoparticles

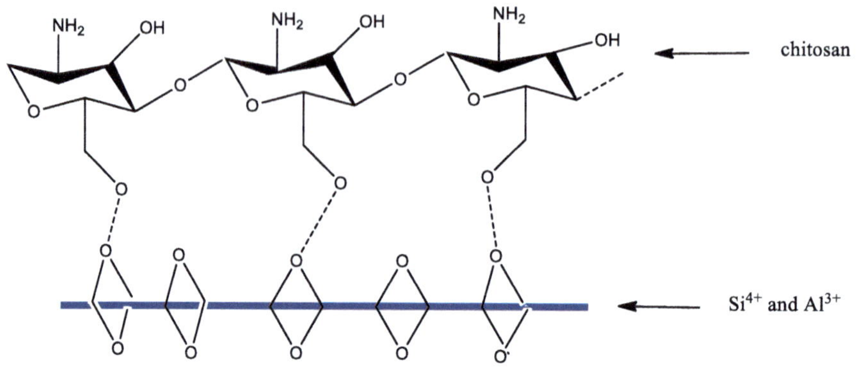

Fig. 4 Depiction of the interaction between chitosan and silicate layers

resulted in enhanced antimicrobial properties as compared to pure chitosan-based nanocomposite films. A strong electrostatic interaction was found between chitosan and silicate layers which is responsible for the significant increase in thermal stability, whereas the layer-layer stacking structure of chitosan nanocomposites enhances the antimicrobial activities (Fig. 4) [6].

4.2 Zinc Oxide (ZnO)

In dental fields, the use of metal-oxide nanoparticles used in dental restoration surgeries is a new and highly used process with its high-antibacterial activities against bacteria growing pathogenic species.

The preparation of ZnO infused chitosan (CTS)-based nanocomposites with size-controlled ZnO nanoparticles is achieved via chemical routes, resulting in stable nanostructures. These chemically extracted ZnO nanoparticles infused CTS nanocomposites proved to be antibacterial, antidiabetic, antioxidant, and cytotoxic agents. These CTS nanocomposite encapsulated ZnO nanoparticles are highly used in restorative dental applications as an antibacterial dental adhesive used during dental surgeries with increased mechanical characteristics and increase in bond strength due to CTS/ZnO interaction Fig. 5 [7].

5 Applications of Chitosan-Based Nanocomposites

As discussed above, chitosan (CTS)-based nanocomposites have proven to be helpful in the field of biopharmaceuticals and biomedicine, adding to that when they get modified with different nano-scaled chemical compounds and chemical films, their bioactivity, biocompatibility, drug delivery, etc., enhance up to significant results,

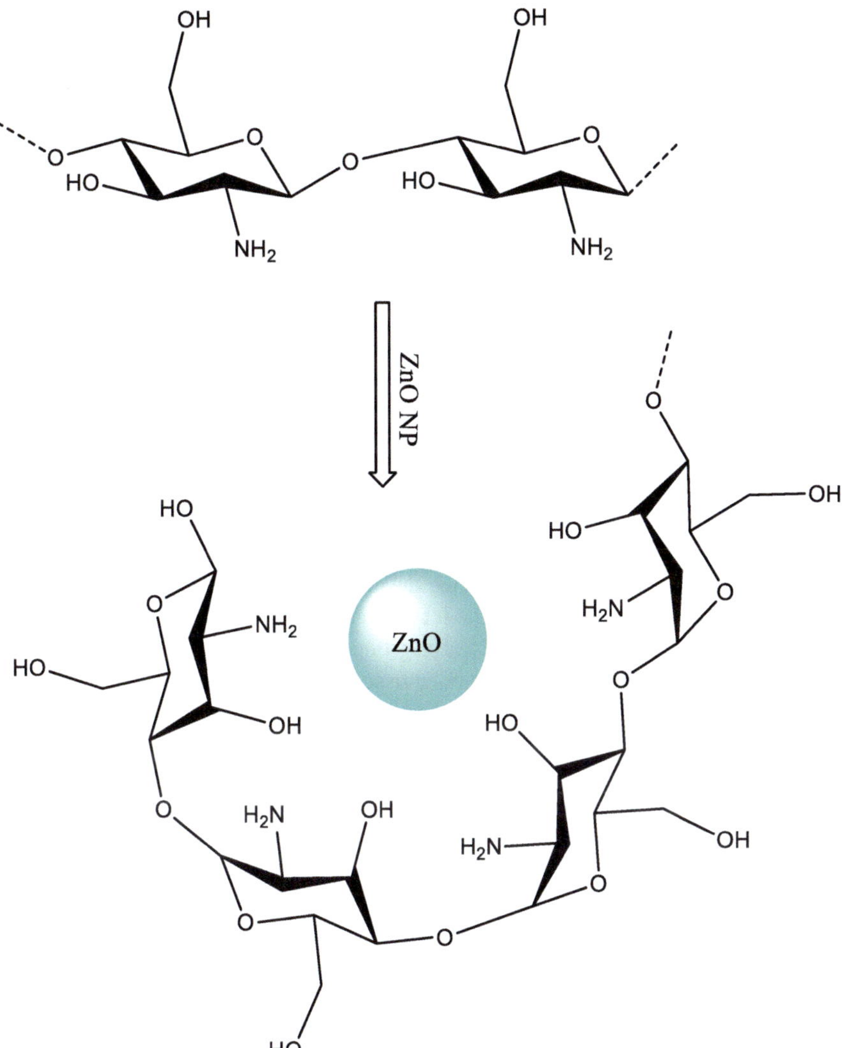

Fig. 5 Interaction of CTS-based nanocomposites and ZnO nanoparticles

this makes these nanocomposites to deliver several biomedical applications. These biomedical applications of chitosan-based nanocomposites (both with CTS alone or modified) have acquired a prominent place in the sectors of orthopaedics and dentistry, in different regions like drug delivery, enamel repair, bone, and dental tissue regeneration.

6 Biomedical Applications of Chitosan-Based Nanocomposites in Orthopaedics

The natural bone comprises of three-dimensional nanocomposite structured matrix. The study of bone structure and matrix comes under the bio-application of bone tissue engineering in the field of orthopaedics.

The field of bone tissue regeneration is comparatively new (started about three decades ago), its applications mainly include the study and modification of 3D (three-dimensional) structures, which serve their part in giving support, reinforcement to the bone material and in certain cases organizing bone replacements and tissue regeneration in a natural way. Bone tissue engineering encourages a full-fledged regeneration of the tissue of a new bone in the human body with the combinations and modifications of different biomolecules, biosynthetic materials, drugs for delivery, growth-enhancing factors, etc. Along with the significant advancement in the field, the selection of material for bone tissue engineering applications becomes crucial to making these applications become a proper clinical practice in the field of orthopaedics. To fulfil the needs of bone tissue engineering as a biomedical application in orthopaedics, researchers have developed biomaterial composites (nano-scaled) that promote osteogenesis (the formation of bone with the help of osteoblast cells) in the bone, and also various research groups have been showing their interest in the significance and advancements in bone tissue engineering resulting in expanding their work associated with the application of modified chitosan-based nanocomposites for orthopaedic procedures regarding bone tissue regeneration, due to the properties of chitosan being biocompatible, provides structural strength, etc.

Also, chitosan-based nanocomposites have been developed under several kinds of research and have been found suitable for the applications of bone tissue engineering because of the non-toxic properties of chitosan along with the ease of getting moulded or modified to give different suitable biomaterials (as discussed in the above section) [8] (Table 2). This makes the function of chitosan-based nanocomposites easier to add side groups to it in order to optimize itself for bone tissue engineering. In recent times, the use of CTS-based nanocomposites at defective bone sites in bone tissue regeneration has been increased [9]. The main focus of these applications is to overcome the problems that occur very often in the field of orthopaedics, like low availability, transfer of different pathogens, donor-site morbidity, immune response causing destruction to the bone, etc., by substituting the treatment methods with suitable ones [8].

7 Biomedical Applications of Chitosan-Based Nanocomposites in Dentistry

Chitosan-based nanocomposites have proved their non-toxicity towards cells in mammals along with different properties like being compatible with the tooth colour

Table 2 Biomedical applications of chitosan-based nanocomposites in orthopaedics

S. No.	Area of study	Biomedical applications of chitosan-based nanocomposites in orthopaedics	References
1	Modification of 3D structure matrix of bone	Giving structure and support to the bone, reinforcement to the bone material, natural tissue regeneration	[8]
2	Bone tissue engineering	Promotes osteogenesis, enhances growth, highly biocompatible, provides structural strength, non-toxic in nature, and increased tissue regeneration at defective bone sites	[8, 9]
3	Orthopaedic surgeries	Easily available, economically favourable, enhanced drug delivery, and enhanced wound healing	[8]

and being economically inexpensive. Their easy availability and wide range of possibilities for modifications make CTS nanocomposites which make them a clear winner for their potential uses. Chitosan-based nanocomposites compatibility and availability for a large range of bio-dental applications are depicted as follows, Fig. 6 [1].

7.1 Oral Drug Delivery

Chitosan-based nanocomposites can be utilized to develop a systematic arrangement of direct drug delivery with the required properties like mechanical strength, enough contact time with a good release profile being in close contact with the mucosa (oral), they also increase the bioavailability for the treatment in different oral pathologies. For the delivery of antibiotics (e.g. nystatin, chlorhexidine, and metronidazole) to periodontal (the structures surrounding and supporting the teeth) tissues, CTS-based nanocomposites in the form of nano-films can be used against fungal infections and mucositis (oral). CTS-based nanocomposites are also useful in the inhibition of plaque pathogens causing dental issues being an effective plaque controller [1].

7.2 Guided Tissue Regeneration

Guided tissue regeneration (GTR) is a developing strategy that involves the regeneration of periodontal cells. The main principle lies in the isolation of the given periodontal defect using a physical membrane that enhances the rate of bone regeneration. Chitosan and its nanocomposites fill this spot with ease, and they prove

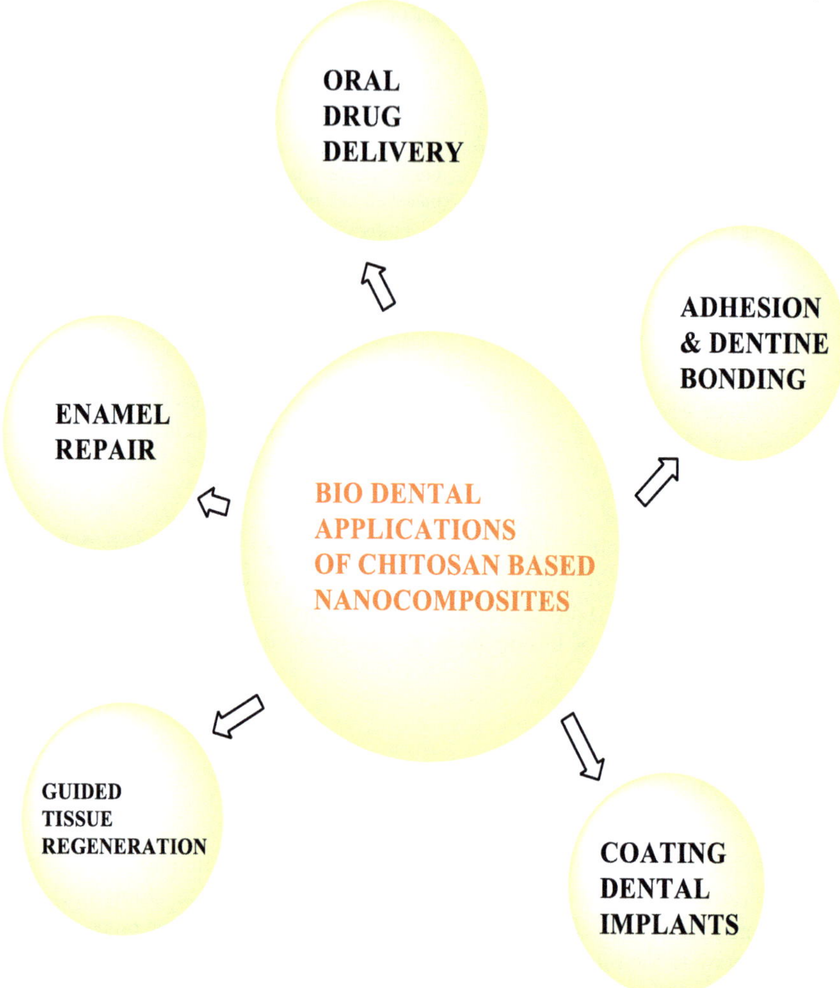

Fig. 6 Bio-dental applications of chitosan-based nanocomposites

to be a favourable substance for GTR because of their optimal size and good biocompatibility (biological behaviour) [1].

7.3 Enamel Repair

Studies have shown that tooth enamel is the hardest non-vascular tissue present in the human body, which makes it difficult to repair or regenerate. An abundant amount of chemical techniques have been developed with the use of chitosan-based nanocomposites to achieve the regeneration of human teeth enamel through the successful delivery of drug agents like organic amelogenin at defect sites. Along with nanocomposites, chitosan-based hydrogels as delivery agents are also being used nowadays for enamel regeneration [1].

7.4 For Coating Dental Implants

The coating of dental implant material in the process depends upon the osseointegration (direct structural and functional connection between ordered, living bone and the surface of a load-carrying implant) value of the material. Several studies have given positive results of chitosan-based nanocomposites being a good coating material in dental implants affecting the surface of the teeth with a change of mechanical, biological, and surface properties [1].

7.5 Adhesion and Dentine Bonding

Research has been conducted to study the durability of dentine bonding and dentine-restoration techniques. This comes under the area of use of bio-adhesive polymers in dental complications, which is fulfilled by chitosan-based nanocomposites resulting in better dentine bonding and a significant increase in shear strength of the bond [1] (Table 3).

8 Conclusion

Chitosan has been studied extensively due to the significant properties of the biopolymer being economically friendly, with excellent biodegradability and biocompatibility, along with non-toxic traits, antimicrobial, and antibacterial properties making the biopolymer a good candidate in the fields of biomedicine and biopharmaceuticals [10]. When chitosan nanocomposites get modified with different nano-scaled materials and films, they show significant properties increase in the rate of bone regeneration, better-targeted drug delivery, enhanced wound healing, and increased osteogenesis. Therefore, research shows a great potential for chitosan-based nanocomposites (both modified and only CTS-based) for biomedical applications in the orthopaedic and dentistry sectors [11].

Chitosan and its nanocomposites offer significantly to the future of biomedicine and biopharmaceutics, and it has also been studied that chitosan-based medical material will be of great use in upcoming times [2].

Table 3 Biomedical applications of chitosan-based nanocomposites in dentistry

S. No.	Area of study	Biomedical applications of chitosan-based nanocomposites in dentistry	Reference
1	Oral drug delivery	Develops a systematic arrangement for drug delivery with mechanical strength, good contact time, and increased bioavailability	[1]
2	Guided tissue regeneration (GTR)	Increased rate of GTR (due to optimal size and good biocompatibility of CTS-based nanocomposites)	[1]
3	Enamel repair	CTS-based nanocomposites works on regenerating human teeth enamel through successful drug delivery	[1]
4	Coating dental implants	Good coating material in dental implants with enhancing mechanical, biological, and surface properties	[1]
5	Adhesion and dentine bonding	Results in better dentine bonding and increase in the bond strength	[1]

References

1. Sanap P et al (2020) Current applications of chitosan nanoparticles in dentistry: a review. Int J Appl Dent Sci 6(4):81–84. https://doi.org/10.22271/oral.2020.v6.i4b.1050
2. Kmiec M et al (2017) Chitosan-properties and applications in dentistry. Adv Tissue Eng Regenerative Med Open Access 2(4):205–211.https://doi.org/10.15406/atroa.2017.02.00035
3. Sharifianjazi F et al (2022) Advancements in fabrication and application of chitosan composites in implants and dentistry: a review. Biomolecules 12(2). https://doi.org/10.3390/biom12020155
4. Venkatesan J, Kim SK (2010) Chitosan composites for bone tissue engineering—an overview. Mar Drugs 8(8):2252–2266. https://doi.org/10.3390/md8082252
5. Zapata MEV et al (2019) Novel bioactive and antibacterial acrylic bone cement nanocomposites modified with graphene oxide and chitosan. Int J Mol Sci 20(12). https://doi.org/10.3390/ijms20122938
6. Issn V-I-M-P, Ranijikanth V, Shukla SS (2019) Original research paper dental science Balaram Naik Tanvi Dugge * Nikolaos Soldatos Mallikarjuna Nadagouda Tejraj M. Aminabhavi Amit Pachlag 3:49–52
7. Javed R et al (2020) Chitosan encapsulated ZnO nanocomposites: fabrication, characterization, and functionalization of bio-dental approaches. Mater Sci Eng, C 116:111184. https://doi.org/10.1016/j.msec.2020.111184
8. Hasnain MS et al (2018) Biodegradable polymer matrix nanocomposites for bone tissue engineering. In: Applications of nanocomposite materials in orthopedics. Elsevier Inc. https://doi.org/10.1016/B978-0-12-813740-6.00001-6
9. Levengood SKL, Zhang M (2014) Chitosan-based scaffolds for bone tissue engineering. J Mater Chem B 2(21):3161–3184. https://doi.org/10.1039/c4tb00027g
10. Aguilar A et al (2019) Application of chitosan in bone and dental engineering. Molecules 24(16):1–17. https://doi.org/10.3390/molecules24163009
11. Murugesan S, Scheibel T (2021) Chitosan-based nanocomposites for medical applications. J Polym Sci 59(15):1610–1642. https://doi.org/10.1002/pol.20210251

Chapter 14
Conclusion and Future Prospects of Chitosan-Based Nanocomposites

Sanjay Kumar, Abhigyan Sarmah Gogoi, Shefali Shukla, Manoj Trivedi, and Shikha Gulati

Abstract Chitosan, as a bionanocomposite has been garnering immense attention and become a center of comprehensive research, owing to its vast structural possibilities for physiochemical alterations to produce novel characteristics, functions, and utilizations, especially in the biomedical field. Chitosan is endowed with numerous properties such as biocompatibility, biodegradability, non-toxicity, antimicrobial activity, low immunogenicity, stimuli sensitivity, adjustable physical strength, and water solubility. This ideal biopolymer can be transformed into nanoparticles, nanocapsules, nano-vehicles, fiber meshes, scaffolds, and 3D printed scaffolds. The last decade is a testimony to the enormous potential of chitosan and chitosan-based nanocomposites, as showcased by the legion of research reports displaying many new applications in the field of targeted drug delivery, modern biomedical instruments, and bioimaging sensors. This chapter unfolds different aspects of chitosan, including its properties and mutations, and focuses on chitosan-based nanocomposites. Emphasis has been laid on the salient biomedical applications of chitosan-based nanocomposites including drug delivery, gene therapy, tissue engineering and regeneration, cancer diagnosis and treatment, and bioimaging among various others.

Keywords Chitosan · Bionanocomposite · Properties · Modifications · Biomedical applications

Abbreviations

AAS	Atomic absorption spectroscopy
BGC	Bioactive glass–ceramic
CG	Gelatin

S. Kumar · S. Shukla · M. Trivedi · S. Gulati (✉)
Sri Venkateswara College, Department of Chemistry, University of Delhi, Delhi 110021, India
e-mail: shikha2gulati@gmail.com

A. S. Gogoi · S. Shukla · M. Trivedi
Department of Chemistry, University of Delhi, Delhi 110007, India

CHO	Chinese hamster ovary
CMC	Carboxymethyl chitin
CNT	Carbon nanotubes
Cryo-SEM/TEM	Cryogenic-scanning electron microscopy and transmission electron microscopy
CS	Chitosan
CTAB	Cetrimonium bromide
DC	Direct current
DDA	Degree of deacetylation
DLS	Dynamic light scattering
DMSO	Dimethyl sulfoxide
DNA	Deoxyribonucleic acid
ECM	Extracellular matrix
EVA	Ethylene–vinyl acetate
FDA	Food and Drug Administration
FTIR	Fourier-transform infrared spectroscopy
GO	Graphene oxide
HAp	Hydroxyapatite
HEK	Human embryonic kidney
HIV	Human immunodeficiency virus
mRNA	Messenger ribonucleic acid
MTT	Microculture tetrazolium
NPs	Nanoparticles
NMs	Nanomaterials
PCL	Polycaprolactone
PEG	Polyethylene glycol
PVA	Polyvinyl alcohol
QDs	Quantum dots
RNAi	Ribonucleic acid interference
RT	Room temperature
SEM	Scanning electron microscopy
TEM	Transmission electron microscopy
UV	Ultraviolet
XPS	X-ray photoelectron spectroscopy
XRD	X-ray diffraction

1 Introduction

Natural materials outperform artificial materials in a variety of industrial applications, from cosmetics to aerospace, due to their diverse qualities and properties. When it comes to biomedical applications, natural materials are frequently found to be preferable to synthetic ones [1]. The development of new standards to overcome technical difficulties and effectively resolve social and ecological challenges is critical for the advancement of sustainable biomedical innovation, particularly in the large-scale production of renewable materials. As a result, it is critical to reduce the use of harmful compounds and instead focus on the creation of environmentally safe nanoscale-based composite materials. In addition, the introduction of nanotechnology opens up more creative and superior options for many pharmaceutical applications, such as vaccination, tissue engineering, diagnostic imaging methods, cancer therapy, and other medicinal, and sustainable drug delivery, among others [2]. In recent years, special attention has been given to biopolymers, such as biodegradable polyesters, polysaccharides, or polypeptides. Bionanocomposites are a type of hybrid materials consisting of biopolymers and inorganic solids, having at least one dimension on the nanometer scale. The inclusion of nano-sized inorganic fillers causes biopolymer characteristics to alter and improve. The increasing interest in bionanocomposites is a global phenomenon now because of their cost-effectiveness, biodegradability, environmental friendliness, and ease of preparation. Also, polysaccharides happen to be the most popular biopolymers due to their sustainable and recyclable nature, as well as their biodegradability. They comprise a bountiful part of the biomass [3].

Chitosan is a linear polysaccharide and is acquired from nature itself. It can be made from chitin by deacetylation in the solid form under alkaline circumstances or through enzymatic hydrolysis by chitin deacetylase [4] (Fig. 1). In 1811, a French scientist named Henni Braconnot was the first to obtain chitin from mushrooms. Crab and shrimp shells are the most abundant hotspots for chitin. Chitin is a polymer made of the 2-acetamido-2-deoxy-D-glucose disaccharide that is joined together by a β bond. Chitosan was discovered in the nineteenth century, after the deacetylation of chitin having numerous acetyl glucosamine units. As a result of the partial deacetylation, chitosan is a heteropolymer with both glucosamine and acetyl glucosamine groups. Since amino groups $(-NH_2)$ allow copolymerization with acrylic, styrene, urethane, and vinyl ester-based monomers to acquire auxiliary features, adjusting the attributes of this polymer is easier [1].

Non-harmfulness, biodegradability, biocompatibility, immune-enhancing, antitumoral, anti-viral, and antimicrobial action are some of the most well-known features of chitosan in the medical context. Chitosan's biodegradability was established in vitro and in vivo, where polymeric chains were disintegrated into sundry tinier segments of monomers. Chitosan and its enzymatic derivatives could harmlessly associate with living cells without any unfavorable impact on the body. Chitosan can diminish cholesterol retention, filter-free radicals from the chain oxidation process, and carry out its antimicrobial and anti-viral action against numerous yeasts and microbial organisms. Chitosan-based NPs can likewise hinder cancer cell development by instigating apoptosis. Chitosan, either singly or blended in with different polymers, active agents, and metallic nanostructures, has been widely utilized in numerous biomedical utilizations, that include wound scaffolding as an antimicrobial specialist, in drug conveyance as a nanoscale transporter to single out tumoral growths while negligibly influencing sites of unaffected tissue, for gene delivery, and

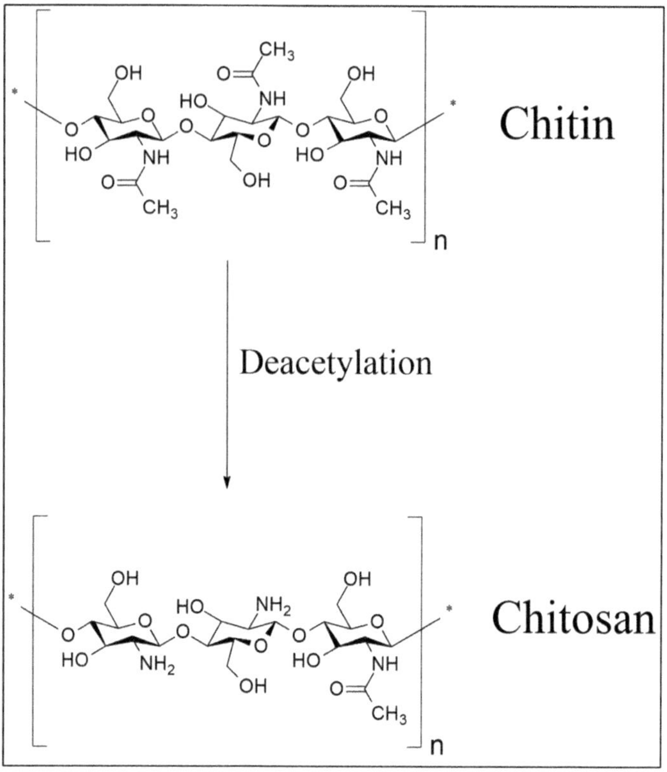

Fig. 1 Preparation of chitosan from chitin by deacetylation

dentistry and also as absorbable structure for immobilizing enzymes [4]. Since the future opportunities of chitosan-based bionanocomposite in different sectors provide us with ample room to maneuver, this chapter presents an itemized examination in regards to these issues. It is coordinated pleasantly by talking about chitosan-based bionanocomposite and their structure. From that point onward, it uncovers crucial data about the origin and attributes, and alteration techniques of chitosan, along with the essential diagrams. It likewise presents the most suitable strategies and methods for creating chitosan-based bionanocomposites. Moreover, the originality of this article is ameliorated by collecting significant data about the various applications along with the advantages and downsides intricately. Prior to closing this article, the future exploration heading is additionally given so as to progress and improve its application for different purposes soon [3].

2 Properties of Chitosan

Chitosan is made up of deacetylated D-glucosamine and N-acetyl-D-glucosamine units. It has a lesser molecular weight and a lesser number of crystalline domains than chitin (whose molecular weight is more than 100 kDa). This incredible biopolymer has also been approved by the FDA. Its design is like cellulose with the exception of the hydroxyl group at the C-2 site in cellulose is supplanted by the amino group in chitosan [6, 7].

2.1 Physiochemical Properties

Chitosan is characterized by a high extent of deacetylation and low crystallinity. This property makes it a compelling possibility for a variety of beneficial applications, such as drugs and biotechnology. Chelation, viscosity, dissolvability in various media, mucoadhesivity, polyelectrolytic behavior, and polyoxysalt formation are only a few of the physical features of chitosan [3]. Each glycosidic residue in chitosan has three receptive positions: one amino group and two hydroxyl groups. The amino group is of incredible importance because it is pH sensitive and answerable for chitosan's cationic nature. The position of the acetyl groups along the chain, deacetylation techniques, and ionic strength all play a role in its solubility. Chitosan has chelating capabilities for a few metal cations at an acidic pH [8].

2.2 Biological Properties

Chitosan has been found to be a non-toxic, biocompatible, and biodegradable polymer in several investigations [9]. Anti-inflammatory, antibacterial, antitumor, anti-fungal, hemostatic, and analgesic characteristics are all present in chitosan and its derivatives [3, 10, 11] (Fig. 2).

3 Fabrication and Chemical Alterations of Chitosan

3.1 Extraction/ Synthesis of Chitosan and Its Based Nanocomposites

The degree of deacetylation of chitin determines the molecular weight of chitosan. Deacetylation is generally done in the presence of an atmosphere of nitrogen gas or by adding sodium borohydride to the sodium hydroxide solution so as to kill any adverse side reactions. Many additional methods for making chitosan, using the exoskeleton

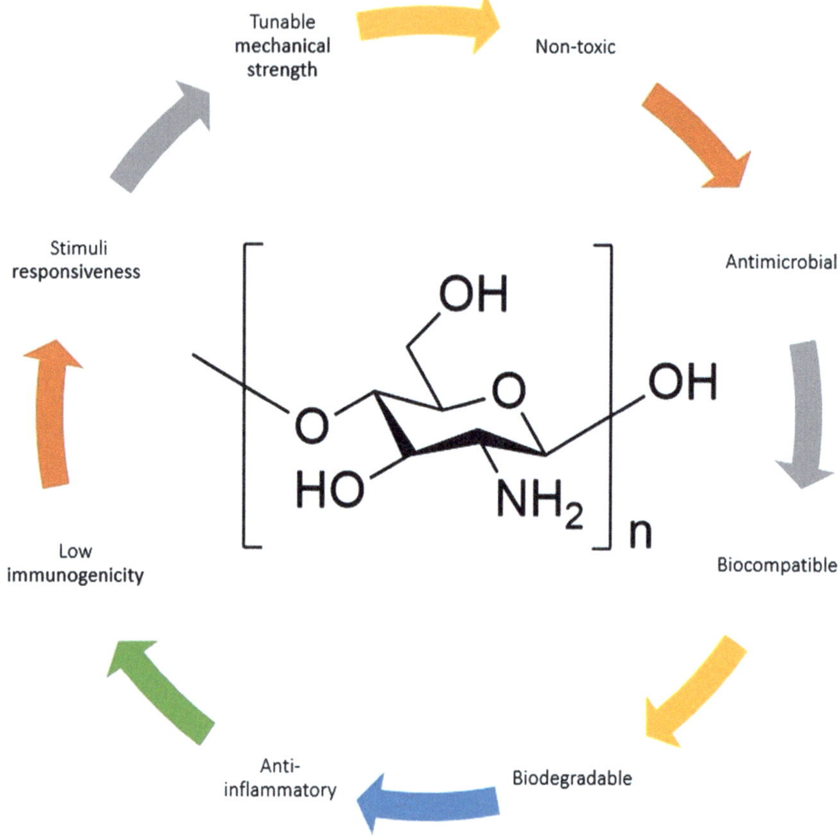

Fig. 2 Exceptional biotic properties permit the utilization of chitosan in numerous biomedical fields

of different crustaceans, fungi, cuttlefish, squid, and insects have been invented, in addition to alkali treatment [12]. To generate chitosan with variable degrees of acetylation from chitin, two techniques are now most widely used. The first approach is the solid phase heterogeneous deacetylation (Fig. 3), whereas the second involves pre-swollen aqueous phase homogeneous deacetylation under vacuum conditions [13]. These procedures employ thermo-mechanical techniques such as a cascade reactor operated at low concentrations of alkali [14]; sequential alkali analyzes with thiophenol in dimethyl sulfoxide [15]; and flash treatment with saturated steam and repeated washing with distilled water [2, 17]. The abstraction of chitosan using irradiation by microwaves has also been proposed as a more advanced approach [16]. Treatment with enzymes, rather than alkaline treatment at high temperatures, is another way to generate chitosan [18]. Following the biological methodology, chitosan is enzymatically created with chitin deacetylase. In the enzymatical fabrication strategy, the N-acetamide bond found in the basic framework of chitin gets

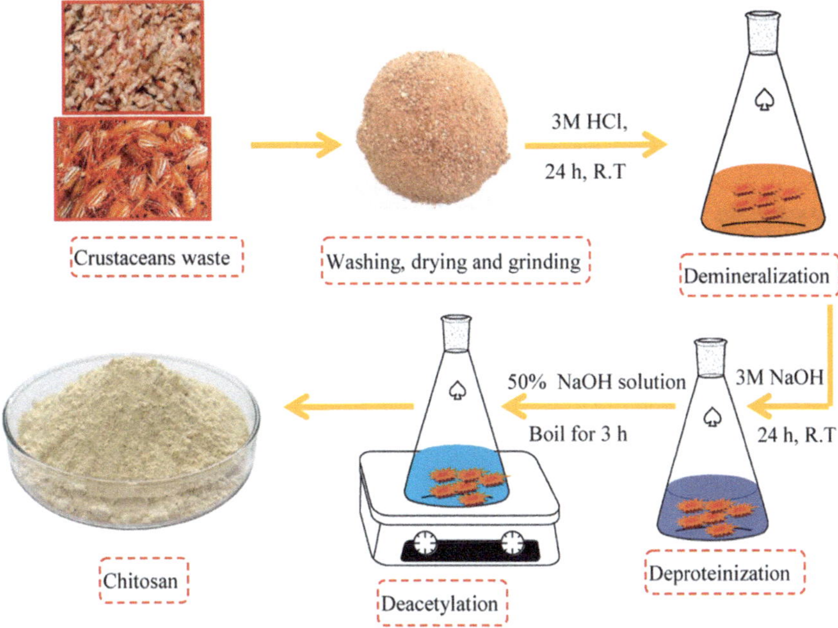

Fig. 3 Traditional method for the extraction of chitosan using the extracellular matrix of crustaceans. Reproduced from Ajahar et al. (2021) with permission from Elsevier

hydrolyzed by the chitin deacetylase enzyme. Prior to hydrolysis, chitin is artificially changed into unrefined chitin, as unrefined chitin is a poor enzyme substrate. As a result, chitin undergoes reprecipitation, glycolation, and depolymerization before being hydrolyzed [2, 19, 20].

Also, typical techniques generally used for the synthesis of chitosan-based bionanocomposites are listed below [3]:

i.   Solution-casting method
ii. In-situ technique
iii. Electrospinning technique
iv. Freeze-drying technique

3.2 Modification of Chitosan

Natural chitosan has lower transfection effectiveness than synthetic chitosan, and it lacks several characteristics that are essential for some applications [2]. In addition to this, their inferior solubility in many polar solvents and water limits their applicability. To overcome this problem, chemical/enzymatic methods (Fig. 4) can be used to modify the chitosan molecule, resulting in new derivatives that not only improve its solvability for prudent use in various biomedical utilizations but also

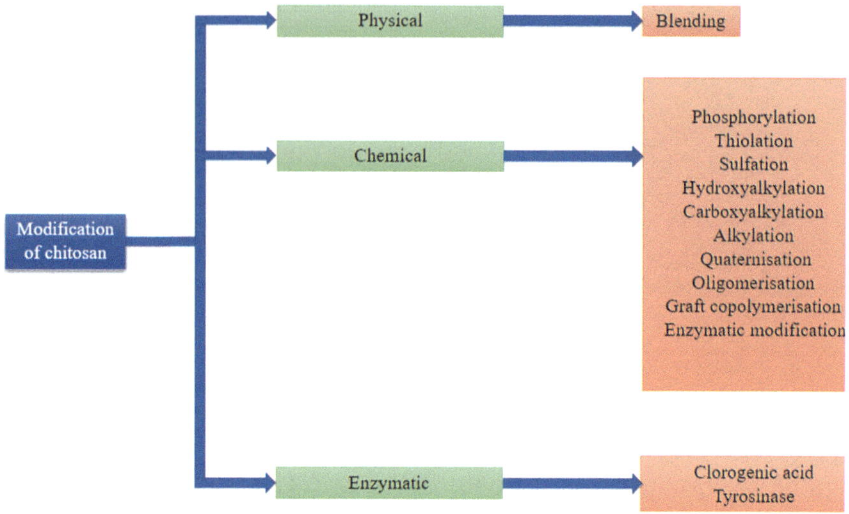

Fig. 4 Techniques to modify chitosan

serve as a powerful resource to promote novel biological actions and change its physical attributes [3]. Surprisingly, structural alterations to chitosan do not modify the molecule's basic structure, but they do give it new characteristics. Chitosan's structural morphology allows it to undergo a wide variety of reactions (like reduction, oxidation, acylation, cross-linking, complexation, halogenation, and phosphorylation) that result in novel derivatives [21]. Chemical or physical processes such as compositing with appropriate fillers, substituent inclusion, cross-linking, blending, curing, graft polymerization, and so on can be used to modify the material. Chemically modified chitosan and its derivatives as biomaterials for antimicrobial action, drug transport, and other biomedical utilizations have been examined extensively by Sashiwa and Alves et al. [22, 23]. Also, Mourya et al. have outlined the various approaches for modifying chitosan in great detail [24].

3.2.1 Physical Modification

Physical modification is commonly accomplished by combining two polymers to create a new material with distinct physical properties. Polymer blending is employed for the creation of materials with the perfect mechanical, chemical, biological, morphological, and structural attributes. When compared to other readily available tactics, this method is considered practicable and judicious, as well as less tedious [25, 26]. Hydrophilic polymers like polyethyl oxide (PEO), polyvinyl alcohol (PVA), and polyvinyl pyrrolidone (PVP) are all known to mix nicely with chitosan [27, 28]. The qualities of the component polymers and the chitosan/PVA blend, which is widely utilized in food applications, are improved through molecular interaction [29].

Furthermore, chitosan combined with PVA has demonstrated excellent mechanical properties for pharmaceutical items and controlled drug conveyance systems.

3.2.2 Chemical Modification

The polymer chain of chitosan possesses reactive amino ($-NH_2$) and hydroxyl (-OH) groups that aid the favorable chemical alteration of the molecule, without amending its fundamental characteristics. The modification of chitosan can be performed by various chemical reactions as listed below [3, 30]:

   i.   Phosphorylation
   ii.   Thiolation
   iii.   Sulfation
   iv.   Hydroxyalkylation
   v.   Carboxyalkylation
   vi.   Alkylation
   vii.   Quaternization
   viii.   Oligomerization
   ix.   Graft copolymerization
   x.   Enzymatic modification

4 Characterization Techniques for Chitosan and Chitosan-Based Nanocomposites

Characterization techniques such as FTIR, DLS, XPS, XRD, TEM, AAS, and SEM give reliable, consistent, and accurate results for improving the chemical and physical properties of nanomaterials. Nanomaterials are characterized in order to gain a better understanding of them, improve already-existing methods, and their applications sustainably.

4.1 Dynamic Light Scattering

The features of chitosan-based nanomaterials, such as particle size, surface charge, and size distribution, can be determined using the dynamic light scattering (DLS) approach. This is the most reliable and profoundly effective method when contrasted with others. DLS is a non-invasive method for estimating the size, size distribution, and particle charge of nanomaterials, especially in the submicron and <1 nm ranges, using cutting-edge technology. Brownian motion of particles (present in suspension) having varied scattering angles (θ) is used to determine particle size in DLS. This technique is primarily focused on the measurement of particles suspended in liquid media. Brownian motion is slowed as the particle size increases. The Stokes–Einstein equation is applied to find the hydrodynamic size by analyzing intensity fluctuations and Brownian motion velocity. The probable size of the dynamic hydrated/solvated particle is determined by the diameter derived from the particle's diffusional characteristics. Using a combo of light scattering and microscopy techniques, a more thorough examination of the size and distribution of NPs can be performed. Chitosan nanoparticles in water have a size range of 40–374 nm, with an average size of ~ 250 nm, as observed by DLS. Furthermore, it was revealed by the study of the number distribution data that 91.6% of chitosan nanoparticles had diameters less than 100 nm [91.28 nm (1.6%), 78.82 nm (6.3%), 68.06 nm (17.6%), 58.77 nm (30.2%), 50.72 (26.8%), and 43.82 (9.1%)] [31–33] .

4.2 Interaction Analysis of Nanomaterials by FTIR Spectroscopy

Interaction evaluation, functional group evaluation, and the types of particular bonds/interactions occurring between various groups are all used to investigate the chemical characteristics of chitosan-based nanomaterials, for which Fourier-transform infrared spectroscopy is employed. Their calculations are crucial when it comes to changing or adjusting the attributes of nanomaterials. The paired peaks of the amino and hydroxyl groups stretching vibration in chitosan are responsible for a band at $3420\,cm^{-1}$. Furthermore, the C–N, $-CONH_2$ stretching vibration of primary amine, and anhydrous glycosidic bond are indicated by bands at 1385, 1649, and $892\,cm^{-1}$, respectively. These peaks are sharper and reposition toward 1315, 1640, and $894\,cm^{-1}$ in Cu-chitosan nanoparticles, indicating boosted interactions. As a result, the synthesis of chitosan NPs is revealed by the relocation of vibrations from greater to smaller wavenumbers [33].

4.3 Elemental Analysis of Nanomaterials

4.3.1 X-Ray Photoelectron Spectroscopy (XPS)

Nanomaterials' surfaces provide sites of interface with other materials and the environment. As a result, a proper understanding of the chemical and physical properties of these nanostructures is a mammoth errand that can be accomplished with the help of XPS. It is a statistical spectroscopic tool for delving into data regarding the chemical and elemental composition of substances found inside nanomaterials [34]. Cu, C, O, N, and P elements are found in Cu-chitosan nanoparticles, according to quintessential elemental research findings. C and O were the most plentiful components found in nanomaterials, while N and P were detected in limited proportions [35].

4.3.2 Atomic Absorption Spectroscopy

AAS is used to evaluate the concentration of elements present in nanomaterials. Photons are absorbed by atoms, which allows them to get excited from lower to higher energy levels. The ionization energy required for the excitation of the electron is specific to the element. As a result, each atom has its own unique wavelength profile, allowing for a qualitative study of chitosan-based nanomaterials. Furthermore, AAS can be used to focus on the entrapped metal ion release profile in chitosan-based NMs at different pH levels, agitation intervals, and rates [33].

4.4 Solid-State Characteristics of Nanomaterials by X-ray Diffraction (XRD)

X-ray diffraction is often employed to study the solid-state characteristics of nanomaterials. The molecular design of nanomaterials can be elucidated using solid-state characteristics data. Understanding the fundamental architecture of nanomaterials can aid in the efficient and precise functionalization and modification of these materials. Furthermore, these features can be used to further create an extraction strategy for chitosan nanoparticles, resulting in a structure that is suitable for a wide range of applications [33]. XRD is a non-destructive technology that studies the structure of materials on an atomic or molecular scale. XRD analysis of chitosan has been reported in a number of research studies. The crystalline peak intensity in the XRD diffractogram was found to be low intensity ($2\theta = 19.5$ and 21.0), which revealed the presence of Cu-chitosan NPs, allowing solid-state characteristics to be assessed [36, 37].

4.5 Internal and Surface Characteristics of Nanomaterials

The properties of the superficial and internal environment of chitosan-based nanomaterials, such as topology, inner architecture, biochemical functionalization, characterization, purity assessment, degree of agglomeration, and dimensions, are studied using electron microscopy and cryogenic electron microscopy [38].

4.5.1 Scanning Electron Microscopy (SEM)

SEM can be considered to be a crucial method for studying geometry, external morphology, and functionalization of diffused and aggregated chitosan-based nanostructures from top to bottom. The exterior of the nanomaterials is examined using a high-energy beam of electrons. Under optimal conditions, many of the state-of-the-art SEM machines can unveil data even in the 2 nm range. Chemical data on various nanomaterials can be obtained using a combination of SEM and energy-dispersive X-ray spectroscopy (SEM–EDX). In the various research publications, SEM data of chitosan-based nanomaterials has been frequently described [31, 39, 40].

4.5.2 Transmission Electron Microscopy (TEM)

Because of its higher spatial resolution relative to SEM, the most popular electron microscopy tool for the evaluation of chitosan-based nanoparticles is TEM. It is a key technique for the characterization process, which includes analysis of size, internal structure, and interactions between nanomaterials [33]. TEM can provide information in the 0.2 nm range under optimum circumstances. TEM images of spherical-shaped chitosan-based nanomaterials have been reported in several publications [39, 41].

4.5.3 Cryogenic-Scanning Electron Microscopy and Transmission Electron Microscopy

Cryo-SEM/TEM has recently emerged as a fundamental method for characterizing nanomaterials. This technique allows researchers to examine nanomaterials in a chemically unaffected and fully hydrated form at cryogenic temperatures (usually $-$100 to $-175\ ^{\circ}C$) [42, 43]. It can be used to investigate the surface profile, geometry, dimension, and inner structural properties of chitosan-based nanomaterials in the same way that SEM and TEM can. The increased adoption of these approaches is due to the speedy pace of specimen processing, efficiency, the convenience of use, and imaging quality. Cryo-SEM/TEM edges out conventional SEM or TEM due to their numerous advantages like excellent resolution, swiftness, and examination in a completely hydrated condition, less dispersed material translocation, and suitability for liquid/semiliquid substances [43].

5 Applications of Chitosan

The creation of efficient and safe chitosan-based biomaterials for diverse biomedical utilizations (Fig. 5) like drug delivery, biosensors, tissue engineering, and wound healing has been the focus of extensive research and development during the last several years. The central emphasis was on prospective biomedical uses based on chitosan nanocomposite, with a specific focus on its upgraded physico-chemical and biological attributes. The customary biological uses of chitosan-based nanocomposites are depicted in the diagram below.

As per the requirement of the desired application, chitosan can be rendered into numerous reactive formulations as shown in Fig. 6.

The next sections focus on the various biomedical application in a summarized manner.

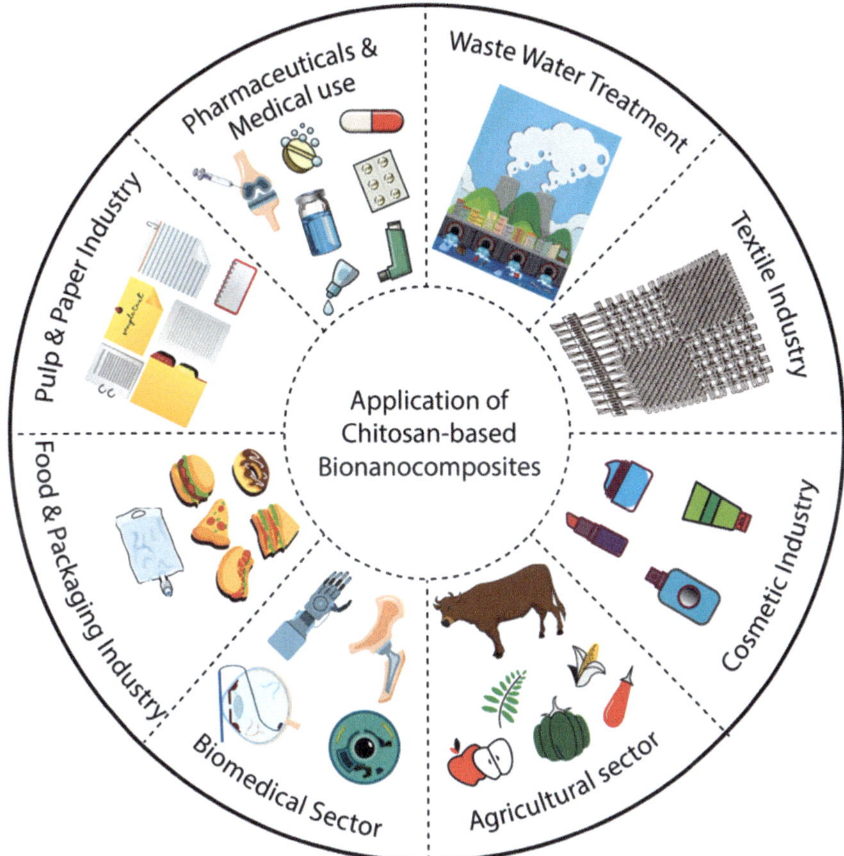

Fig. 5 Broadened application of chitosan-based bionanocomposites. Reproduced from Motia et al. (2021) with permission from Elsevier

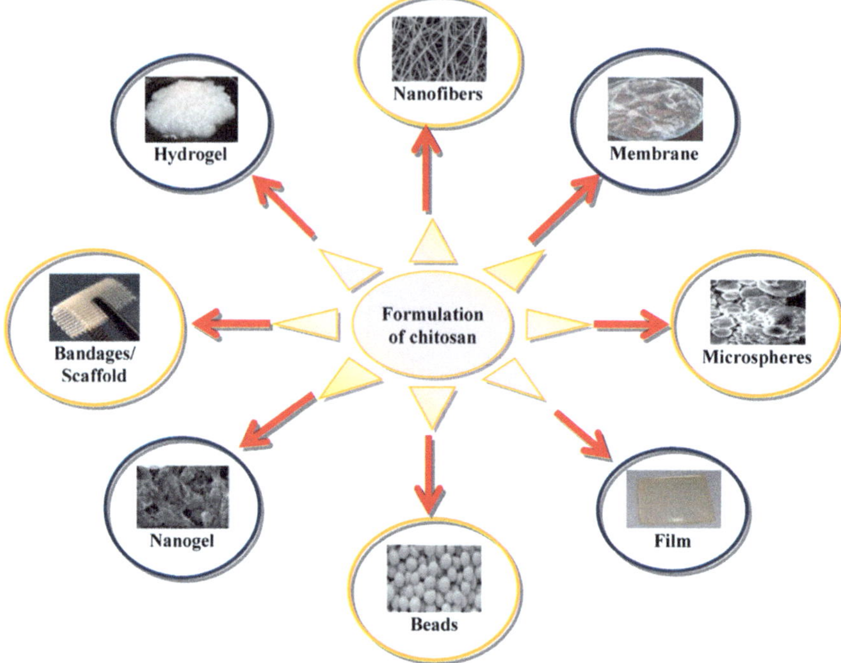

Fig. 6 Different formulations of chitosan. Reproduced from Ajahar et al. (2021) with permission from Elsevier

5.1 Drug Delivery

The use of nanomaterials in pharmaceuticals is becoming increasingly popular. Because of their tiny dimensions, nanoparticles can permeate through various biological membranes and convey drugs to specific areas with greater effectiveness [44]. Chitosan is perhaps the most essential polysaccharide for drug delivery due to its cationic characteristics and the existence of primary amino groups. These are in charge of many of their properties, including permeation augmentation, in-situ gelations, mucoadhesion, and calculated release of drugs [45]. Premeditated chitosan bio-decomposition allows for the controlled and steady discharge of loaded moieties while decreasing dosing frequency, which is beneficial for improving patient medication adherence [46].

Numerous research publications have identified chitosan-based nanoparticles to be enticing candidates for oral drug conveyance, owing to their unique perks like enhanced solvation of loaded hydrophobic medications, regulated drug delivery, minimal cytotoxicity, and improved therapeutic performance. Furthermore, nanoparticles in the intestinal system prevent unstable medications from deteriorating enzymatically. Nanoclays, reduced graphene oxide, gold NPs, layered double hydroxide, hydroxyapatite, mesoporous zeolites, and Fe and SiO_2 NPs are among the various nanoparticles employed in the fabrication of chitosan-based drug carrier systems, according to the literature [1, 47–49]. Also, because of the creation of grueling routes for clays, 2D layered nanosheets such as carbon nanotubes (CNT), MXenes (Ti_3C_2, Nb_2C) nanoclay, or graphene exhibit better drug release profiles [50]. Moreover, other researchers have sought to incorporate mesoporous silica and other nanoparticles, such as hydroxyapatite, into chitosan frameworks in order to fine-tune drug-loading efficiency and release properties [51, 52].

Dev et al. [53] adopted an emulsion approach to produce poly(lactic acid) (PLA)/CS nanoparticles for anti-HIV medication delivery. Lamivudine, a hydrophilic antiretroviral, was incorporated into PLA/CS nanoparticles. In vitro, drug release tests revealed that when the medium's pH changed from basic to acidic to neutral, the rate of drug release from PLA/CS nanoparticles reduced. When comparing acidic and basic pH, the rate of drug release decreased in acidic pH. This could be owing to the fact that H^+ ions repel cationic groups found in polymeric NPs. These findings suggest that PLA/CS NPs are a potential delivery system for controlled anti-HIV and cancer medication delivery.

Sandhyarani and Chandran [54] used the solution-casting method to synthesize electric field-sensitive nanocomposite thin films comprised of chitosan/Au NPs and incorporated 5-Fluorouracil. In an electrolyte solution, the produced nanocomposites had a greater drug delivery efficacy (63%) and consistent-release regulated by an externally manifested electric field (DC). Shah et al. [55] used in-situ co-precipitation to create chitosan/Ag NP composite films with integrated moxifloxacin medications. The drug-encased nanocomposite films have superior mechanical qualities and antibacterial activity against a variety of microorganisms. These achievements in the realm of drug delivery are piquing attention and sending a clear message to produce sophisticated and upgraded nanomaterials/nanodevices for enhancing drug delivery efficiency by employing chitosan-based bionanocomposites (Table 1).

5.2 Gene Therapy

Chitosan-based bionanopolymers are strongly cationic, making them ideal choices for intracellular transport of nucleic acid biomolecules. Due to the cationic property of chitosan, it produces a complex of polyelectrolytes with oppositely charged nucleic acid moieties, which shields nucleic acids from nuclease disintegration [56].

Multiple illnesses such as cystic fibrosis and Parkinson's can be treated employing these genetic components (ribonucleic acid and deoxyribonucleic acid). The advantage of chitosan carriers is that they minimize the cytotoxicity problem that plagues the plurality of the artificial polymeric devices, in addition to having the unique capability of transcellular movement. Furthermore, because the amine groups are positively charged, they can transfer plasmid DNA (pDNA) into cells via membrane invagination and endocytosis [57].

Different techniques for constructing chitosan-based protein-embedded nanoparticles have been documented. Emulsification along with cross-linking is the most

Table 1 Some examples of chitosan and its derivatives-based nanocomposites in diversified functional formulations for drug delivery [2]

Serial no	Nanocomposites	Formulation	Receiver	Characteristics
1	Glycol chitosan-o-nitrobenzyl succinate	Nano-micelles	Mouse	Biologically compatible and superior drug carrier system for antitumor treatment
2	Chitosan-Mg–Al-PO$_4$-nanoclay	Nanohybrid hydrogel, scaffold	Mouse	Benign, more biologically compatible, and protracted distribution with site-specific cellular proliferation
3	Chitosan cross-linked-6-phosphogluconic trisodium	Hydrogel (ionic)	Swine	Benign, causes no skin itching, may be used as a medicine carrier and also for the dressing of wounds
4	Interferon-alpha embedded chitosan	NPs	Mouse	Oral delivery, harmless, intrusive, and more patient-friendly organic medication can be rendered as cytokines and protein therapeutics

popular method of preparation, although the use of organic solvents and cross-linkers can have a deleterious impact on protein expression [58]. To improve chitosan's effectiveness for gene delivery, researchers tried grafting polyethyleneimine into 1-butyl-imidazolium acetate [59] or fabricating chitosan derivatives in 1-ethyl-3-methylimidazolium chloride [57]. Consolidation of nucleic acids, shielding against breakdown, preservation in physiological settings, cellular incorporation, endolyso-somal discharge, unloading, and transportation of the genetic material to the nucleus are all essential steps in the gene delivery process [60]. A range of methodologies has been attempted to increase chitosan's protein-sponging capacity, including the insertion of various functionalities (imidazole, histidine, etc.). In vitro and in vivo, chitosan has been exploited to transfect an array of cell varieties. A549, HeLa, HEK293, and COS-1 seem to be the most widely transfected cell varieties with chitosan-pDNA complexes under in-vitro conditions [61]. RNAi was also effectively implanted into HEK293, H1299, CHO-K1, and HepG2 cells using chitosan-RNAi complexes [61]. For pharmaceutical application, gene medications can be administered in a multitude of ways, but one of the more inventive strategies involved incorporating complexes into a scaffold and propagating them to the problematic location [4].

In an exclusively built-in-situ reactor, Chen et al. [57] detailed the production of O-alkylated chitosan derivatives in 1-ethyl-3-methylimidazolium chloride solvent with N, N' -carbonyldiimidazole as a binder. It is claimed that the ionic liquid solvent's unique properties are responsible for the selective alkylation of hydroxyl groups without protecting the amino ($-NH_2$) groups in chitosan. Furthermore, the chitosan derivatives' increased solubility in the presence of organic solvent may facilitate their future use in gene delivery studies [2].

Bionanocomposites based on chitosan offer hope for genome treatment and nucleic acid transport so as to cure genetic diseases. Future studies on improving the formulation of nucleic acid and gene delivery to targeted cells will hopefully upgrade the features and enhance the conveyance of nucleic acid and gene therapy on target sites.

5.3 Tissue Engineering and Regenerative Medicine

Tissue engineering is the process of leveraging live cells to create biological replacements for insertion into the body and/or to encourage active tissue reformation by manipulation of their extracellular space or genetics. It is an effective technique for repairing, replacing, preserving, or augmenting the performance of a particular organ/tissue [62].

A detailed grasp of bone architecture and the process of healing is required to select suitable biological materials for tissue engineering. In terms of selecting appropriate biomaterials, chitosan nanocomposites containing nanofillers such as bioactive glass, zeolite, hydroxyapatite, Cu NPs, carbon filler, etc., are widely used in tissue engineering applications in the form of fiber meshes, scaffolds, thin films, and hydrogels [1]. Chitosan possesses a hydrophilic surface that stimulates cell adhesion and growth, unlike many artificial polymeric materials. Chitosan molecules are very adaptable, allowing them to be easily changed into thin films and scaffolds, with a wide range of uses in tissue regeneration and cell transplantation. Chitosan may be treated in a variety of ways to create 3D scaffolds with a range of porous structures for bone-tissue regeneration. These functionalized scaffolds induce bone-forming osteoblast cell growth as well as the generation of a mineralized bone framework [3].

Multiple physiologically functional moieties, extracellular matrix (ECM) constituents, and cells are now well understood to coordinate at the nanometer scale. Electrospun nanofiber mat has a structure that is highly comparable to human native ECM, making it suitable scaffolding equipment for tissue engineering and cell culture [63]. Water-soluble carboxymethyl chitin (CMC)/PVA blend, electrospun for tissue engineering applications was described by Shalumon and colleagues [64]. To produce nanofibers, the proportion of PVA (8%) and CMC (7%) was tuned, then combined in various ratios (0–100%) and electrospun. By cross-linking fibers with glutaraldehyde fumes and then heating them, the fibers became insoluble in water. This led to the creation of bioactive and biocompatible nanofibers. Cells were able to adhere and proliferate in nanofibrous scaffolds, according to cell attachment experiments. These findings show that the nanofibrous CMC/PVA scaffold promotes cell proliferation and adherence, indicating that it could be a great pick for the purpose of tissue engineering.

Nanosurfaces have a significant impact on cell behavior. Nanophase ceramics, compared to microphase ceramics, are known to have better cell–material interconnections [65]. The development of chitosan/nBGC and chitin/nBGC composite scaffolds for the purpose of tissue engineering is a good example of this [62]. The lyophilization procedure was used to create chitosan or chitin /nBGC composite scaffolds. When the nBGC were evenly distributed on the pore walls, the composite scaffolds displayed satisfactory porosity. Apart from their potential to become bioactive, the generated nanocomposite scaffolds exhibited acceptable swelling and disintegration characteristics. Direct contact test, MTT assay, and cell attachment tests were used to analyze the chitin/nBGC and chitosan/nBGC scaffolds' cytocompatibility. There was no evidence of toxic effects, and the cells were discovered to adhere to the scaffolds' pore walls. These findings indicated that the composite scaffolds generated may be employed in the field of tissue engineering [62].

5.3.1 Cartilage

Because of its similarities to glycosaminoglycans found in the extracellular matrix, chitosan is the most commonly employed biopolymer for cartilage tissue engineering [66]. Due to the biocompatible and non-poisonous characteristics of biopolymers like gelatin, silk fibroin, alginate, and collagen, their chitosan-based nanocomposites have garnered significant research in the area of cartilage tissue engineering [67]. By using an in-situ precipitation approach, Zhi-Sen et al. [68] were able to create a robust, porous, and resistant chitosan–gelatin-based hydrogel with Young's modulus of 3.25 MPa and a tensile rigidity of 2.15 MPa. The hydrogel was able to demonstrate multiplication and wonderful adhesion of human thyroid cartilage cells due to its incredibly porous architecture, which aided the cells' growth and transit of nutrients. Furthermore, it was observed that 65.9% of the produced hydrogel was decomposed after a period of 70 days.

5.3.2 Bone

Chitin and chitosan have minimal mechanical capabilities by nature. As a result, chitin can be employed as a bone-substituent material for bone healing and rebuilding, only if the mechanical characteristics can be enhanced by adding biomaterials such as bioactive glass–ceramic (BGC), hydroxyapatite (HAp), or other biomaterials. BGCs are a class of bone-healing materials consisting of osteoconductive silicates. Chitosan

(CS)-gelatin (CG) composite scaffolds with nBGC were developed by integrating gelatin and chitosan with nBGC [69]. According to the findings, the scaffold has a macroporous internal structure with pore sizes varying from 150 to 300 μm. With the addition of nBGC, protein adsorption improved and the nanocomposite scaffolds' disintegration and swelling tendency decreased. As incubation time rose, a higher degree of mineral deposition on the nanocomposite scaffold was revealed by biomineralization investigations. The performance of these nanocomposite scaffolds in the direct contact test, MTT assay, and cell attachment experiments showed that they are more favorable choices for cell spreading and attachment. Hence, these scaffolds can be employed to regenerate alveolar bone successfully [69] (Table 2).

5.4 Cancer Diagnosis and Treatment

Cancer is an exceptional disease portrayed by abnormal cell multiplication, which in the long run prompts the development of a cluster of cells called malignant cancer. Chemotherapy is usually utilized for the therapy of various diseases like lung cancer, breast cancer, prostate cancer, and so on. Despite the fact that chemotherapy showed strong anti-cancer action, it has numerous limits which at times hinder its application in medical settings. Chemotherapy is limited by three factors: (1) brief half-life, (2) non-specific anti-cancer activity, and (3) extreme aftereffects. To defeat these impediments, a targeted drug transportation system has emerged as a promising option for traditional chemotherapy. Deliberate preying on tumor cells, extending medication half-life, and regulating the release of the drug have all been demonstrated with the designated drug carrier framework. Due to the sheer diminutive dimensions of nano-transporters and their ability to preserve the drug from enzymatic defilement along with renal sieving, target-tailored drug carrier devices prolong the half-life of chemotherapeutic circulation. Even though these drug-loaded nanocarrier frameworks can evade enzymatic degradation and renal sieving, they were discovered to be eliminated through absorption of the reticuloendothelial framework. PEG-covered nanocarriers and PEGylated nanocarriers [70, 71] have been developed in the recent years and are regarded as remarkable enhancements due to their cloak and dagger effect, which allows drug-loaded nanocarriers to avoid renal sieving and enzymatic degradation and thus disseminate freely in the bloodstream, resulting in increased bioavailability of the drug at the assigned location and broadening its restorative competency.

The pharmacodynamics and pharmacokinetics of the drug-encased chemotherapeutic medications are substantially influenced by the nanocarriers used. Several chitosan nanoparticle-based chemotherapeutic formulations have been designed and tested in vivo and in vitro on various tumors in an attempt to identify a superior tumor treatment [70]. Bionanocomposites based on chitosan can preferentially permeate the membranes of cancerous cells and provide an antitumor effect through a variety of antiangiogenic, enzymatic, apoptotic, and immune-enhancing mechanisms. They are hidden from non-tumor cells and attack cancerous cells with increased bioavailability

Table 2 Summarizes a few popular chitosan nanocomposites for the purpose of bone engineering [1]

Nanocomposites additives	Chitosan information/synthesis technique	Attributes compared to original chitosan materials	Study of cell lineage and cytotoxicity when compared to original chitosan materials
Tobermorite type nanoclay (chitosan film)	Molecular mass low/solvent casting	Cell compatible; bioactive and biodegradable	MG-63 human osteosarcoma cells; viability of cells improved up to 30%
Hydroxyapatite (chitosan film)	DDA of chitosan >95%/solution casting	Alkaline phosphatase level rose by 377%, collagen I grew by 479%, and osteopontin climbed by 597%: all outstanding osteodifferentiation features	Mesenchymal stem cells; viability of cells enhanced by 52%
Silicon dioxide and zirconia NPs (chitosan scaffold)	75–85% DDA and low molecular mass/freeze-drying technique	Protein adsorption, deswelling, and biodegradation rates, and accelerated biomineralization abilities are significantly improved	Osteoprogenitor cells; zero toxicity shown by nanocomposites scaffold at fewer proportions of SiO_2/Zirconia NPs encapsulation
Nano-crystalline calcium phosphate (chitosan scaffold)	DDA = 92.3%/solution approach	Enhanced proliferation, mechanical characteristics, fibronectin, and cell adhesion	Human embryonic palatal mesenchymal cells; viability of cells > 65%
Bioglass (chitosan scaffold)	Needle punching technique	Porosity was raised by around 86% without impacting mechanical integrity; cell adherence and multiplication were enhanced	Human bone marrow stromal cells; viability of cells raised by four times after seven days of culture

over time. The magnetic sensitivity of bionanocomposites speeds up the antitumor agent's intravenous dispersion while reducing cytotoxicity [72]. Chitosan coating significantly enhanced the selectivity of magnetic NPs under conditions of hyperthermia. Thus, magnetic hyperthermia is indeed a potential tumor treatment method. The use of chitosan coating increased the targeting of magnetic nanoparticles in hyperthermia. Anti-cancer action of the complexes of chitosan/metal is owing to their interactions with cellular DNA and antioxidative capacity. The multiplication of cancerous cells has been reported to be suppressed by the more dissoluble variants with lower molecular weight chitosan oligosaccharides [73]. Thus, drug transport devices combined with the bioavailability, and antioxidative attributes of chitosan-based nanocomposites can create a new vista for creating novel medications and techniques for the treatment of cancer.

For the treatment of bladder tumors, chitosan hydrogels embedded with β-glycerophosphate and magnetic NPs (for the steady and long-term administration of bacillus Calmette Guérin) were synthesized by Zhang and colleagues [74]. The apoptosis of cancerous cell clusters was studied by Bae and team [75] using chitosan-g-PEG/heparin bio-nanocomplexes. These complexes showed superior internalization of cells in comparison with unbound heparin exclusively. Furthermore, after the internalization of cells, heparin discharges induced apoptosis of the tumorous growth via activation of caspase. Sasidharan and colleagues [62] applied an aqueous chemistry approach at RT to synthesize a breakthrough nanomaterial structure centered on mannosylated ZnS with robust fluorescence intensity and extensive durability. Under this investigation, d-Mannose was utilized to effectively functionalize chitosan entrapped zinc sulfide NPs, thereby generating mannosylated ZnS with a dimension of approximately 120 nm. According to in-vitro cytotoxicity assessment using MTT assay, the mannosylated zinc sulfide NPs had negligible toxicity for normal and tumor cells alike. The mannosylated nanostructures were used to specifically attack tumor cells. The target specificity of mannosylated zinc sulfide nanostructures toward mannose-containing KB cells was revealed by fluorescence microscopic investigations, with no particular adhesion on normal cells. These studies greatly demonstrate the importance of chitosan-centric nanomedicine in cancer treatment (Table 3).

Table 3 Summary of a few chitosan-based nanocomposites in diversified formulations having anti-cancer action [2]

Bionanocomposites	Formulations	Receiver	Characteristics
Glycol chitosan-o-nitrobenzyl succinate	Nano micelles	Mouse	Biocompatibility; superior drug carrier for tumor treatment
GO/dimethylmaleic anhydride-altered chitosan	NPs	HepG2 cells	Smart antitumor nanocomposite having enhanced curative activity
Chitosan/dextran sulfate/chitosan	NPs	HepG2 cell	Nanoscale drug carrier with enhanced impediment toward cancerous clusters
Disulfide-bridged chitosan-Eudragit S-100	NPs	Mouse	Steady release of drugs at the specified site; tumors in the colorectal area are affected
GO-functionalized chitosan polyelectrolyte	Nanocomposite	MCF7, HeLa and L929 cells	Antitumor medication delivered in a regulated manner to the desired site

5.5 *Bioimaging*

Bioimaging has been widely utilized for both scientific surveys and primary diagnostic trials [76, 77]. This approach aids in the investigation of physiological processes spanning from cells at the nanoscopic level to the mammalian degree. Because of its ability to automatically measure, recognize, and profile phenotypic variations, this method is regarded as a viable technique for learning more about complex physiological phenomena [78, 79]. Altered chitosan and its hybrid nano polymers have evolved as interesting bioimaging materials of late. Chitosan's photoluminescent property was studied for bioimaging, and it showed photoluminescence in the aqueous phase.

Salehizadeh et al. [80] created a magnetically susceptible core–shell nano polymer for bioimaging by mixing Fe_3O_4-AuNPs with chitosan as the stabilizer. The superior magnetic character of magnetite ensured magnetic resonance for imaging, while gold NPs served the purpose of photothermal conversion, and stimulate optical qualities. So as to create exceptionally robust covalently cross-polymerized fusion nanogels, Wu's research group [81] demonstrated in-situ confinement of cadmium selenide QDs in chitosan–poly (methacrylic acid) nanogels. They came to the conclusion that the synthesized bionanocomposite can be used for efficient bioimaging and biosensing. The nanogels created established the likelihood of concurrent surveillance and assessment responsiveness during therapies. Lin's team [82] used an ethyl alcohol-mediated counter-ion complex formation technique in an aqueous medium to fabricate CdSe/ZnS quantum dot-embedded chitosan fusion nanospheres. It was discovered that produced fusion nanostructures may well be absorbed by cancer cells and hence act as cell imaging tagging tools. It might be used to image malignancies in rodents having tumors, utilizing intra-tumoral injection of a substance that would concentrate at the cancer site through the circulatory system.

Due to the diverse functions of chitosan-centric biomaterials, the bioimaging domain appears to be a viable arena for assisting and speeding up the curative procedure. In any event, a slew of testing difficulties persists to this day. These can be answered by large-scale medical tests and further analysis of the reconfigured nanocomposites.

5.6 *Wound Healing*

Healing of wound progressive mechanism that involves the controlled synthesis of inflammatory, vascular, and connective tissues. The injury dressing biomaterials must create a moist condition surrounding the wound, soak secretions, mechanically maintain the tissue, prevent drying of the wound, stimulate cellular proliferation, and function as a barrier against microbial pathogens while allowing for the exchange of gases. To accelerate wound regeneration, the perfect wound patch should not be allergenic or virulent and have a high antibacterial and cytocompatibility rating [83,

84]. A variety of biomaterials have been evaluated as wound coverings for both acute and persistent wounds. Inflammation, hemostasis, motility, proliferation, and restructuring are all coordinated activities in the wound repair mechanism, and the tempo of regeneration fluctuates according to the biopolymers utilized.

Even though numerous artificial or organic materials are being used, chitosan-centric nanopolymers (Fig. 7) have received much interest for a variety of reasons, including superior microbiological resistance, oxygen penetration, photothermal consequences, stimuli sensitivity, and simplicity of usage. Antibiotics like sulfadiazine, ciprofloxacin, or tetracycline, metallic antibacterial particulates (nCu, nAg, or nZnO), and natural substances/extracts (Salix alba leaves, Juglena regia, aloe vera, and so on) can all be incorporated into chitosan for the production of enhanced antimicrobial wound dressing [85–88]. Nanofibers made of chitosan trigger macrophages and speed up tissue repair. Furthermore, chitosan-induced re-epithelialization and restoration of the granular surface of the skin have also been documented [89].

Via electrospinning followed by photo-polymerization, Zhou's research team [90] effectively created a bio-nanofibrous scaffold exhibiting increased stability in water. They used photo-cross-linked maleilated chitosan/methacrylated poly(vinyl alcohol) in the aqueous phase. The developed product has great biocompatible features, according to in-vitro cytotoxicity evaluation and could be used as a potential bandaging material for wounds. A micro-porous, elastic, and chitosan hydrogel/nano ZnO-based bandage for wound treatment was created by Sudheesh Kumar and colleagues [91]. It works as a barrier against microbes and offers a moist atmosphere and cooling effect. These synthesized chitosan-based bandages improved edema, blood coagulation, and antimicrobial properties. Bionanocomposite scaffolds are cytocompatible with the dermal fibroblast cells of human beings, which promotes their usage for diabetic foot ulcers, burn injuries, and chronic injuries.

With the addition of bio-nanofillers, it is anticipated that researchers will soon create dramatically improved chitosan-centric bionanocomposites that will accelerate re-epithelialization and optimize the efficiency to eliminate microbes from the injured area.

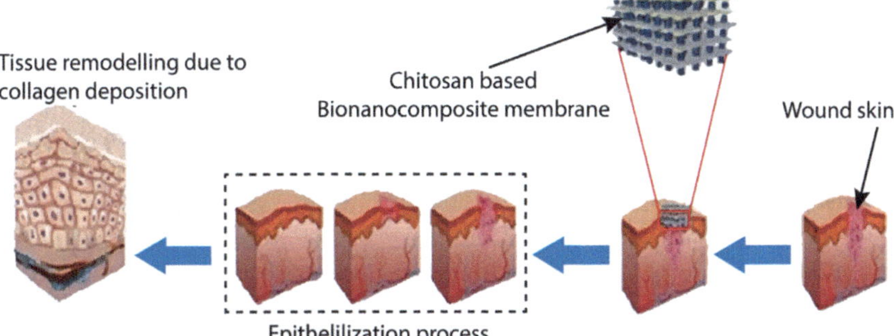

Fig. 7 Mechanism of wound healing with bionanocomposites based on chitosan. Reproduced from Motia et al. (2021) with permission from Elsevier

5.7 Antibacterial and Anti-fungal Action

Chitosan provides a variety of antagonistic efficacies against various fungi and bacteria. Attributed to variations in cellular framework, the mechanism of antibacterial action differs for both Gram-positive and Gram-negative bacteria. Chitosan is actually believed to be fungistatic instead of fungicidal, similar to bacterial activity [4]. Chitosan does not require chemical manipulation to trigger its antimicrobial effects. Nevertheless, chitosan NPs considerably boost their anti-fungal action.

Multiple concepts on the mechanism of antimicrobial activity, including the following, have lately been suggested and endorsed, despite the fact that the precise mechanism for the antibacterial and anti-fungal activity of chitosan and its derivatives has not yet been adequately elucidated.

- Polycationic nature of chitosan [92, 93]
- Interaction with bacterial DNA (mRNA suppression) [4, 94]
- Chelation agent (nutrients and essential metals) [95, 96]
- Chitosan has the ability to envelop the bacterial cell surface which blocks the uptake of nutrients. Additionally, it obstructs the O_2 pathway and prevents the development of aerobic microorganisms [97, 98].

For the enhancement and modification of the antibacterial action of both artificial and organic compounds, chitosan was used as a nanodevice for their transportation. Antibiotics, natural chemicals, antimicrobial peptides, and proteins are some of these entities. To boost the efficiency of the medication against bacterial development, various kinds of antibiotics including penicillins, tetracycline, vancomycin, cephalosporins, and aminoglycosides were encased in chitosan nanocarriers [99]. These nanocarriers were used, especially to enhance the administration of antibiotics into cells infiltrated by intracellular bacteria or to strengthen their potency against pathogens with multiple resistances.

The most popular chitosan-metal complexes for research studies are Ag-based nanostructures. According to studies done by Du's team [100], chitosan solution and nanoparticles comprising of metals like Cu(II), Mn(II), Zn(II), and Fe(II), all had reduced antibacterial effects against Gram-negative and Gram-positive bacteria (MIC varying from 3.0 to 6.0 g mL^{-1}) in comparison with Ag$^+$/chitosan complex NPs. The effectiveness of cefazolin-embedded chitosan NPs against multi-resistant Gram-negative bacteria like P. aeruginosa, E. coli and K. pneumonia was assessed by Jamil and colleagues [101]. Data obtained from the agar well diffusion assay and broth microdilution method showed that the drug-embedded chitosan NPs boosted antibacterial activity against the three pathogens, relative to simple cefazolin solution. Furthermore, Vancomycin's effectiveness against drug-resistive S. aureus, as proven by Chakraborty and others [102], firmly established the potency of drug-embedded chitosan NPs against antibiotic-resistive strains of bacteria. In a different investigation by Zhang's team [103], catechin and catechin/zinc complex were incorporated into chitosan NPs to increase the bioavailability and antibacterial impact of catechin. *L. innocua & E. coli* were used to test the antibacterial effect, and the catechin-zinc complex encased chitosan NPs showed a greater impact than the normal catechin-encased kind. Additionally, all specimens, along with the unloaded chitosan NPs, showed greater antibacterial action against *L. innocua*. A study done by Panwar's group [104] showed that the surface of chitosan NPs has a positive charge which was critical in promoting their engagement with the fungal cells' oppositely charged plasma membrane. This prevented Candida albicans from forming a biofilm by compromising its physical stability. These findings were consistent with research done by Ing and colleagues [105], which revealed that Candida albicans were successfully eliminated by chitosan NPs at a dosage of 1 mg/mL. For enhanced anti-fungal action, Fawzya et al. [3, 106] created chitosan oligosaccharide polymers by hydrolyzing chitosan with an enzyme from Aeromonas media KLU 11.16. In another study [107], EVA/chitosan-based PEG-PCL micelles nanocomposite films were produced by solution-casting technique. These nanocomposites possessed improved anti-fungal efficacy and temperature-responsive drug deliverance. Through pre-harvest fertilization, they led to the production of grape-fruit of superior standards.

5.8 Anti-viral Action

Living beings frequently suffer from a variety of viral diseases, some of which can be fatal. Although standard anti-viral medications are effective in curing viral infections, they can have negative consequences or even be toxic to the body. Several methods emphasizing the physical methodology have been used to create Ag nanoparticle/chitosan blends [108]. In order to create Ag/chitosan bionanocomposites with a range of reducing and/or stabilizing components, thermal techniques like the freeze-drying process and also specialized materials like polyvinylpyrrolidone and carboxymethyl chitosan have been used [108]. Ghosh and his team [109] applied a layer-by-layer synthesizing approach, which involved a thin layer of Ag-chitosan composite being poured onto a quartz and stainless-steel strip. More recently, applying the principles of green chemistry, chitosan-based bionanocomposites with integrated silver nanoparticles have been created using a natural polymer that performs double duty as a stabilizing agent and a reductant [110].

Silver nanoparticles with chitosan coating were developed by Mori and colleagues, and their efficacy against viruses like H1N1 was examined. The findings showed that the composites with finer silver nanoparticles had a considerable inhibitory action, whereas chitosan in the pure form alone showed little anti-viral action [111]. Additionally, a new study by Loutfy et al. demonstrated a chitosan-curcumin nanohybrid that can cure human hepatoma cells of the genotype 4a of hepatitis C virus [112]. In a different research, antimicrobial and anti-viral films for the safe packaging of foods were created by integrating green tea extracts with eatable coatings of chitosan. The nano-polymeric film synthesized successfully suppressed and neutralized the murine norovirus and Escherichia coli bacterial colonies [113].

5.9   *Orthopedics and Dentistry*

The most significant and common dental remedy for preventing oral caries is fluoride. However, the substantial prevalence of dental fluorosis raises alarm [56].

Chitosan and bioactive glass–ceramic (BGC) blended nano scaffold has been created and is being researched for a variety of orthopedic applications, including bone regeneration and engineering [70]. BGC is an assortment of osteoconductive bone-repairing agents which are based on silicates. Hench synthesized BGC for the first time in 1991, and it was employed as a biocompatible substance for bone healing. Owing to its prospective capacity to bind to hard as well as soft tissues, BGC is now an important material in the world of dentistry and orthopedics. A nanophase version of bioactive glass–ceramic has recently been created using the sol–gel method. When compared to micro-BGC, it has demonstrated improved interactions between the material and the cells [62]. A chitosan/macroporous nano-BGC scaffolding frame with dimensions of pores spanning from 150 to 500 μm was created by Peter's team [69]. This bandaging nanostructure enhances adherence, dispersion, and proliferation of cells, according to in-vitro research on osteoblast-like cells (MG-63) [114]. A biofilm for laminating orthopedic implants, consisting of chitosan and multi-walled carbon nanotubes, was developed by Ahmed and colleagues [56]. Titania and zirconia are two of the most used materials for orthopedic laminations and bone implantations. For laminating 316L stainless-steel substratum, Clavijo et al. [115]. examined the bioglass/chitosan/TiO_2 matrix and achieved superior biomechanical characteristics for implantation purposes. In addition, Bartma'nski and his team achieved excellent biocompatibility in a Ti13Zr13Nb alloy nanocomposite wrapped with chitosan/nAg [116].

Chitosan and propolis together showed a synergistic relationship for enhancing antibacterial action. Chitosan incorporated with dentifrice significantly inhibited the demineralization of enamel around the dental braces during orthodontic therapy. Both gypsum-based chitosan and calcium phosphate carboxymethyl-chitosan bionanocomposites can be used as pulp capping agents to enhance cytocompatibility and proliferation of human dental pulp stem cells. The calcium phosphate carboxymethyl-chitosan composite has the odontogenic capability, enhanced physical qualities, accelerated curing, is biocompatible, and solidifies quickly upon freezing, thereby satisfying the fundamental requirements of a prospective agent for pulp capping and regeneration [117, 118]. Using the freeze-drying technique, a porous chitosan-based scaffolding nanostructure was produced. The surface of the fabricated scaffold was then loaded with human dental pulp stem cells. The observations indicated improved gene delivery and that the dental pulp stem cells developed into an odontoblast-like phenotype [56]. Moreover, Costa's team suggested that chitosan NPs can be used to fight plaque on the enamel surface, which is linked to gingivitis, periodontitis, and cavities in human beings [99]. The antibacterial effects of chitosan NPs with varying degrees of acetylation and molecular weight on Streptococcus mutans biofilm were studied by Chavez de Paz and colleagues [119]. The chitosan nanoparticle formulations with low molecular weights (up to 150 kDa) effectively engaged with bacterial cells (more than 95% of afflicted cells) and disrupted the cell membrane stability in Streptococcus mutans throughout the whole region affected by plaque formation. The bactericidal efficacy of chitosan NPs against four distinct species of cariogenic streptococci (Streptococcus sanguis, Streptococcus mutans, Streptococcus sobrinus, and Streptococcus salivarius) was assessed in the research paper published by Aliasghari's team [120]. According to the findings, these compounds can inhibit the formation of biofilms in vitro, in addition to having antibacterial and anti-adhesion properties.

6 Drawbacks and Challenges

Deliberations about the drawbacks enable us to improvise material-based investigations and explore novel utilizations. Chitosan has promising prospects for various applications in the biomedical field, as evidenced by the rising number of research reports on chitosan-based nanocomposites. But despite having unprecedented biological and physiochemical attributes, chitosan suffers from certain pitfalls that pose significant hurdles and restricts its utilization in some foremost sectors. In order to fully exploit the benefits of this promising biopolymer and guarantee its availability for any application in distinctive sectors, the following noteworthy limitations must be addressed imperatively:

- Inferior solubility in physiological pH is a constraint in the biomedical sector [121].
- The Food and Drug Administration (FDA) has deemed it as food-contact material, but the European Food Safety Authority says otherwise. So, the safety of this material is still questionable [122].
- Low colloidal stability inhibits drug delivery [123].
- Economic practicality of eco-friendly bio-nanopolymer in the market of the real world [3].
- Efficacy of drug discharge, drug-loading capability, rate of decomposition of used nanocomposites, and delivery time period are other challenges faced by researchers [2, 124].

7 Future Prospects

The future holds great prospects for the innovation of chitosan and chitosan-based nanocomposites.

- *Development in agriculture*

    Recent in-vivo and in-vitro studies showcase chitosan-based nanocomposites as a game-changer in agri-economics to lessen the negative environmental consequences of traditional chemical fertilizers and deliver the upcoming generation of sustainable, compostable, and environment-friendly fertilizers. Primary usages include pisciculture improvement, pest control, meat production, seed preservation, and enhancement of plant immune by gradual, regulated, and selective nutrient conveyance to plants. Chitosan-based nanomaterials serve as a rich supply of nutrients for plants, providing them with C (~54–48 wt%), O (~42–30 wt%), N (~8–6 wt%), and P (~6–3 wt%) [125]. Additionally, via its functional groups, chitosan nanomaterials can be further modified to provide additional nutrient content. Thus, extensive research can be committed to developing this efficient and realistic approach to safeguard farming operations and also our ecosystem.

- *Development of chitosan carrier for brain drug delivery*

Chitosan-based nanocarriers have received a lot of attention for the delivery of drugs to the brain, and extensive research has been done on them with respectable outcomes. But to dismay, it was discovered that the chitosan-based nano delivery device interacted with the antibodies. Following this, in-vivo studies were conducted which showed their nanocarriers conjugating with antibodies, and their cytotoxicity, in the long run, was also assessed [126]. Therefore, the scholar can focus on these topics in their subsequent research investigation.

- *Improving the properties for medical application*

Studies have shown that the tissue sensitivity for chitosan-based biopolymers is discernably different when contrasted with lone materials. Cellular feedback, adherence, differentiation, and proliferation can be modulated through surface functionalization of nanoparticles and modifications of chitosan by moieties such as peptides, proteins, small biomolecules, and polymers, which in turn provides for improved bio-sensitivity, thereby facilitating enhanced cell-substrate reciprocation, and ameliorate physical properties of the substrate. Chitosan-based nanomaterials with bespoke characteristics show encouraging end results, such as selective drug/gene conveyance, faster bone rebuilding, superior wound regeneration, boosted osteogenesis, and angiogenesis. Nonetheless, it is of the utmost importance to perform research and surveys before starting clinical trials, since chitosan-based nanomaterials may elute as time goes by and also, the study of biodistribution in long-term animal trials should be done exhaustively. To be more precise, cells may absorb the nanoparticles released from the nanocomposites through endocytotic pathways, as exhibited by nanographene sheets that have been proven to infiltrate cells through phagocytotic uptake or clathrin-mediated endocytosis. The molecular mass and degree of deacetylation of chitosan are some additional disadvantages that severely affect its use. One recent study found that chitosan promotes the rupture of lysosomes that causes macrophage cytokines to be produced when glucosamine concentrations are more than 30,000 g/mol [127]. Moreover, the nanoparticulate concentration (such as CNTs, silver sulfadiazine, and bioactive glass), employed as fillers beyond a particular wt% leads to an appreciable decline in cell viability which hampers tissue engineering applications. Other nanoparticles did not enhance any properties. However, regardless of these blemishes of chitosan-based nanomaterials, they have a lot of potential for further study in the biomedical field.

- *Wound healing*

UV irradiated chitosan/ZnO nanocomposite is a superb research topic in the near future, owing to the distinctive optical and semiconducting attributes of ZnO combined with UV's ability to promote photocatalysis. All these factors improve the bacteriostatic characteristics of the synthesized nanocomposite [128].

- *Heterogeneous nanocatalysis*

Inefficiency, impotent and low yield of conventional catalysts like iBr, CTAB, CuBr, salicylic acid, and L-proline are huge barriers in the synthesis of essential derivatives having antidiabetic, antitumor, geroprotective, antitubercular, and pain-killing activities. This is where chitosan and chitosan-based bionanocomposites come in as excellent future research options for heterogeneous bio-nanocatalysis because lately, Asgharnasl and his team achieved encouraging results from magnetically susceptible chitosan-terephthaloyl-creatine bionanocomposite [129].

- *Discovering substitute sources*

The exoskeleton of crustaceans is primarily made of chitin, which is also present in several other species, including fungi, mollusks, and insects. The a-chitosan from crustacean chitin is produced from recycled shrimp and crab shells and is the most widely exploited variant of chitosan. An enzyme known as chitosanases that is found in the species itself can also be used to extract chitosan from bacteria and fungus. But with rising popularity comes the fear that these natural resources will disappear. To restore the ecological balance, researchers must, therefore, hunt for alternate supplies.

- *Eco-friendly and cost-efficient production*

The traditional chemical technique of obtaining chitin poses several environmental concerns. Ionic liquids, microbial fermentation, deep eutectic solvents, microwave-mediated extraction, enzyme-mediated extraction, ultrasonic-mediated extraction, subcritical water extraction, and electrochemical extraction are a few methods that have been fruitfully established recently, using green harvesting methods for retrieving chitosan and chitin from different sources.

8   Conclusion

The indiscriminate exploitation of non-renewable resources and the harmful effects of non-biodegradable materials have led to the rise of a new era in the study and development of environmentally safe materials. Bionanocomposites are the leading contenders among the possible candidates with the intrinsic qualities of biocompatibility, biodegradability, and non-toxicity, along with their improved structural and functional features. They possess numerous extraordinary properties which can be considered to be a boon to the human race.

This chapter has expounded on the chitosan-based bionanocomposites and the apprehension of their exceptional qualities. Different sources of chitosan, alteration techniques, and manufacturing methods have also been discussed. Global applications, advantages, and limitations of the chitosan-based bionanocomposites are described in depth. This chapter also draws attention to certain areas that have scope for further scientific study and investigation and provides insights to the scientists to work on ways to augment the physiochemical properties of this biopolymer. This chapter eventually concludes that even though multiple chitosan bionanocomposite-based utilizations are still in infancy, a booming future lies ahead of it with enhanced distinctive attributes like thermal, mechanical, protective properties, crystallization, and decomposition rate, which reflects their métier in the biomedical field, industrial field, agronomic field, and so on.

References

1. Murugesan S, Scheibel T (2021) Chitosan-based nanocomposites for medical applications. J Polym Sci 59(15):1610–1642, Wiley. https://doi.org/10.1002/pol.20210251
2. Khan A, Alamry KA (2021) Recent advances of emerging green chitosan-based biomaterials with potential biomedical applications: a review. Carbohydrate Res 506. Elsevier Ltd. https://doi.org/10.1016/j.carres.2021.108368
3. Azmana M, Mahmood S, Hilles AR, Rahman A, Bin Arifin MA, Ahmed S (2021) A review on chitosan and chitosan-based bionanocomposites: promising material for combatting global issues and its applications. Int J Biol Macromolecules 185:832–848, Elsevier B.V. https://doi.org/10.1016/j.ijbiomac.2021.07.023
4. Kravanja G, Primožič M, Knez Ž, Leitgeb M (2019) Chitosan-based (Nano)materials for novel biomedical applications. Molecules 24(10). MDPI AG. https://doi.org/10.3390/molecules24101960
5. Bagheri-Khoulenjani S, Taghizadeh SM, Mirzadeh H (2009) An investigation on the short-term biodegradability of chitosan with various molecular weights and degrees of deacetylation. Carbohydr Polym 78(4):773–778. https://doi.org/10.1016/j.carbpol.2009.06.020
6. Moreno JAS et al (2018) Development of electrosprayed mucoadhesive chitosan microparticles. Carbohydr Polym 190:240–247. https://doi.org/10.1016/j.carbpol.2018.02.062
7. Husain S et al (2017) Chitosan biomaterials for current and potential dental applications. Materials 10(6). MDPI AG. https://doi.org/10.3390/ma10060602
8. Kumirska J, Weinhold MX, Thöming J, Stepnowski P (2011) Biomedical activity of chitin/chitosan based materials-influence of physicochemical properties apart from molecular weight and degree of N-Acetylation. Polymers 3(4):1875–1901. https://doi.org/10.3390/polym3041875
9. Ghanbari B, Shahhoseini L, Hosseini H, Bagherzadeh M, Owczarzak A, Kubicki M (2018) Synthesis of Pd(II) large dinuclear macrocyclic complex tethered through two dipyridine-bridged aza-crowns as an efficient copper-and phosphine-free Sonogashira catalytic reaction. J Organomet Chem 866(Ii):72–78. https://doi.org/10.1016/j.jorganchem.2018.04.009
10. Kim S (2018) Competitive biological activities of Chitosan and its derivatives: antimicrobial, antioxidant, anticancer, and anti-inflammatory activities. Int J Polym Sci. https://doi.org/10.1155/2018/1708172

11. Ardila N, Daigle F, Heuzey MC, Ajji A (2017) Effect of Chitosan physical form on its antibacterial activity against pathogenic bacteria. J Food Sci 82(3):679–686. https://doi.org/10.1111/1750-3841.13635

12. Rashid TU, Rahman MM, Kabir S, Shamsuddin SM, Khan MA (2012) A new approach for the preparation of chitosan from γ-irradiation of prawn shell: effects of radiation on the characteristics of chitosan. Polym Int 61(8):1302–1308. https://doi.org/10.1002/pi.4207

13. Methacanon P, Prasitsilp M, Pothsree T, Pattaraarchachai J (2022) Heterogeneous N-deacetylation of squid chitin in alkaline solution. [Online]. Available: www.elsevier.com/locate/carbpol

14. Focher B, Beltrame PL, Naggi A, Torri G (1990) Alkaline N-Deacetylation of Chitin enhanced by flash treatments. Reaction Kinetics and Struct Modif

15. Pelletier A, Sygusch J, Chornet E, Overend RP (1990) Chitin/Chitosan transformation by thermo-mechano-chemical treatment including characterization by enzymatic depolymerization

16. Lertwattanaseri T, Ichikawa N, Mizoguchi T, Tanaka Y, Chirachanchai S (2009) Microwave technique for efficient deacetylation of chitin nanowhiskers to a chitosan nanoscaffold. Carbohydr Res 344(3):331–335. https://doi.org/10.1016/j.carres.2008.10.018

17. Aranaz I et al (2009) Functional characterization of Chitin and Chitosan

18. Cai J et al (2006) Enzymatic preparation of chitosan from the waste Aspergillus niger mycelium of citric acid production plant. Carbohydr Polym 64(2):151–157. https://doi.org/10.1016/J.CARBPOL.2005.11.004

19. Zhao Y, Park R-D, Muzzarelli RAA (2010) Chitin deacetylases: properties and applications. Mar Drugs 8:24–46. https://doi.org/10.3390/md8010024

20. Yamada M, Kurano M, Inatomi S, Taguchi G, Okazaki M, Shimosaka M (2008) Isolation and characterization of a gene coding for chitin deacetylase specifically expressed during fruiting body development in the basidiomycete Flammulina velutipes and its expression in the yeast Pichia pastoris. https://doi.org/10.1111/j.1574-6968.2008.01361.x

21. Qin Y, Li P, Guo Z (2020) Cationic chitosan derivatives as potential antifungals: a review of structural optimization and applications. Carbohydrate Polym 236, Elsevier Ltd https://doi.org/10.1016/j.carbpol.2020.116002

22. Sashiwa H, Aiba SI (2004) Chemically modified chitin and chitosan as biomaterials. Prog Polym Sci 29(9):887–908. https://doi.org/10.1016/J.PROGPOLYMSCI.2004.04.001

23. Alves NM, Mano JF (2008) Chitosan derivatives obtained by chemical modifications for biomedical and environmental applications. Int J Biol Macromol 43(5):401–414. https://doi.org/10.1016/J.IJBIOMAC.2008.09.007

24. Mourya VK, Inamdar NN (2008) Chitosan-modifications and applications: opportunities galore. React Funct Polym 68(6):1013–1051. https://doi.org/10.1016/J.REACTFUNCTPOLYM.2008.03.002

25. Meredith JC et al (2003) Combinatorial characterization of cell interactions with polymer surfaces

26. Abolhasani MM, Arefazar A, Mozdianfard M (2010) Effect of dispersed phase composition on morphological and mechanical properties of PET/EVA/PP ternary blends. J Polym Sci Part B Polym Phys 48(3):251–259. https://doi.org/10.1002/polb.21857

27. Khor E, Lim LY (2003) Implantable applications of chitin and chitosan. Biomaterials 24(13):2339–2349. https://doi.org/10.1016/S0142-9612(03)00026-7

28. Khoo CGL, Frantzich S, Rosinski A, Sjöström M, Hoogstraate J (2003) Oral gingival delivery systems from chitosan blends with hydrophilic polymers. Eur J Pharm Biopharm 55(1):47–56. https://doi.org/10.1016/S0939-6411(02)00155-8

29. Park SY, Jun ST, Marsh KS (2001) Physical properties of PVOH/chitosan-blended films cast from different solvents. Food Hydrocoll 15(4–6):499–502. https://doi.org/10.1016/S0268-005X(01)00055-8

30. Kumari S, Singh RP (2012) Glycolic acid-g-chitosan-gold nanoflower nanocomposite scaffolds for drug delivery and tissue engineering. Int J Biol Macromol 50(3):878–883. https://doi.org/10.1016/J.IJBIOMAC.2011.10.014

31. Choudhary RC et al (2017) Cu-chitosan nanoparticle boost defense responses and plant growth in maize (Zea mays L.) OPEN. https://doi.org/10.1038/s41598-017-08571-0

32. Saharan V, Pal A (2016) Properties and types of Chitosan-based nanomaterials. pp 23–32. https://doi.org/10.1007/978-81-322-3601-6_3

33. Choudhary RC et al (2019) Characterization methods for Chitosan-based nanomaterials. In: Nanotechnology in the life sciences, Springer Science and Business Media B.V., pp 103–116. https://doi.org/10.1007/978-3-030-12496-0_5

34. Korin E, Froumin N, Cohen S (2017) Surface analysis of nanocomplexes by X-ray photo-electron spectroscopy (XPS). ACS Biomater Sci Eng 3(6):882–889. https://doi.org/10.1021/acsbiomaterials.7b00040

35. Matienzo LJ, Winnacker SK (2002) Dry processes for surface modification of a biopolymer: Chitosan

36. Ieva E, Trapani A, Cioffi N, Ditaranto N, Monopoli A, Sabbatini L (2009) Analytical characterization of chitosan nanoparticles for peptide drug delivery applications. Anal Bioanal Chem 393(1):207–215. https://doi.org/10.1007/s00216-008-2463-4

37. Li LH, Deng JC, Deng HR, Liu ZL, Xin L (2010) Synthesis and characterization of chitosan/ZnO nanoparticle composite membranes. Carbohydr Res 345(8):994–998. https://doi.org/10.1016/j.carres.2010.03.019

38. Han J, Zhou C, Wu Y, Liu F, Wu Q (2013) Self-assembling behavior of cellulose nanoparticles during freeze-drying: Effect of suspension concentration, particle size, crystal structure, and surface charge. Biomacromol 14(5):1529–1540. https://doi.org/10.1021/bm4001734

39. Saharan V, Mehrotra A, Khatik R, Rawal P, Sharma SS, Pal A (2013) Synthesis of chitosan based nanoparticles and their in vitro evaluation against phytopathogenic fungi. Int J Biol Macromol 62:677–683. https://doi.org/10.1016/J.IJBIOMAC.2013.10.012

40. Nguyen Van S, Dinh Minh H, Nguyen Anh D (2013) Study on chitosan nanoparticles on biophysical characteristics and growth of Robusta coffee in green house. Biocatal Agric Biotechnol 2(4):289–294. https://doi.org/10.1016/j.bcab.2013.06.001

41. Liu H, Gao C (2009) Preparation and properties of ionically cross-linked chitosan nanoparticles. Polym Adv Technol 20(7):613–619. https://doi.org/10.1002/pat.1306

42. Sriamornsak P, Thirawong N, Cheewatanakornkool K, Burapapadh K, Sae-Ngow W (2008) Cryo-scanning electron microscopy (cryo-SEM) as a tool for studying the ultrastructure during bead formation by ionotropic gelation of calcium pectinate. Int J Pharm 352(1–2):115–122. https://doi.org/10.1016/j.ijpharm.2007.10.038

43. Kuntsche J, Horst JC, Bunjes H (2011) Cryogenic transmission electron microscopy (cryo-TEM) for studying the morphology of colloidal drug delivery systems. Int J Pharmaceutics 417(1–2):120–137, Elsevier B.V. https://doi.org/10.1016/j.ijpharm.2011.02.001

44. Wang JJ et al (2011) Recent advances of chitosan nanoparticles as drug carriers. Int J Nanomed 6:765–774. https://doi.org/10.2147/ijn.s17296

45. Barman M, Mahmood S, Augustine R, Hasan A, Thomas S, Ghosal K (2020) Natural halloysite nanotubes /chitosan based bio-nanocomposite for delivering norfloxacin, an anti-microbial agent in sustained release manner. Int J Biol Macromol 162:1849–1861. https://doi.org/10.1016/J.IJBIOMAC.2020.08.060

46. Rajitha P, Gopinath D, Biswas R, Sabitha M, Jayakumar R (2016) Chitosan nanoparticles in drug therapy of infectious and inflammatory diseases. Expert Opinion on Drug Delivery 13(8):177–1194, Taylor and Francis Ltd. https://doi.org/10.1080/17425247.2016.1178232

47. Liu W et al (2018) Molecules Galactosylated Chitosan-functionalized mesoporous silica nanoparticle loading by Calcium Leucovorin for colon cancer cell-targeted drug delivery. https://doi.org/10.3390/molecules23123082

48. Baktash MS, Zarrabi A, Avazverdi E, Reis NM (2021) Development and optimization of a new hybrid chitosan-grafted graphene oxide/magnetic nanoparticle system for theranostic applications. J Mol Liq 322:114515. https://doi.org/10.1016/J.MOLLIQ.2020.114515

49. Gowda BHJ et al (2022) Current trends in bio-waste mediated metal/metal oxide nanoparticles for drug delivery. J Drug Deliv Sci Technol 71:103305. https://doi.org/10.1016/J.JDDST.2022.103305

50. Murugesan S, Scheibel T (2020) Copolymer/clay nanocomposites for biomedical applications. https://doi.org/10.1002/adfm.201908101
51. Ahmed A, Hearn J, Abdelmagid W, Zhang H (2012) Dual-tuned drug release by nanofibrous scaffolds of chitosan and mesoporous silica microspheres. J Mater Chem 22(48):25027–25035. https://doi.org/10.1039/c2jm35569h
52. Pourshahrestani S et al (2017) Potency and cytotoxicity of a novel gallium-containing mesoporous bioactive glass/chitosan composite scaffold as hemostatic agents. ACS Appl Mater Interf 9(37):31381–31392. https://doi.org/10.1021/acsami.7b07769
53. Dev A et al (2010) Preparation of poly(lactic acid)/chitosan nanoparticles for anti-HIV drug delivery applications. Carbohydr Polym 80(3):833–838. https://doi.org/10.1016/J.CARBPOL.2009.12.040
54. Chandran PR, Sandhyarani N (2014) An electric field responsive drug delivery system based on chitosan-gold nanocomposites for site specific and controlled delivery of 5-fluorouracil. RSC Adv 4(85):44922–44929. https://doi.org/10.1039/c4ra07551j
55. Shah A, Ali Buabeid M, Arafa ESA, Hussain I, Li L, Murtaza G, (2019) The wound healing and antibacterial potential of triple-component nanocomposite (chitosan-silver-sericin) films loaded with moxifloxacin. Int J Pharm 564:22–38. https://doi.org/10.1016/j.ijpharm.2019.04.046
56. Sharifianjazi F et al (2022) Advancements in fabrication and application of Chitosan composites in implants and dentistry: a review. Biomolecule 12(2). MDPI. https://doi.org/10.3390/biom12020155
57. Chen H et al (2012) O-Alkylation of Chitosan for gene delivery by using ionic liquid in an in-situ reactor. Engineering 04(10):114–117. https://doi.org/10.4236/eng.2012.410b029
58. Amidi M, Mastrobattista E, Jiskoot W, Hennink WE (2010) Chitosan-based delivery systems for protein therapeutics and antigens. Adv Drug Deliv Rev 62(1):59–82. https://doi.org/10.1016/J.ADDR.2009.11.009
59. Chen H, Cui S, Zhao Y, Zhang C, Zhang S, Peng X (2015) Grafting Chitosan with polyethylenimine in an ionic liquid for efficient gene delivery. https://doi.org/10.1371/journal.pone.0121817
60. Ahsan SM, Thomas M, Reddy KK, Sooraparaju SG, Asthana A, Bhatnagar I (2018) Chitosan as biomaterial in drug delivery and tissue engineering. Int J Biol Macromol 110:97–109. https://doi.org/10.1016/J.IJBIOMAC.2017.08.140
61. Raftery R, O'brien FJ, Cryan S-A (2013) Molecules Chitosan for gene delivery and orthopedic tissue engineering applications. Molecules 18:5611–5647. https://doi.org/10.3390/molecules18055611
62. Jayakumar R, Menon D, Manzoor K, Nair SV, Tamura H (2010) Biomedical applications of chitin and chitosan based nanomaterials—a short review. Carbohydrate Polym 82(2):227–232. https://doi.org/10.1016/j.carbpol.2010.04.074
63. Jayakumar R, Prabaharan M, Nair SV, Tokura S, Tamura H, Selvamurugan N (2010) Novel carboxymethyl derivatives of chitin and chitosan materials and their biomedical applications. Prog Mater Sci 55(7):675–709. https://doi.org/10.1016/J.PMATSCI.2010.03.001
64. Shalumon KT et al (2009) Electrospinning of carboxymethyl chitin/poly(vinyl alcohol) nanofibrous scaffolds for tissue engineering applications. Carbohydr Polym 77(4):863–869. https://doi.org/10.1016/j.carbpol.2009.03.009
65. Webster TJ, Ergun C, Doremus RH, Siegel RW, Bizios R (2000) Enhanced functions of osteoblasts on nanophase ceramics. Biomaterials 21(17):1803–1810. https://doi.org/10.1016/S0142-9612(00)00075-2
66. Francis Suh JK, Matthew HWT (2000) Application of chitosan-based polysaccharide biomaterials in cartilage tissue engineering: a review. Biomaterials 21(24):2589–2598. https://doi.org/10.1016/S0142-9612(00)00126-5
67. Kuo CY, Chen CH, Hsiao CY, Chen JP (2015) Incorporation of chitosan in biomimetic gelatin/chondroitin-6-sulfate/hyaluronan cryogel for cartilage tissue engineering. Carbohydr Polym 117:722–730. https://doi.org/10.1016/J.CARBPOL.2014.10.056

68. Sen Shen Z, Cui X, Hou RX, Li Q, Deng HX, Fu J (2015) Tough biodegradable chitosan-gelatin hydrogels via in situ precipitation for potential cartilage tissue engineering. RSC Adv 5(69):55640–55647. https://doi.org/10.1039/c5ra06835e

69. Peter M, Binulal NS, Nair SV, Selvamurugan N, Tamura H, Jayakumar R (2010) Novel biodegradable chitosan-gelatin/nano-bioactive glass ceramic composite scaffolds for alveolar bone tissue engineering. Chem Eng J 158(2):353–361. https://doi.org/10.1016/j.cej.2010.02.003

70. El-Sherbiny IM, El-Baz NM (2015) A review on bionanocomposites based on chitosan and its derivatives for biomedical applications. Adv Struct Mater 74:173–208. https://doi.org/10.1007/978-81-322-2473-0_6

71. Hu C-MJ, Zhang L (2009) Therapeutic nanoparticles to combat cancer drug resistance

72. Arias JL, Reddy LH, Couvreur P (2012) Fe_3O_4/chitosan nanocomposite for magnetic drug targeting to cancer. J Mater Chem 22(15):7622–7632. https://doi.org/10.1039/c2jm15339d

73. Sharan Adhikari H, Nath Yadav P (2018) Anticancer activity of Chitosan, Chitosan derivatives, and their mechanism of action. https://doi.org/10.1155/2018/2952085

74. Zhang D et al (2013) A magnetic chitosan hydrogel for sustained and prolonged delivery of Bacillus Calmette-Guérin in the treatment of bladder cancer. Biomaterials 34(38):10258–10266. https://doi.org/10.1016/J.BIOMATERIALS.2013.09.027

75. Bae KH, Moon CW, Lee Y, Park TG (2009) Intracellular delivery of heparin complexed with chitosan-g-poly(ethylene glycol) for inducing apoptosis. Pharm Res 26(1):93–100. https://doi.org/10.1007/s11095-008-9713-1

76. Lin J, Chen X, Huang P (2016) Graphene-based nanomaterials for bioimaging. Adv Drug Deliv Rev 105:242–254. https://doi.org/10.1016/J.ADDR.2016.05.013

77. Chen M, Yin M (2014) Design and development of fluorescent nanostructures for bioimaging. Prog Polym Sci 39(2):365–395. https://doi.org/10.1016/J.PROGPOLYMSCI.2013.11.001

78. Chen W, Xia X, Huang Y, Chen X, Han JDJ (2016) Bioimaging for quantitative phenotype analysis. Methods 102:20–25. https://doi.org/10.1016/J.YMETH.2016.01.017

79. Yang J et al (2014) Multifunctional quantum dot-polypeptide hybrid nanogel for targeted imaging and drug delivery. Nanoscale 6(19):11282–11292. https://doi.org/10.1039/c4nr03058c

80. Salehizadeh H, Hekmatian E, Sadeghi M, Kennedy K, (2012) Synthesis and characterization of core-shell Fe_3O_4-gold-chitosan nanostructure. J Nanobiotechnol 10. https://doi.org/10.1186/1477-3155-10-3

81. Wu W, Shen J, Banerjee P, Zhou S (2010) Chitosan-based responsive hybrid nanogels for integration of optical pH-sensing, tumor cell imaging and controlled drug delivery. Biomaterials 31(32):8371–8381. https://doi.org/10.1016/J.BIOMATERIALS.2010.07.061

82. Lin Y et al (2011) Water-soluble chitosan-quantum dot hybrid nanospheres toward bioimaging and biolabeling. ACS Appl Mater Interfaces 3(4):995–1002. https://doi.org/10.1021/am100982p

83. Gomes SR et al (2015) In vitro and in vivo evaluation of electrospun nanofibers of PCL, chitosan and gelatin: a comparative study. Mater Sci Eng C 46:348–358. https://doi.org/10.1016/J.MSEC.2014.10.051

84. Khorasani MT, Joorabloo A, Moghaddam A, Shamsi H, MansooriMoghadam Z (2018) Incorporation of ZnO nanoparticles into heparinised polyvinyl alcohol/chitosan hydrogels for wound dressing application. Int J Biol Macromol 114:1203–1215. https://doi.org/10.1016/J.IJBIOMAC.2018.04.010

85. Simões D, Miguel SP, Ribeiro MP, Coutinho P, Mendonça AG, Correia IJ (2018) Recent advances on antimicrobial wound dressing: a review. Eur J Pharm Biopharm 127:130–141. https://doi.org/10.1016/J.EJPB.2018.02.022

86. Yang L et al (2016) Fabrication of a novel chitosan scaffold with asymmetric structure for guided tissue regeneration. RSC Adv 6(75):71567–71573. https://doi.org/10.1039/c6ra12370h

87. Huang X et al (2015) Using absorbable chitosan hemostatic sponges as a promising surgical dressing. Int J Biol Macromol 75:322–329. https://doi.org/10.1016/J.IJBIOMAC.2015.01.049

88. Mi FL, Shyu SS, Wu YB, Lee ST, Shyong JY, Huang RN (2001) Fabrication and characterization of a sponge-like asymmetric chitosan membrane as a wound dressing. Biomaterials 22(2):165–173. https://doi.org/10.1016/S0142-9612(00)00167-8

89. Jayakumar R, Prabaharan M, Sudheesh Kumar PT, Nair SV, Tamura H (2011) Biomaterials based on chitin and chitosan in wound dressing applications. Biotechnol Adv 29(3):322–337. https://doi.org/10.1016/J.BIOTECHADV.2011.01.005

90. Zhou Y et al (2017) Photocrosslinked maleilated chitosan/methacrylated poly (vinyl alcohol) bicomponent nanofibrous scaffolds for use as potential wound dressings. Carbohydr Polym 168:220–226. https://doi.org/10.1016/J.CARBPOL.2017.03.044

91. Sudheesh Kumar PT et al (2012) Flexible and microporous chitosan hydrogel/nano ZnO composite bandages for wound dressing: In vitro and in vivo evaluation. ACS Appl Mater Interf 4(5):2618–2629, American Chemical Society. https://doi.org/10.1021/am300292v

92. Severino R, Ferrari G, Vu KD, Donsì F, Salmieri S, Lacroix M (2015) Antimicrobial effects of modified chitosan based coating containing nanoemulsion of essential oils, modified atmosphere packaging and gamma irradiation against *Escherichia coli* O157:H7 and Salmonella Typhimurium on green beans. Food Control 50:215–222. https://doi.org/10.1016/J.FOODCONT.2014.08.029

93. Rabea EI, Badawy MET, Stevens CV, Smagghe G, Steurbaut W (2003) Chitosan as antimicrobial agent: applications and mode of action. Biomacromol 4(6):1457–1465. https://doi.org/10.1021/bm034130m

94. Sudarshan NR, Hoover DG, Knorr D (1992) Antibacterial action of Chitosan. Food Biotechnol 6(3):257–272. https://doi.org/10.1080/08905439209549838

95. Wang X, Du Y, Fan L, Liu H, Hu Y (2005) Chitosan- metal complexes as antimicrobial agent: synthesis, characterization and Structure-activity study. Polym Bull 55(1–2):105–113. https://doi.org/10.1007/s00289-005-0414-1

96. Dutta PK, Tripathi S, Mehrotra GK, Dutta J (2009) Perspectives for chitosan based antimicrobial films in food applications. Food Chem 114(4):1173–1182. https://doi.org/10.1016/J.FOODCHEM.2008.11.047

97. Devlieghere F, Vermeulen A, Debevere J (2004) Chitosan: antimicrobial activity, interactions with food components and applicability as a coating on fruit and vegetables. Food Microbiol 21(6):703–714. https://doi.org/10.1016/J.FM.2004.02.008

98. Yuan G, Lv H, Tang W, Zhang X, Sun H (2016) Effect of chitosan coating combined with pomegranate peel extract on the quality of Pacific white shrimp during iced storage. Food Control 59:818–823. https://doi.org/10.1016/J.FOODCONT.2015.07.011

99. Perinelli DR et al (2018) Chitosan-based nanosystems and their exploited antimicrobial activity. European J Pharmaceutical Sci 117:8–20. Elsevier B.V. https://doi.org/10.1016/j.ejps.2018.01.046

100. Du WL, Niu SS, Xu YL, Xu ZR, Fan CL (2009) Antibacterial activity of chitosan tripolyphosphate nanoparticles loaded with various metal ions. Carbohydr Polym 75(3):385–389. https://doi.org/10.1016/j.carbpol.2008.07.039

101. Jamil B et al (2016) Cefazolin loaded chitosan nanoparticles to cure multi drug resistant Gram-negative pathogens. Carbohydr Polym 136:682–691. https://doi.org/10.1016/J.CARBPOL.2015.09.078

102. Chakraborty SP, Sahu SK, Pramanik P, Roy S (2012) In vitro antimicrobial activity of nanoconjugated vancomycin against drug resistant Staphylococcus aureus. Int J Pharm 436(1–2):659–676. https://doi.org/10.1016/J.IJPHARM.2012.07.033

103. Zhang H, Jung J, Zhao Y (2016) Preparation, characterization and evaluation of antibacterial activity of catechins and catechins-Zn complex loaded β-chitosan nanoparticles of different particle sizes. Carbohydr Polym 137:82–91. https://doi.org/10.1016/j.carbpol.2015.10.036

104. Panwar R, Pemmaraju SC, Sharma AK, Pruthi V (2016) Efficacy of ferulic acid encapsulated chitosan nanoparticles against Candida albicans biofilm. Microb Pathog 95:21–31. https://doi.org/10.1016/J.MICPATH.2016.02.007

105. Ing LY, Zin NM, Sarwar A, Katas H (2012) Antifungal activity of Chitosan nanoparticles and correlation with their physical properties. Int J Biomater 2012. https://doi.org/10.1155/2012/632698

106. Jeon YJ, Shahidi F, Kim SK (2000) Preparation of chitin and chitosan oligomers and their applications in physiological functional foods. Food Rev Int 16(2):159–176. https://doi.org/10.1081/FRI-100100286

107. Xiao NY et al (2020) Construction of EVA/chitosan based PEG-PCL micelles nanocomposite films with controlled release of iprodione and its application in pre-harvest treatment of grapes. Food Chem 331:127277. https://doi.org/10.1016/J.FOODCHEM.2020.127277

108. Jeevanandam J et al (2022) Synthesis approach-dependent antiviral properties of silver nanoparticles and nanocomposites. J Nanostruct Chem. https://doi.org/10.1007/s40097-021-00465-y

109. Ghosh S, Ranebennur TK, Vasan HN (2011) Study of antibacterial efficacy of hybrid chitosan-silver nanoparticles for prevention of specific biofilm and water purification. Int J Carbohydr Chem 2011:1–11. https://doi.org/10.1155/2011/693759

110. Sharma S, Sanpui P, Chattopadhyay A, Ghosh SS (2012) Fabrication of antibacterial silver nanoparticle—sodium alginate-chitosan composite films. RSC Adv 2(13):5837–5843. https://doi.org/10.1039/c2ra00006g

111. Mori Y et al (2011) Simple and environmentally friendly preparation and size control of silver nanoparticles using an inhomogeneous system with silver-containing glass powder. J Nanoparticle Res 13(7):2799–2806. https://doi.org/10.1007/s11051-010-0168-z

112. Loutfy SA et al (2022) Antiviral activity of chitosan nanoparticles encapsulating curcumin against hepatitis C virus genotype 4a in human hepatoma cell lines. Int J Nanomed 2020:15–2699. https://doi.org/10.2147/IJN.S241702

113. Mallakpour S, Azadi E, Hussain CM (2021) Recent breakthroughs of antibacterial and antiviral protective polymeric materials during COVID-19 pandemic and after pandemic: coating, packaging, and textile applications. Current Opinion in Colloid and Interf Sci 55. Elsevier Ltd. https://doi.org/10.1016/j.cocis.2021.101480

114. Liu X-P et al (2010) Open access research anti-tumor activity of N-trimethyl chitosan-encapsulated camptothecin in a mouse melanoma model. https://doi.org/10.1186/1756-9966-29-76

115. Clavijo S, Membrives F, Quiroga G, Boccaccini AR, Santillán MJ (2016) Electrophoretic deposition of chitosan/Bioglass® and chitosan/Bioglass®/TiO$_2$ composite coatings for bioimplants. Ceram Int 42(12):14206–14213. https://doi.org/10.1016/J.CERAMINT.2016.05.178

116. Bartmá Nski M et al (2020) Electrophoretic deposition and characteristics of Chitosan-nanosilver composite coatings on a nanotubular TiO$_2$ layer. https://doi.org/10.3390/coatings10030245

117. Muˌsat VM et al (2021) A Chitosan-Agarose polysaccharide-based hydrogel for biomimetic remineralization of dental enamel. https://doi.org/10.3390/biom11081137

118. Zhang C et al (2021) Preparation and application of chitosan biomaterials in dentistry. Int J Biol Macromol 167:1198–1210. https://doi.org/10.1016/J.IJBIOMAC.2020.11.073

119. de Paz LEC, Resin A, Howard KA, Sutherland DS, Wejse PL (2011) Antimicrobial effect of chitosan nanoparticles on Streptococcus mutans biofilms. Appl Environ Microbiol 77(11):3892–3895. https://doi.org/10.1128/AEM.02941-10

120. Aliasghari A, Khorasgani MR, Vaezifar S, Younesi H (2022) Evaluation of antibacterial efficiency of chitosan and chitosan nanoparticles on cariogenic streptococci: an in vitro study antimicrobial effects of *Camellia sinensis* and *Achillea millefolium* on several antibiotic resistant strains view project biodegradable PLGA implants containing anti cancer drug-loaded chitosan nanoparticles view project. [Online]. Available: https://www.researchgate.net/publication/303788651

121. Kumari S, Singh RP, Chavan NN, Annamalai PK (2016) Chitosan-based bionanocomposites for biomedical application. Bioinspired Biomim Nanobiomater 7(4):219–227. https://doi.org/10.1680/jbibn.15.00015

122. Butnaru E, Stoleru E, Brebu A, Nicoleta Darie-Nita R, Bargan A, Vasile C (2019) Materials Chitosan-based bionanocomposite films prepared by emulsion technique for food preservation. https://doi.org/10.3390/ma12030373

123. Onnainty R, Granero G, Onnainty R (2017) In: Onnainty R, Granero G (eds) Chitosan-Clays based nanocomposites: promising materials for drug delivery applications
124. Barra A et al (2020) Biocompatible chitosan-based composites with properties suitable for hyperthermia therapy. J Mater Chem B 8(6):1256–1265. https://doi.org/10.1039/c9tb02067e
125. Prajapati D et al (2022) Chitosan nanomaterials: a prelim of next-generation fertilizers; existing and future prospects. Carbohydr Polym 288:119356. https://doi.org/10.1016/J.CARBPOL.2022.119356
126. Aderibigbe BA, Naki T (2019) Chitosan-based nanocarriers for nose to brain delivery. https://doi.org/10.3390/app9112219
127. Fong D et al (2017) Lysosomal rupture induced by structurally distinct chitosans either promotes a type 1 IFN response or activates the inflammasome in macrophages. Biomaterials 129:127–138. https://doi.org/10.1016/J.BIOMATERIALS.2017.03.022
128. Wang K et al (2020) Review article recent advances in Chitosan-based metal nanocomposites. https://doi.org/10.1155/2020/3827912
129. Asgharnasl S, Eivazzadeh-Keihan R, Radinekiyan F, Maleki A (2020) Preparation of a novel magnetic bionanocomposite based on factionalized chitosan by creatine and its application in the synthesis of polyhydroquinoline, 1,4-dyhdropyridine and 1,8-dioxo-decahydroacridine derivatives. Int J Biol Macromol 144:29–46. https://doi.org/10.1016/J.IJBIOMAC.2019.12.059

Printed by Printforce, the Netherlands